Ham Radio DX
A Complete Guide

How to go from Karaoke to a DXCC Rockstar

Lucas L. Ford, W6AER

COPYRIGHT © 2022, Lucas L. Ford
FIRST EDITION – December 1st, 2022

Publisher's Cataloging-in-Publication Data

Names: Ford, Lucas L., author.
Title: Ham radio DX : a complete guide : how to go from karaoke to a DXCC rockstar / Lucas L. Ford.
Description: Pacifica, CA: Vitamin DX Publishing, 2022.
Identifiers: LCCN: 2022918726 | ISBN: 979-8-9869499-2-5 (hardcover) | 979-8-9869499-1-8 (paperback) | 979-8-9869499-0-1 (Kindle)
Subjects: LCSH Radio--Amateurs' manuals. | Amateur radio stations. | BISAC TECHNOLOGY & ENGINEERING / Radio TECHNOLOGY & ENGINEERING / Telecommunications
Classification: LCC TK9956 .F67 2022 | DDC 621.3841/6--dc23

Copyright © 2022 by Lucas L. Ford

All Rights Reserved. No part of this book may be reproduced in any form or used in any manner except with written permission of the Author. Exception to the above is to be used in a quotation in a book review, for non-commercial or educational purposes as permitted by United States copyright law. For permission, please contact the author. Lucas@w6aer.com

Vitamin DX Publishing, Pacifica, CA - vitamindx.com / w6aer.com

Printed in the United States of America

Library of Congress Control Number: 2022918726

ISBN: 979-8-9869499-1-8

FIRST EDITION – Rev. 1.03
10 9 8 7 6 5 4 3 2 1

Dedications

To my wonderful wife, Jill, who has put up with my *"wires"* as she refers to everything ham radio related as these.

To all my family and friends who have supported me in this hobby, or over the years. And of course, to those who have helped me during my journey of learning more about ham radio and improving my ham radio skills.

And last, but not least, to all those who made this book possible.

Photo Credits & Notes of Appreciation

All photos are provided by and are the property of the author, **Lucas L. Ford, W6AER** unless otherwise indicated below. These may be used with permission and proper credit. Contact the author for more information.

DXpedition photos from Conway Reef, South Georgia and South Sandwich were provide by **Paul Ewing, N6PSE** with an additional photo by **Mike Flowers, K6MKF** as credited.

Dual hexbeam, tower stack and hardline vs. LMR400 comparison photos generously provided by **Stan "Stax" Schwartz, KE5EE** as credited.

Vatican City, HV0A photo provided by **Kären Wiggins-Dowler, KM6IND**.

Navassa Island, K1N photo provided by **John Miller, K6MM**.

Thanks to **Dr. Antonis Papatsaras, AA6PP** for photos of his shack setup as well as his SOTA/POTA photo as credited.

An **extra special thanks** to the other "*beta readers*" for providing feedback to make this book more enjoyable for readers.

Dr. Bruce Bern, MD, K3NQ who also contributed "*The most important piece of equipment in your shack…*" and for providing fantastic beta reader feedback.

John Miller, K6MM who in addition to being a beta reader also assisted with chapter title ideas and content arrangement suggestions.

Dr. Antonis Papatsaras, AA6PP for the beta reader feedback and suggestions.

Dr. Andrew Calman, MD, K6ZP for the many great recommendations and ideas on which topics to elaborate on.

Paul Ewing, N6PSE for his valuable reader feedback and passionate support.

Special thanks to **Charles "Rusty" Epps, W6OAT** for brainstorming ideas with me as well as suggestions for additional topics of interest to DXers.

Thanks to **Tony Dowler, K6BV** for assisting with some ideas for the final title for the book as well as providing photos from his shack and outside antennas as credited.

QSL card images are used with permission and are courtesy of the following:

- **Savvas Pavlidis, 9H1AE**
- **Mohammed Al Saie, A92EE**
- **Nasir Khan, AP2NK**
- **Helmut Heindl, HS0ZIV**
- NU5DE, **Naturalist ARC**, photo by **Steven Bosbach**
- TL7M, **Kenneth Opskar, LA7GIA**
- WH6EYY, **Eric Brundage, KH6EB**

Author photo by **Jill Ford,** XYL (Wife)

The NCDXC logo is the property of the **Northern California DX Club**.

The NCDXF logo is the property of the **Northern California DX Foundation**.

The "**DX code of conduct**" text and logo have been identified as free of known copyright restrictions.

Cover design via **Canva**, main cover sun image courtesy of **NASA** from their royalty free archives.

For disclaimers, warnings and other legalese please see the end of the book.

Table of Contents

Dedications	3
Photo Credits & Notes of Appreciation	5
Table of Contents	7
About the Author	15
CHAPTER 1: Welcome to the World of DX	**17**
Why I Wrote a DX Book	17
What to Expect from this Book	17
My Weird DX Journey	20
So…What Exactly is DX?	21
Where is the DX?	21
International Operators	22
Visiting Operators aka the "*DX Tourists*"	24
DXpeditions	25
Where does the Money come from?	27
Beware of Pirates, Slims and Bootleggers	27
Getting DX News	28
CHAPTER 2: Overcoming Obstacles	**31**
An Arm and a Leg?	31
How Much Space Do I Need to DX?	32
HOA and Tower Issues	34
Remote Operating	35
License and Registration Please	35
Microphone Fright and Other Phobias	36
Age is Just a Number	37
CHAPTER 3: Blueprint for a DX Station – Hardware	**39**
Your Operating Position	39
The Most Important Piece of Equipment in your Shack…	43
HF Rigs aka Your Radio	44
Specifications (Over) Simplified	47
Boat Anchors, SDRs and in Between	49

Buying Used	51
A Second Rig	52
The Bare Minimums you will Need	52
Radio Accessories	53
Computers – MacIntosh vs. Windows vs. Linux	53
Radio Audio Interface	61
Headphones, Microphones and Speakers	63
Antenna Tuners & SWR	66
Power & SWR Meters	69
Amplifiers	69
Solid-state vs Glowing Tubes	71
Power Supplies, Protection & More	72
Antenna Analyzers	76
Dummy Loads	76
Rack Mounting	77
Optional "Nice to Have" Hardware	78
One Final Step	79
CHAPTER 4: Metal in the Sky - Antennas	**80**
Watch for Power Lines	80
Can you do it with Just one Antenna?	81
Does size Really Matter?	82
What About Height and Location?	83
A Quick Word on Loss and Gain	84
Indoor vs. Outdoor Antennas	85
Resonant Dipoles - G5RV, Bazooka, Fan Dipoles	86
Non-resonant Dipoles - OFC & End Fed	90
Verticals – Elevated Radials vs. Ground Radials	90
The Yagi and Other Multi-Element Antennas	95
Horizontal Loops	97
Vertical Loops	97
Low Band Antennas	98
Magnetic Loop Antennas	98

Beverage Antennas	100
Inverted V Antennas	100
Inverted L Antennas	101
RF Exposure	101
Ham Radio Towers & Masts	102
Other Things to Consider	105
Rotators / Rotors	106
Transmission Lines	108
Baluns & Chokes	112
Ugly Baluns	113
Connecting it all Together	114
Grounding & Bonding - Fighting Mother Nature	117
Keeping Everyone Happy	120
CHAPTER 5: Getting Binary - DX Software	**122**
Logging Software	122
TQSL & LOTW	125
Automation & Remote Access Software	125
Radio Control Software / CAT	127
CW Decoders & Skimmers	128
Clock Sync / GPS Timekeeping	128
DX Clusters	131
Award Tracker Software	134
Optional *Nice to Have* Software	134
Linux and Raspberry Pi	135
Extra Software for FlexRadio Users	136
Troubleshooting Ham Radio Software	137
Save Time with Scripts	138
Backup your Logs and Settings!	141
CHAPTER 6: Bagging the DX - Preparation	**142**
Working Split	142
Don't be a Dupe	143
Your Callsign is not Written in Stone	144

The Never-Ending Q-codes	145
Your Grid Square	147
Know your Radio	147
RF Gain / AGC-T	148
Preamplifier	148
Attenuator	149
Signal Width	149
DSP & Noise Blanker	149
Notch Filter	150
Roofing Filters	150
AGC	150
RIT and XIT	151
Signal Overload, work your RF Gain	152
ESP and Insurance Contacts	153
QSB – Fading	153
Use What you Paid for	153
Low Band Noise Issues	154
RFI and EMI Should be a 4-Letter Word	155
Finding Reception Noise Sources	157
Powerline Noise	158
Antenna Phasing & Noise Canceling	159
Ferrites Are Your Friends	159
Get Some Fresh Air	162
DQRM & Other Jammers	164
Join a Wolfpack	164
CHAPTER 7: Nice to Hear Your Voice - Mode	**166**
Speech Processors / Compressing	167
Equalizers	167
To VOX or not to VOX	169
ALC & Microphone Gain	169
Phonetic Alphabet, relearning your ABCs	171
Working the Phone Pileup	172

Timing, Tailgaters and Tail-enders	174
Rule Breakers	175
The HF Blue, Meet the Badge	176
CHAPTER 8: Dits & Dahs - CW & Morse Code	**178**
CW Keys & Paddles	181
Using Memory	183
CW Filters and Bandwidth	184
Do you Really need to Master CW?	185
Learning to Pound the Brass	186
Working the CW Pileup	188
What is QSK	190
CHAPTER 9: Digital Modes	**191**
The Ingenious FTs	192
Setting up your WSJT-X Software	193
Often Forgotten but a Must for DXers	197
Multiple WSJT-X Profiles	199
Using the WSJT-X Software	200
Tweak the WSJT-X Waterfall/Spectrum Display	203
How Much Power to Use	204
Working Split with WSJT-X	205
Fox & Hound, No Animals were Harmed	205
Adding Working Frequencies	208
Check your ALC…Again and Often	208
Watch your Duty Cycle	210
Digital Etiquette	211
Final Adjustments & Checks	212
Digital Mode DX Clock Trick	212
JTAlert – A Must for the Successful DXer	213
Alternatives & Other FT8 Software	216
RTTY	218
AFSK vs. FSK	219
RTTY Software	220

Working RTTY and RTTY Pileups	221
The PSK Modes	222
SSTV	224
Other Digital Modes	226
What is a "LID" and how not to be One	227
CHAPTER 10: On the Air – Navigating DX Challenges	**229**
The Solar Cycle	230
Day vs. Night	231
Enter the Ionospheric Layer Alphabet	231
Propagation Reports Demystified	232
Solar Flux Index	232
Sunspot Count and the Wolf Number	233
Sunspots: The High, The Low and the Ugly	233
Geomagnetic Storms	234
Get to Know your K and A	234
Maximum and Minimum Usable Frequencies	235
Sporadic E	235
Trans Equatorial Propagation	236
The Grayline Phenomenon	236
Long path, Short Path and Weird Path?	237
Hops are not Just for Beer	238
Ground Wave Propagation	239
NVIS	239
Check the CW Beacons	239
Other Propagation Resources & Alerts	240
Band Characteristics	242
The Magic Band(s) – 6m (and 4m)	242
The Upper Bands – 10m to 15m	243
The Middle Bands – 17m to 30m	244
The Lower Bands – 40m to 160m	245
DX Nets	246
Holidays and Soccer	247

Scheduling a Contact	248
Frequencies to monitor	248
Finding Information on Active DX	249
DX in Contests	249
The Sixes – Mark of the Beast?	251
On-Air Etiquette	252

CHAPTER 11: After the QSO — **253**

Setup your QRZ page	253
Design your own QSL Card	253
Print your QSL Card	257
Proper Card QSL Fill out Procedures	258
Storing your QSL Card Collection	258
QSL and Getting your Contacts Confirmed	258
Send Some Green Stamps & SASE	259
The Bureau System	260
Upload your Logs Regularly	261
LOTW	261
eQSL	262
ClubLog	262
HRDlog	263
QRZ	263
QSL Managers	263
OQRS Systems	263
SWL Reports	264
Tips for Fixing Logs, Missing QSOs	264
Card Checkers	266
Meet Other Hams / Conferences	266
Keep Learning for Life	268
Give Back if You Can	269

CHAPTER 12: Available DX Awards — **271**

The Prestigious DXCC Award	271
The DXCC Challenge	273

When Contacts Don't Count?	274
The Odd Bands	274
Going Fishing? Will not Count for DXCC!	275
No Permission Granted	276
What Else is Out?	276
Worked All Continents Award	277
CQ DX Award	277
Worked All Zones Award	277
Collecting Prefixes	277
Hunting for Islands	278
Other Notable Awards	278
Final Thoughts	**281**
Glossary	**282**
Disclaimers, Warnings, and other Legalese	**290**

About the Author

I became interested in electronics at a very early age. One of my first electronics experiences was playing with the RadioShack 160 in 1 electronic kit in the 1980s, which included the ability to build a simple AM receiver. This was followed by many AM crystal radio receiver units, which always picked up the local radio station, KGO.

While working my first ever job, a Realistic TRC-219 CB radio became my very first purchase. As with many hams, CB radio was the gateway. While in high school, I took several years of analog and digital electronics courses. This turned out to become tremendously helpful later in both career and hobby. As a regular reader of QST and 73 magazines around this time, ham radio became an instant interest of mine.

As for many hams, life, school, and work got in the way. I have obtained a dual degree in biology/biochemistry and followed up with a master's degree in teaching. This ironically led to a career in high tech, where I have been working for several decades now.

I currently hold a 5-band DXCC with endorsements for 12m, 17m, 30m bands (8 Band DXCC) and have confirmed contacts with around 300 entities in under a decade. Most of which I worked during the bottom of the solar cycle and from a tiny lot near San Francisco. I have also earned VUCC awards for satellite and 6m with endorsements and numerous other DX awards.

I also hold the *"WAS Triple Play"* for having worked all 50 US states on phone, CW, and digital mode with WAS endorsements for 9 bands and mode-specific digital endorsements for RTTY, JT65, JT9, PSK31, FT4 and FT8.

Life memberships include: ARRL (American Radio Relay League), AMSAT (Radio Amateur Satellite Corporation), NCDXC (Northern California DX Club), NCCC (Northern California Contest Club), San Francisco Radio Club, Ten-Ten International, and QRZ. I am also a member of the Intrepid DX Group.

Over the years, I have also held several board member positions at various local clubs and have given dozens of presentations on various ham radio related topics.

My current interests include ham radio antenna design and construction, exotic digital modes, SDR technology, solid-state amplifiers, transverters, satellite radio operations, VHF/UHF DX and of course HF DX. The current station can operate on all Ham bands from 1.8MHz (160m) to 1.2GHz (23cm) including 220 and 900Mhz bands with occasional, seasonal setups for 630/2200m and above 1.2GHz.

I am also an avid collector of QSL cards, and I have accumulated several dozen overflowing albums worth now, organized by continents and sometimes entities.

CHAPTER 1: Welcome to the World of DX

Why I Wrote a DX Book

While it seems long ago, it all started with a simple website in the early 90s. The internet was just taking off. Although a lot slower and, for many, it involved a dial tone still. Not to mention the use of CDs which used to fall out of magazines. Remember those? If you recall this, you were around too. If you don't, then just trust me when I say things are a lot easier now.

On this website, I wrote and posted articles on scanner and ham radio modifications, provided frequency lists for various things along with other technical content, mainly about computers and programming. I removed the site at some point due to time and costs involved. Hosting used to cost a lot more and online advertising was not as big yet to offset costs. Soon after, I started getting emails regularly about why I took the site down. Some also asked me how they could get copies of the site, its articles, and lists. I happily provided my old content when requested. This went on for a while. A lot has changed since, of course.

Skip forward a decade or so, the w6aer.com website came about. Here I have published many brand-new articles over the years on various ham radio related topics. The contents included presentations I have given to local clubs, articles, guides for new users, and much more.

I always felt there was no straightforward way on a website to connect all the content I had written. Also, I felt like I needed to fill in many missing blanks and address commonly asked questions I heard regularly. I was getting questions from new and even seasoned hams who wanted to try something new, like digital modes or DX, for example. These were some of the same questions I used to ask at one point. The answers to these questions, in part, form the main contents of this book. Along with many other technological advancements in DXing since most DX books I have ran across, it was time.

And so here we are. Thanks to the pandemic, I am spending more time at home instead of elsewhere. I have finally pulled all this information together into one place, filled in the blanks, and then some. This book results from combining research, my selected web content, experiences (good and bad), common questions and my past presentations. I guess one could say this book came from the spirit of sharing and ultimately, getting organized. But to answer the above question simply, I wrote this book for you!

What to Expect from this Book

I like to think of this book as a guide to hunting DX, but it is a lot more than that. While the focus of this book is DX, it is for everyone! **Beginner to experienced**. There is no specific intended audience. From those just getting started in ham radio with an interest in HF and/or DX to helping those more seasoned work the rare ones as they come up. Perhaps via a mode new to them. Maybe this book will inspire some to try something they have not done before. Either due to fear of getting started, being afraid to ask, not knowing who to ask or not even being aware of its existence. If this rings true for you, this book was certainly written for you.

Even if you consider yourself a seasoned DXer, there might be things you have never attempted. And vice versa is likely also true. It is quite possible some things a veteran DXer has done or tried, I may have not. Therefore, we all try to learn more by reading, watching and especially by talking to other hams. Think of this book as yet another tool in your DX hunting arsenal. I started with the basics for a reason, working up to the more complex topics.

If you are looking for math formulas and theory, this is not the book for you. There are plenty of those and they serve their purpose, though most of us will not be caught reading these on a sunny beach. The goal here is to simplify complex ideas into something that a non-engineer or newly minted ham can both easily understand.

If you are looking for great DX *"fishing stories"*, you're likely also out of luck. I will keep these to a bare minimum, or as needed, to prove a point. Again, many DX books like these are out there, and they are a good read. These are acceptable for beach use, by the way. I own many of them and love them. But these books rarely cover the topics for many DXers or really dive into what you really need to know.

There are also numerous books out there covering ham radio topics, usually focused on the 2m band or DMR. There have also been some books written for preppers which touch on ham radio. Those usually do a great job as well. Many of the above have not much to do, if anything, with DX. Specifically, HF DX, which is where this book focuses.

In this book, I tried to get to the point quickly, and want to be sincere about what I learned. Some of these may contradict things you have been taught, heard, and others may not agree with some of what I have to say. And that is perfectly OK. I speak from my experience, my experimenting, and my errors. If you are interested in the end results, read on!

The focus here will be mainly on "*little pistols*" as we are called. Yes, I certainly fall into this category. Those of us who do not have acres of property or maybe even have HOA restrictions but want to compete with the "*big guns*". Those with multi-element mono-banders on 85ft (26m) towers. For most, including me, that is just a dream. Maybe one day. But as you will see from this book, it can be done with less and, frankly, it is even more rewarding when you get to challenge yourself a little. One is certainly forced to get creative and learn more about this wonderful hobby when you deal with certain limitations. I would know! Getting creative because of limitations and using technology to overcome obstacles is half the fun. So, if you are wondering if it can be done. **Yes, it can!**

I have talked to many new hams over the years when I was elmering at local radio clubs, as well as chatting with those who are getting back into ham radio before I started this book. Think of an Elmer as a ham radio mentor by the way. Someone who can answer some questions or assist with issues you may come across. Some folks are experts in a given area and most are usually very open to helping. Though nobody knows it all and that includes me. If they tell you they do, run! I certainly have a list of folks I turn to for help and have done so regularly when I run into issues, including when I was writing this book. I bounce ideas off them for a sanity check and they do the same. In the process over time, I learned a lot. It is time to return the favors and pass on the knowledge the best I know how.

Lucky for me, and hopefully you, I took good notes during my DX journey, the website certainly helped in documenting. Although I must admit, the initial organization for this book was like herding feral cats. You will find some overlap and I may address some things from a different perspective in various chapters. There was no other way of completely separating out sections without doing so.

There are things which may often get overlooked or have been forgotten by some. Especially those who perhaps have been licensed for a while. Think of it as a refresher if you fall into this category. I did this to get everyone up to the same level, making this book the "*great equalizer*".

When I was asked once what it takes to work DX, I had to think about it a bit as many things came to mind. Then I narrowed it down to patience, constant learning and improving skills. Also, losing sleep. If you work low band DX, you know what I mean. Working from home sometimes certainly helps with DXing. Our recent pandemic has undoubtedly proved this to those of us who are still in the workforce. If you are lucky enough to be retired, you certainly have a significant advantage.

I recall when I was getting started in DX, I struggled to find things to read which were current or found things where I was not the target audience. Setting up a station is an adventure for beginners. I sure struggled and hit many hurdles on the way. Digital modes are rarely covered in detail nor are computer hardware or software. If they are, the information is usually a bit dated and, frankly, also inaccurate at times. As someone who has worked in IT for over 30 years, this sometimes makes my hair stand up.

There are some DX books written by other hams from W6/7 land, the US West coast. There is a good reason for this. Many would agree that this part of the US is one of the most challenging to work DX from. This region had many coined nicknames for this reason. These include the "*Forgotten coast*" or even the "*Suffering sixes*" but you will also hear about the "*Suffering sevens*". In the Pacific Northwest **(W7)** and Alaska **(KL7)**, as well as the West coast of Canada **(VE7)**, tend to not get as good openings on the higher frequencies as their counterparts a little further South.

This does not mean that this book is only for those on the West Coast, but since I live in the San Francisco Bay Area, I will certainly share my experiences based on my challenging location. This DX tricky location, combined with tiny lot sizes and difficult terrain (and soil), has many hams having to work much harder. If you are on the East coast of the US for example, you may find that working Asia is more of a challenge for you. There are certainly some international locations where hams have their own challenges, such as valleys, auroras and so on. Techniques mentioned in this book can be used anywhere. Hopefully, you will find them useful.

I would like to give you a **new perspective and a new spin on things** in some cases with some insight instead of just being a parrot. Of course, I will cover things you likely heard before if you are seasoned but will do my best to add some value in each area.

Keep in mind that some topics will have two, or more, schools of thought. Sometimes I feel the "*truth*" is in the middle. Certain things work for some hams may not work for others. I will discuss what I have done and seen others do successfully, the rest is up to you.

I am also **hoping to save you some money**. I have made some mistakes in getting things I thought I needed, doing things in a certain way that looking back did not make much sense. Sometimes because of listening to others sadly. Falling for marketing hype and all the other good stuff I am sure you have also encountered. Hopefully, I can help you avoid these pitfalls from lessons I have learned. What worked, what didn't. Some may work for you; some you may already be doing though some may not apply. What you do with the information is up to you.

Lastly, in the back of this book, you will find a ham radio glossary. This may come in useful if you run into a term I have not explained yet or just to clear things up for you as needed. Hopefully, you will find this useful.

My Weird DX Journey

…is far from finished. While around the writing of this book, I am hovering around 300 entities, I still have a few dozen left to work, though they are getting harder and harder. I am now mainly waiting for new DX activations. I certainly have many band fillers to get on some bands, mainly on 6 and 160m. This is really a lifelong hobby for a reason.

Joining clubs, especially those with a DX focus, has been tremendously helpful for me and I would highly recommend it to others if you are serious about DXing. If you want to find DX clubs in your area, check out the list from **The Daily DX**. It can be found at dailydx.com/dx-clubs. Might also want to consider subscribing while you are at it, well worth it. This has helped me tremendously.

I owe much of my learning and experience to the **Northern California DX Club** (NCDXC) and its many prominent members who help each other out and share the knowledge. I also made some great, lifelong friends in clubs and that is just one extra added social benefit of being part of this wonderful hobby. The NCDXC can be visited at ncdxc.org.

You can belong to more than one DX club, just as you can join more than one radio club. You can even join many regional DX clubs if you are outside the area as an associate member. This will usually be like being a full member, though without voting rights in most cases. Something to consider!

Finally, you are welcome to visit my current ham radio website at w6aer.com as I keep this site current with new content and includes several searchable databases. You can also scan this QR code to visit. I tend to document my DX journey here as time allows. This is  not just my DX journey, as I am involved in many other aspects of the hobby. Some tiny parts of the content of this book may also exist there, though likely very much expanded upon. As I had mentioned, it was in part the motivation for this book. There are also many topics on this site related to ham radio, some even outside of ham radio, which were not appropriate for this book as I tried to stick with HF DX. If interested, swing by.

So…What Exactly is DX?

To put it simply, **working a station outside of your general geographical area**. Sometimes it is just defined as "*distance*". When on the HF bands, DX is going to be characterized somewhat differently than on VHF and above. Distances achievable point to point on the higher bands, such as VHF, are much shorter on earth than on the HF bands. But this is misleading. Technically, if there is a line of sight, this limitation does not apply. Just think about satellite communications using ham radio or even moon bounce EME (**E**arth **M**oon **E**arth) where we are talking about some serious distances via VHF and UHF frequencies. Now that is DX! But the problem, at least as far as ham radio goes, is the earth is round. Yep, don't let the flat paper maps fool you.

But in all seriousness, an HF signal is more likely to hop (sometimes called skip) when conditions are right. The distances involved in these skips make DX by definition be pretty much anywhere on earth. However, I get excited when I get to work a neighboring state on the 2m band. Mind you, I am on the coast, a 3-hour drive from Nevada on a good day for perspective. So, it is all relative.

Our main focus in this book will be the HF bands. To be more specific, the 160m to 6m bands. But even this is not all that black and white.

Technically, the 160m ham band (1.8-2.0MHz in the US) falls into the medium wave frequency range, right above the AM broadcast band and 6m (50-54MHz) is technically VHF since it falls above 30MHz. Don't let this trip you up, the term HF is generalized in this case and since most ham radio gear out there covers these bands, it is a lot easier on the tongue than saying I am going to hop on my upper MF, HF and lower VHF rig. I think you get the point; hams often generalize. You will see this again and again in this book.

Where is the DX?

DX comes about in various ways. It could even be you as **you are DX to someone else**. It is all relative. But for now, let's get the basics down. Some DX are rather common, some are not so much and some are quite rare! These latter ones are the ones DXers get the most thrilled about. In fact, some of the rarest entities may not even get activated for a decade or longer. And when they do is when things get exciting! Serious DXers have been known to schedule time off around known activation dates and I know quite a few who "*caught Covid*" recently during a rare one…if you know what I mean.

In this book, I will use the country prefix next to the name of the entity. This will hopefully help you to recognize entities by their prefixes more easily when you encounter them in the wild. Frankly, it is also good practice for me. When ham radio operators use the word "*work*" or "*worked*" in reference to an entity, this has nothing to do with employment. In fact, it refers to making contact with an operator in that locale. This is also known as a QSO, or simply a "Q" as some hams like to call it. Because you know, three letters were just too much! Non-ham friends of mine are quite often confused by my contextual usage of this word.

Next, let's look at an overview based on how I like to classify different types of DX. Some overlap, but there is a reason I presented them this way. Hopefully, it will be clear by the end of the section.

International Operators

Just as here in the US **(K/W)**, many countries around the world give out ham radio licenses. Notably, Japan **(JA)** has the most operators in the world after the United States. This becomes obvious quickly as there is no shortage of Japanese stations to work for those of us on the West Coast **(W6/7)**.

The Hawaiian Islands **(KH6)**, along with Alaska **(KL7)**, are part of the 50 US states. Since I know I have readers outside the US, I figured I will point that out as most only think of the lower 48. When I talk to relatives outside the US, they also rarely realize this. These two states do, however, count as separate entities when it comes to ham radio due to geographical separation.

In North America, besides the US, the number of licensed hams in Canada **(VE)** is also rather high. Most live near the US border, so if you are hunting Canadian provinces (equivalent to US states) this can be hard. Especially in the Northern regions of Canada where there is a less dense population. I always make sure I work the rarer provinces when they show up.

Mexico **(XE)** is also relatively easy to work from the US with some well-known operators there, especially near the US border and spots popular with tourists and ex-pats. Moving further West, ham radio is becoming more popular in the rest of Asia and mainly in China **(BY)**, although most contacts are still being made from radio club stations there. Indonesia **(YB)** and Thailand **(HS)** can be easily and regularly worked from the West Coast as well. Others like Laos **(XW)**, Cambodia **(XU)**, and even Vietnam **(3W)** are harder to come by due to lack of local operators.

The Eastern side of Asia, commonly referred to as the middle East, is another story. Sometimes the difficulty of working these stations is due to war, politics, and lack of operators. Also, for those on the West Coast, it is just a bit harder than from the East Coast **(W1/2/3)**, or even the US South **(W4/5)**. This was one of the regional West coast challenges I was referring to earlier.

Russia also has many operators throughout the country. Something to keep in mind with Russia though, it is rather large. It sprawls over two continents, not to mention the insane number of time zones. There is a European Russia **(R)**, there is also an Asiatic Russia **(UA)** as an entity, and each are counted separately.

There is no shortage of stations there to work, especially during contests. Kaliningrad **(R2)** is also part of European Russia though geographically separated. You know…just to complicate things a bit more. The good news is it counts as a separate entity. Former Soviet states, which are now independent, can vary in frequency of showing up on the air. Ukraine **(UR)** is a lot more common for example then Tajikistan **(EY)**.

The European Union and the rest of Europe, mostly, also have no shortage of operators, with some notable minor exceptions such as Mount Athos **(SV2A)** which has one operator or Monaco **(3A)**, to name a few. Common countries are Germany **(DL)**, England **(G)**, Spain **(EA)**, France **(F)**, Netherlands **(PA)**, and Italy **(I)**. Pictured here is a QSL card from Malta **(9H)** which tends to be a bit rare from the US West coast.

South America also has a healthy number of local operators, with minor exceptions. The countries with numerous operators include Brazil **(PY)**, Argentina **(LU)**, and Venezuela **(YV)**, which is a surprise to some, including me.

The continent of Africa has most of its operators on the Southern tip of the continent in South Africa **(ZS)**. Sadly, a large portion of the continent is quite a challenge with very few operators, if any, in some entities. The Canary Islands **(EA8)** is often forgotten and

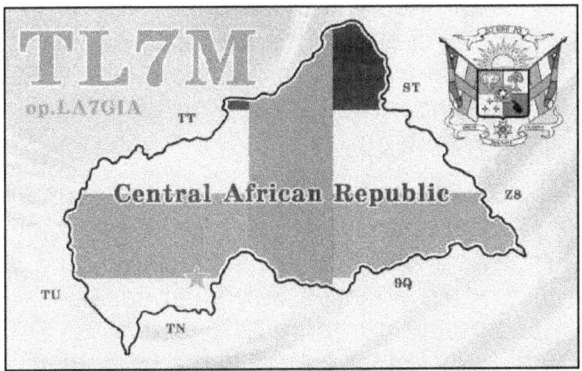

is in fact part of Africa as well. You can often hear contest stations operating here. For many, this is their first African DX. It can be difficult to work parts of Africa for many. I find this the second hardest region, after the Middle East for me. If I see any stations with an African or Middle Eastern callsign, I will try to work them before any other stations. I suggest you do the same, especially if you are on the US West coast.

Switching gears to much cooler places, don't forget about arctic entities like Jan Mayen **(JX)**, Greenland **(OX)** and on the flip side of the planet, Heard Island **(VK0)**. I still recall the excitement when I finally got to work the 2016 DXpedition, VK0EK. That was a tough one for me and I am sure it was no picnic for the DXpedition either given the conditions there.

Australia and Oceania are largely very active from Australia **(VK)** and New Zealand **(ZL)**. Though there are many smaller islands around here, there is not too much activity. For some of these, you will need to wait for activations due to lack of operators. In some cases, there are no permanent operators at all or even human inhabitants. Usually, there are many birds and crabs though! Based on my experience, wherever there are engineers, there are hams!

Visiting Operators aka the "*DX Tourists*"

One way to solve this operator shortage issue in some regions is by becoming a visiting operator. Sometimes these are also called "*Mini DXpeditions*" but they can generate pileups that are anything but mini. This is somewhat common for those places that are touristy. Especially if they have great year-round weather and other enticements as well. Operators often combine vacations in the Caribbean with operating, for example. Fun in the sun, beaches and a little ham radio as time allows.

Some islands even have well established "*super stations*" often used during contests. It is not unusual to hear the Cayman Islands **(ZF)** as well as Puerto Rico **(KP4)** on the air during these. Of course, Cuba **(CO)** also has many local hams who do an amazing job with what they must work with.

There are exchange programs with certain countries where someone licensed in the US automatically is permitted to operate in given entities covered by the agreement. **These programs include CEPT and CITEL.**

Some operators combine their vacations with operating. As with Tony, K6BV who took a side trip to Vatican City while vacationing in Italy. He got to be HV0A for a few hours, generating quite a pileup on sideband. *Photo provided by Kären Wiggins-Dowler, KM6IND.*

CEPT is the European Conference of Postal & Telecommunications Administrations. If you hold a US Advanced (grandfathered in) or US Extra class license, you may be able to operate via this program in some European countries. You can find out more at arrl.org/cept.

CITEL is the Inter-American Telecommunications Commission, which works to allow inter-operating in certain counties in the Americas. You can find out more at arrl.org/citel. You do not need to get a license in Canada if you already hold a license in the US and vice versa. There is a reciprocal agreement in place. You can get more information on that above by reading arrl.org/international-operating from the ARRL or by visiting the International Amateur Radio Permit (IARP) website if you are in the US at arrl.org/iarp.

Sometimes we get lucky and get an operator who stays in a location for an extended period. This could be due to work, humanitarian, or religious missions. Keep your eyes open for these operators!

However, some places may need not only a license for the specific entities to operate but landing permits as well. This is mainly the case for protected islands and areas, not to be confused with a travel visa. This, of course, greatly decreases the chance of these entities being activated and tends to place them much higher on the ham radio DXCC wanted list. This is where DXpeditions come into the picture!

DXpeditions

You likely realize by now if you been a ham for a while, that there are groups of folks out there who like to be on the "*other side of the pileup*" as they say. Although this is somewhat oversimplified. Essentially, that is what we are talking about here, but to the extreme. A group of hams who raise money and often put in their own money to activate a location for other hams is referred to as a **DXpedition. This is a combination of the words DX (Distance) and Expedition.** Sometimes written as "*DX-pedition*" and varies by preference and even the spell checks attempt at correction.

The use of special tents is illustrated above as seen in this photo from the VP8STI/VP8SGI 2016 DXpedition. Curious guests showed up to investigate the tents and the "metal in the sky". *Photo provided by Paul Ewing, N6PSE.*

Certain entities/countries are incredibly hard to work currently due to lack of operators or even visitors there, so this fills the gap. Though decades could go by before some entities are visited or at least get on the air. When the chance comes up for those looking to work rare locations, it is true madness during this time in the world of DX.

Some of these places are nearly impossible to activate. This can even be due to politics or banned ham radio. Examples of these places are North Korea **(P5)** and Yemen **(7O)** (as of 2022). A few are politically unique places where ham radio operators are not always welcome, even if legal, and can even be accused of being spies. Yes, this has happened…more than once! This can potentially end badly for them. Needless to say, not many are willing to risk this.

My estimate is that **about half of your worked entities will require a visiting operator or even a full on DXpedition to get in the log.** In other words, of the 340 entities, only about 180-200 have licensed and regularly active operators at all!

Geography can be another limiting factor. Location and/or climate are, at times, the deal-breakers as on Bouvet Island **(3Y0)** for example, which is the #2 most wanted entity at the moment this book is being written. Not to mention the expense of getting there is yet another factor not to be taken lightly. They estimated it to hover close to a million US Dollars for a small team for a few weeks.

Behind the team is the famous *"Braveheart"* which has been used by many well known DXpeditions for transportation to remote destinations, including the VP8STI/VP8SGI 2016 trip pictured above. *Photo provided by Paul Ewing, N6PSE.*

Don't wait too long to work a DXpedition once they get on the air. What I usually do is I make sure I get at least one contact in the log early and then wait for the crowds to die down if it is in high demand. For most it will be a full-on mayhem likely once activated.

Paul Ewing, N6PSE is shown operating from 3D2D Conway Reef in 2012. A very different climate from the ones mentioned earlier! This was also one of those operations which required a skilled team and some serious funding to make happen. *Photo by Michael Flowers, K6MKF.*

The reason I make sure I get at least one contact in the log early is due to possible problems that can occur while the DXpedition is active. Things happen where DXpeditions need to leave early, such as political uprisings, acts of nature (earthquakes, volcanoes, etc.) or unexpected weather-related issues. I can think of all the above happening at least once since I have been DXing.

Where does the Money come from?

The simple answer is, mostly from you. But it is way more involved than that. One way to show your appreciation for these folks like above is to donate to their cause directly or the **NCDXF** (**N**orthern **C**alifornia **DX** **F**oundation) for example, who help make much of these possible. You can find them at ncdxf.org. Additionally, **INDEXA** (**In**ternational **DX** **A**ssociation) is also a major contributor to many of these DXpeditions. We can find them at indexa.org. The **GDFX** (**G**erman **DX** **F**oundation) found at gdxf.de and **EUDXF** (**E**uropean **DX** **F**oundation) found at eudxf.eu are also major contributors. These foundations get funded by hams and in turn grant funds to DXpedition groups as they see fit.

Some operators like to say that they will donate once they worked an entity. Well, I get that, but only sort of. Many of these operations have high upfront costs and need this money before they get on the air. If they have to cancel, most likely you will be refunded if the operation is reputable. Most are!

Therefore, I encourage donating to DXpeditions before they get there to assist them. If you worked them and want to donate more to offset their costs, you can always do so later. These donations can even be tax deductible but check with the group (usually stated on the website) and with your tax accountant to be sure this applies to you. This may also vary from country to country but is certainly the case here in the US in many cases. Foundation donations are generally tax deductible as well.

Beware of Pirates, Slims and Bootleggers

Pirates are still around and not just those off the coast of Somalia **(T5)** but also the ham radio kind. I cannot for the life of me comprehend why folks do this. Certain people just like to pretend to be someone else I suppose. Basically, they take over a callsign and just use it like their own. When this happens, they become known as pirates.

I have even seen several US **(K/W)** and Japanese **(JA)** hams being impersonated as in actually **hijacking the callsign**. The **pirates** do not get award credit for this (nor do you) and, just frankly, it is a waste of everyone's time. I guess it will be just one of those things I will never understand.

Sometimes you hear pirates referred to as *"slim"* and technically, these are similar. Some would argue they are the same. They are theoretically not the same though. Rare or semi-rare DX are the ones spoofed most often and when this occurs, is when the name is used…mostly. Though you will hear the terms used interchangeably. Does not always involve a callsign hijacking, rather just **pretending to operate from a different location than they are in**.

Lastly, a **bootlegger is someone who makes up a callsign and starts operating**. Nothing to do with moonshining. Sometimes these are easy to pick out as they do not follow the convention of standard callsigns, but at other times, requires a lookup. This is not too common though. I see about 1 to 2 a year. One time I worked a station in Monaco **(3A)** just to realize it was a bootlegger. Yes, this example overlaps with the above "slim" to a degree.

There have been cases where hams had their license expire and continued operating. I suppose that sort of falls into one of these above categories as well. I do know of one instance where a ham did not renew for some reason. About 6 or so years later decided to get on the air again just to find that his call is now someone else's vanity call. Go figure! Basically, **none of the above contacts count towards your total entity count**.

The rule of thumb in DX is, if not sure, **just work them first and worry about it later**. Why? Because sometimes we get pleasantly surprised as did those who heard a call from North Korea **(P5)** station a few years ago here on the US West coast **(W6)**. Without giving it much thought, they jumped on it, worked the station, and turned out to be a legit ham doing some brief operating from the hermit kingdom. This is the #1 most wanted entity (as of 2022) so needless to say there was some serious celebrating for a select few. Sadly, I missed all the fun, but loved to hear the stories. There is always next time! Champaigne is on ice, cigar in the humidor.

Getting DX News

When you first start out DXing, almost anything new, even if somewhat *"common"* is exciting. To a degree, it always will be for a true DXer. However, things get harder and harder as your DX entity count goes up. This is where information starts to be one of the main keys to your <u>continued</u> success.

There are many sources out there for getting news on DX. I was pleasantly surprised at this when I first got serious about DXing. Some are better than others, or perhaps just different. I will go over what I recommend and use. There are other resources out there, some in other languages as well, so I encourage you to look around especially if you are multilingual.

A fellow ham, **Bill Feidt, NG3K** maintains a simple and easy to read page where it shows a calendar of active and upcoming DXpeditions at ng3k.com/Misc/adxo.html. I have this not only bookmarked but is set as one of my startup pages when I turn on my browser in the morning. The page is to the point and that is how I like things personally. Very cool and I thank him for maintaining this!

Speaking of mornings, I am always looking forward to my daily edition of "*The Daily DX*" from Bernie McClenny, W3UR. I have already touched his DX club list. **This service is a must-have in my opinion**. It is a paid service but very inexpensive for what you get and well worth it. You will get the inside scoop and all the latest information you need for your DXing. Kind of like a one stop shop for DX news. There are a couple of ways you can get news from Bernie. There is a weekly summary edition called the "*The Weekly DX*" appropriately, as well as the daily edition per above. You can find out more here: dailydx.com.

Another website I like is **DX Heat** which can be found at dxheat.com/dxc. The main thing I like is the activity meters where I can quickly assess where the action is as well as the "*sun's mood*" for the day. The **color-coded SFI meter alongside the K and A meter is very helpful**. We will talk more about these measurements as well as propagation later in the book. This site also has a nice filter for DX spots by band and mode, amongst other features such as continent filters and lookups.

Also keep your eyes on **QRZ** found at qrz.com as this site often also has DX related news. Here you can also lookup stations by callsign. Often the pages will direct you to websites for the DXpedition if applicable, as well as information on how to QSL (confirm your contact).

One place to also check when you have time is **DX News**, which can be found at dxnews.com. This site gets a lot of visitors and often is one of the first to report DX news. Yes, the name fits well. The very popular **OPDX Bulletin** edited by Tedd Mirgliotta, KB8NW can be found at papays.com/opdx.html but sadly has ceased publication at the end of 2022. You can still view the online archive at the above address if interested.

DX-World is at dx-world.net and is maintained by Col McGowan, MM0NDX and provides some exceptional content. One of the many excellent resources, he is based in Europe. Perhaps my favorite is the expedition timeline.

One of the few things I print monthly and pin near my desk. We can find this at hamradiotimeline.com/timeline/dxw_timeline_1_1.php.

Speaking of Europe, another free resource online is **DX Coffee** from Italy. We can find this at dxcoffee.com and is definitely one to bookmark as well. Staying in the same country, yet another DX news source from Italy is the **425 DX News** found at 425dxn.org and they also publish a monthly magazine. There is also a very handy DX calendar at 425dxn.org/index.php?op=wcal.

DX Zone lives at dxzone.com and is a nice complement to the above, with great articles to round things out. **DX Maps** also has a DX calendar at dxmaps.com/dxcalendar.html which is a great way to visualize upcoming operations. This is something I even print at times to remind me of what is coming up.

Spotting networks can also provide valuable news for your DX hunt. I will cover these later in more detail. Some of what you find above may or rather likely will have overlaps, as there is only so much news to go around in the ham radio world. **You will likely find that paying for a service will save you a lot of time and effort** as it not only provides a nice summary but also tends to have information that the other sources often do not. Something to seriously consider. Your valuable time you could be used to chase DX!

CHAPTER 2: Overcoming Obstacles

The limitations mentioned in this chapter could be imaginary or perhaps very real. This varies with each individual and of course, their situation. Regardless, keep an open mind, as this chapter may surprise you and may make you think outside the box. If you are new to DX or potentially interested in it, reading this book might just be your first step to doing so. If nothing else, you already made it this far. As they say, the hardest step is always the first step.

You will find that many new hams, as well as those who have been at it for a while, find reasons not to try certain things. For some, this is DX. I was one of those hams. It took me a while after getting licensed to get serious about HF DXing as well. I have a small lot, on a hillside to top it all off. This is far from exactly ideal. I was told it cannot be done. Well, spoiler alert: It can be, and I have now written a book about it. Due to this misconception, I initially focused on ATV (Amateur Television), ham satellites and VHF/UHF operating. Obviously, this changed after I got creative, talked to folks in similar shoes and most importantly the right people.

Hams of all levels of experience at times shy away from DXing or feel like they cannot be competitive. Sometimes this is due to myths or lack of information. At times even wrong information, as above. So, let's get started!

An Arm and a Leg?

No, it does not have to be expensive. Far from it! Hams often joke: "*Get your kids into ham radio and they will not have money for booze and drugs.*" There is some truth to this but does not mean it has to be an expensive hobby. It is what you make of it, and you take it only as far as you want to take it. I have seen plenty of examples of both extremes. I know hams who got a DXCC from a college dorm room, in one semester mind you, and I also know hams who have a tower and are struggling. I will get to that shortly…

You can get started on a very limited budget and there is nothing wrong with buying used gear with some exceptions. I started with all used gear, mainly from eBay, some from local hams upgrading. Sadly, also some from hams who have become silent keys (SK). This is a reference to a ham who has passed away. From my first HF rig to my first antennas and even my coax was all second hand. As time went on, I slowly replaced my equipment as needed. But I often still look for "*gently used*" gear.

If you do your homework, or even better, have an Elmer to help you, it can be a cheap and fun process building a station. If you do it right, you will have plenty of money left over in most cases, than you had budgeted. You can always upgrade later as you gain more experience…or don't! You may find that you are perfectly happy with what you have and what works for you. There is absolutely nothing wrong with that. It is a hobby, after all. Treat it as such!

Pictured here is my very first HF radio, the Yaesu FT450. I used this with a simple vertical for many years and got my first 100 entities with it. The total for the radio, power supply, vertical and coax was under $1500. All were obtained used.

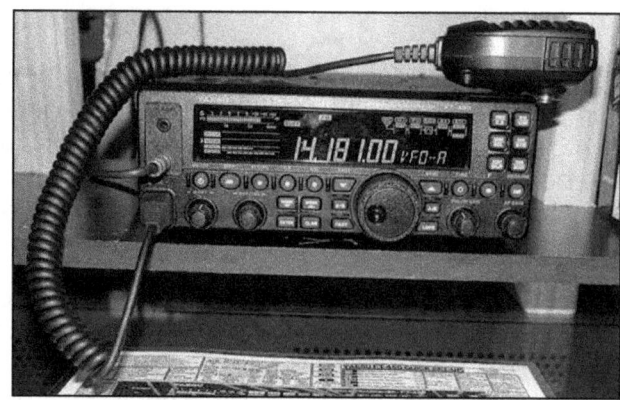

I am going to take a gamble here and say if you already caught the DX bug, there will be some gear upgrades in your future. The good news is that, unlike the computer market, you may find a buyer for your ham gear who will often give you somewhat close to what you had paid initially.

Some used radios, which are highly desirable, may sell for more than when they were new. I sold my Yaesu FT-736r radio for way more than I paid for it several years earlier. I almost felt guilty about it. The demand was, and still is, high. This popular radio is no longer made and there is nothing quite equal to it on the market. This was considered one of the best VHF/UHF rigs made by many, including me. Some boat anchors (older rigs) also fall into this category, where you may get more for it later. Though for different, often nostalgic reasons.

With old computers, of course, they may ask you for a recycling fee instead of giving you money when you are ready for your upgrade. Stay away from used computers in general, there is always a reason they are being sold. Folks rarely sell a used computer, "*just because*". Is it worth losing your logbook or having FT8 issues due to an old clunker? But more on this later.

If you do what I do and **only keep the minimum amount needed and get rid of gear you no longer use**, there might be several positives which come from this. For one, you offset the cost of the newer gear. This is usually more than you think. Good places to sell gear are local electronics flea markets if you have one in your area and of course ham fests. Online options include QRZ at qrz.com, eHam at eham.net and QTH classifieds at swap.qth.com. However, sometimes keeping a spare is a good idea so don't go too minimalistic.

Second, more space in your house, garage, storage or wherever you normally hoard. I do not believe in keeping a radio museum in my house. If you have the room, and the money, more power to you! Many hams seem to have a hard time parting with their old gear. I know a few hams who have dozens of radios (that I know of), yet only 2 ears. Most have not been on in decades. **My rule of thumb is, if I do not touch a piece of gear in a year, on sale it goes.** And lastly, your spouse or partner will likely also appreciate not having old gear all over the house or the garage. So will your wallet!

How Much Space Do I Need to DX?

One of the biggest complaints I hear is "*I have no space*". I hear you! This is especially true in urban areas and even some suburban areas depending on the part of the country you live in. This "*space issue*" can be broken down into two areas. Space for a shack (the easier one) and space for antennas....yes, this can be an issue for many, including me.

Living in a crowded city like San Francisco, for example, many find it hard to erect an antenna, even a small one. Yet, dozens of San Franciscans have a DXCC award and they managed to obtain it within city limits. How could this be? What kind of witchcraft must they be practicing? As I said, it can be done!

Shacks do not need to be huge. They can be a very simple, minimalist setup. To be clear on the term shack, does not literally refer to separate shed like building your radio gear sits in. For most hams, this is a room, part of a room or a basement, for example. Many also find themselves in the attic, the garage, their man cave or as in my case, a spare bedroom which doubles as my home office. It is also where everything else goes which does not "*fit*" with the rest of the place. I will let you figure that one out for yourself.

My radio desk doubles as my home office desk. I purposely purchased an oversized one. A server rack is used for overflow gear. I will talk more about this later in the book. While, yes, I have a lot of things stacked on shelves and had to get creative, the square footage used is comparatively small.

My backyard is also relatively small, as I had mentioned. For California, I suppose it is regarded as average or maybe even a touch above average for the Bay Area, as I cannot actually touch my neighbor's walls from my property. But for those in many other US states, I would be considered living at the poverty level with such a small lot. So how did I, and others, do it? Careful planning and many adjustments over the years.

Antennas can get very large for HF, especially if you are going to work the lower bands. This would be 40m and below. If you are coming from the VHF world, 2m antennas are under 2ft (½ m) wide at ¼ λ (wavelength). 6m antennas move up to just under 5ft (1.5m) in element width. Therefore, an HF Yagi antenna can be expected to be 10ft (3m) across at least, plus however many elements it has. They can get frighteningly large, especially for a 40m Yagi. A 160m or even an 80m dipole is no joke either. It can be done though.

The 10m antenna ¼ λ is still just over 8ft (2.5m), but then this jumps to 33ft (10m) on the 40m band. Therefore, you see antennas start to go vertical in limited spaces. Many hams manage to put up antennas covering 6-20m bands but below the 20m band, it gets harder due to increased element lengths. Do not let this deter you, just do what you can, start with what can get you on the air.

I have managed to install a 160m antenna on my small lot, but it had to be a vertical. To give you an idea, if I was to install a trap free 160m dipole It would cross nearly 6 lots on my street. Not a joke, I did the math! Nearly half a block. That means I would have to get an OK from 5 neighbors. Not happening! Verticals antennas, trapped dipole antennas or even a trapped vertical can solve this problem. See the antenna section for more on this. If you were to run a ½ λ dipole for 160m you are now talking about 130ft (40m) of wire on each side if not using traps. Therefore, a total of 260ft (80m).

HOA and Tower Issues

Yes, the dreaded Homeowners Association or HOA for short. Some hams firmly believe it should be a 4-letter word. HOAs can be a major headache for some hams. Depending on where you live, you may not have a way around this. Condominiums are not the only units which may have HOA. Some housing developments, gated communities and so on also may. If you are moving, be sure to check on this and make sure you know what you are getting into.

On the flip side, this may not be an issue even if there is an HOA. Some are quite reasonable and down to earth. Regardless, be sure to check and maybe even get something in writing before moving if you plan on having anything publicly visible. Also, be careful with what you sign! The same goes for apartments (or houses) if you are renting them. You may have additional restrictions from the homeowner. Sometimes these can be negotiated if you explain to them what your hobby is all about.

Luckily, I do not have to deal with an HOA, but I know many who do and manage to DX with a strict HOA rules. Sometimes they call these "*guidelines*" but if so, I am not sure why they enforce them so strictly. I know of one ham in nearby San Jose who throws a random wire out of their condo window at night and gets on the air. Sounds like something out of a spy movie, but it works. While I would not run legal limit power using this method, it does get the job done.

Some cities can make it a real pain if you want to put up a tower. Not that you need a tower technically. I do not have one, or even really the space to install one properly. In parts of Silicon Valley, it is nearly impossible to erect anything resembling a tower anymore. In some parts of the country, I am told when going in for a permit, they wonder why you are even bothering them with such a "*ridiculous thing*". It all comes down to your QTH and mainly your region. Some places are over regulated, like California and New York. Others are a complete 180 degrees in what goes on. Somewhere in the middle, I feel is ideal, as with most things in life.

Flagpoles antennas are often used as verticals. This does not mean you cannot load up an actual flagpole. I do recall one person operating on one of these from a trailer park. He described this to me in detail over an SSB QSO I had with him many years ago. I bet his neighbors have no idea he is on HF based on his QRZ photos. He just appears to be very patriotic. Good for him!

Simple wires antennas can work great as well for these situations. QRP operators have even been known to operate using very thin speaker wires strung between trees. I have heard of folks loading up rain gutters and other metal objects. There is even an urban legend about a mattress spring box being used on the 20m ham band. Who knows, maybe it is true. Though doubt it did well unless raised up high.

I have even heard of a company now making antennas that look like yard art. Do a little looking around online. It is amazing the workarounds people come up with. There are also many excellent books covering this topic specifically. I will cover additional antenna options as well as towers and push-up masts in more detail in the antennas section.

Remote Operating

There are two ways of doing this and often folks confuse the two. One is to operate your own or someone else's radio remotely while away from home. The other is to operate via a service, such as remotehams or remotehamradio for a small fee at times. All the above are legitimate ways of operating if you follow the rules.

So why would you do this? Sometimes it is just not feasible to operate any other way. For example, some hams find that when they move into a nursing home or move in with relatives, antennas are not an option. They do not want to give up the hobby. Remote operating fills this void. We already talked about HOAs, so need I say more. Sometimes remote operating is the only option. You might be tight on money as well and cannot afford to spend even the minimal amount for gear. This, again, is a perfectly viable option.

Operating a remote station, other than your own, is a controversial topic but I am going to cover it anyway. Some feel like it's cheating. It's generally OK and it is not considered cheating. It is perfectly fine to use these if you follow the required rules set forth for awards. Also, if you follow the rules for operating and for station identification. Meaning, don't operate somewhere you are not licensed and pay attention to the award requirements if you are planning on applying for any.

In other words, for the DXCC award for example you must operate from US soil, so use only a US based station. You cannot log into your friend's station in Croatia **(9A)** and make a QSO with Monaco **(3A)** using US callsign and have it count towards your DXCC count. Not to mention you would be breaking the law. You must make the call from US soil. For WAS and VUCC, this is a different story. For these you must be within a certain distance from your station and the same grid, respectively. In other words, be very careful. Your choices might be limited. For general rag chewing, checking into nets and such, go for it! You can check out remotehams at remotehams.com and alternatively check out, remotehamradio at remotehamradio.com.

License and Registration Please

While it is certainly possible to get many, if not all DX awards with a US General license (or equivalent overseas), sometimes you will miss some great opportunities. There are some operators, both US and DX, who only work in the US Extra segment. I am told by some; this is mainly to avoid oversized pileups and bad operating habits. This is especially true for CW. The Technician license is a good start to get your feet wet regardless. Everyone must start somewhere.

Keep in mind that with a technician license here in the US you are restricted to only tiny portions on the 10,15,40 and 80m bands for CW and some very limited privileges on 10m for data and SSB phone. Therefore, at least a General license is highly recommended.

There are folks who have gotten DXCC though with only a Technician license at the top of the solar cycle and using the 10m band only. So, yes it can be done but this is not for everyone. You will likely miss out on a lot, will take you a lot longer and during the bottom of the cycle you will be bored out of your skull on most days.

If you are only planning on operating FT8 for example, and only on the designated frequencies, you are likely going to be able to get away with a US General license or overseas equivalent. The DXCC desk will note your license class and operating frequency shows up in LOTW logs when you upload. You will likely not get credit for contacts if you operated outside the band portion, an area you are allowed to operate in. Possibly receive a firm warning or may even be disqualified from the DXCC program if you operated outside of your license class on a regular basis. **Always make sure you are operating where you're allowed to do so based on your license class.**

Of course, if you upgrade to Extra (or equivalent outside the US), this is no longer an issue as all areas and power levels are now permitted for you. Now you just need to make sure you follow the band plan if your country requires you to do so. The US does and frankly I feel it is a bit dated. I think a gentleman's agreement would be enough as most other entities do not enforce band plans, so kind of defeats the point. No? Band plans refers to operating CW only in the assigned CW segment and phone in designed phone segment, and so on. The flip side of the argument is that the US band restrictions provide welcome relief to DX operators. Therefore, allowing them to use parts of the band where they can relax and ragchew with other operators without having to worry about a huge pileup.

Upgrading is relatively easy and certainly not expensive. If you are serious about becoming a DXer don't limit yourself and possibly miss great opportunities. Go for it! While you are on the roll, may even consider getting a Commercial FCC (**F**ederal **C**ommunications **C**ommission) license as well. This is practically the same test as the US extra class license. The cost is a bit higher to take the exam, but it is for life and may open some doors to some new employment opportunities as well. Just think of how much fun it would be to work with telecommunications gear for a living not just to play with as a hobby. Plus, it may give you access to bigger and better "*toys*". You can read up on these more at fcc.gov/licensing.

Microphone Fright and Other Phobias

Some folks just can't get themselves to key down on the microphone once licensed. And this is not just on phone mode, but there is also key fright in CW which I did go through. I will be the first to admit I am not the best or fastest CW operator. There is even digital fright by some, even in cases where most of the lifting is done by a computer such as FT8. A bit of caution is good, uncertainty and hesitation are normal.

I started my ham journey mainly on amateur television (ATV) as I had mentioned and in the pre-FT8 digital era. Phone and digital mode contacts were not a big deal, at least for me. I was also somewhat active on the 2m band before getting more serious about HF and DX. Many folks start out with a US Technician license or equivalent and may have experience on 2m, like using FM or one of the digital voice modes like DMR, likely via a local repeater with limited reach. On HF, with the conditions being right, the world can hear you and this is very intimidating for some. I can certainly understand this. Though the same can be true for DMR depending on where you key up.

My first HF QSO once getting back to ham radio was on PSK31 which is also where I worked my first "*real*" DX outside of North America, the Czech Republic (**OK**). Yes, I worked this from the West coast even before Japan (**JA**). Go figure! The funny part is,

I did not realize it was as DX at the time. PSK31 is a mode which I will cover later. This mode does not perform error correction and I thought it was a US station with a calling, starting with the letter K. I assumed the O was just a glitch in the decode as it often happens with weak PSK31 contacts. To my surprise, it was not a glitch. Long story short, this contact perhaps marks the beginning of my DX career. The person on the other end was Karel, OK1EP.

Truthfully, it is rather simple. It really is just like public speaking. Those who fear it know, hopefully, that the way to get past it is by just doing it. Just go for it! I know this is a generic and perhaps non-helpful sounding statement, but you will not trigger a tsunami or the apocalypse. Make sure you understand the basics. Ask for help if needed. Know your callsign phonetically and listen to some conversations to see how they normally flow. Listen and listen some more, next thing you know, you will have a good feel for it. Then get everything setup right and key up. The more you do it, the more natural and second nature it will become. Soon you will realize that it is not such a big deal. Believe me, DXers don't think twice about keying down that microphone. In fact, some you can't even shut up once they do. Yes, I had to go there!

A good place for the hesitant is perhaps checking into a local HF net and working your way up from there. Think about this. When working DX all you really need to say is 2-3 short things. Your callsign in phonetics, a report *"five nine"* (we all lie...more on this later) and maybe *"73"* or a *"thank you"*. The latter, if you were raised right, you have said thousands of times in your everyday life. Right?

Key fright, as I had mentioned earlier, is what I personally dealt with as well to some degree. CW is certainly a skill that is often, and sadly, judged by some harshly. I am not saying this to scare you, but I told you I would be very honest. There are CW snobs out there. Don't worry about them! We all need to start somewhere. And the somewhere with CW is slow. Likely very slow! So just start with the speed you are comfortable with, others will probably work with you and slow down to your speed. I always do, even during contests. You may be the one slowing down for me if you get pretty good. I initially got through my CW hesitance by using a memory keyer (which I still often use) and this got me more comfortable in just a few weeks of being on the air.

To summarize, just get started. Like I said before, the first step is always the hardest. You will get better at it with time and practice. The next thing you know, it is second nature, and you will be helping others. Trust me!

Age is Just a Number

If you are a younger ham or perhaps just a newly minted ham, possibly looking for something to do in retirement, the world of ham radio, especially DX, can be intimidating. Don't let it be and especially do not get scared off by the number of hams who have been at it for decades. I have heard these concerns many times. I have also felt this at times. Then I realized that it was all in my head, like most things.

Age really is just a number, and everyone must start somewhere. I have certainly been to some club meetings where I walked in, and I felt like I was one of the youngest ones there. I have no issues with this though, nor should you. The same is true if you are 90 and got your license yesterday. Welcome to ham radio! I hope to work you soon!

Hams are largely very welcoming. Not to mention, many older hams have been licensed for a long time. And guess what, logically this means they were younger at one time also and just starting out. Nobody is born a ham.

If you are a younger ham, don't be afraid to ask for help from more seasoned hams. In fact, you may find you can help each other out. Remember when I said that nobody is an expert on everything? Well, if you are under 40, there is a good chance you are much better at trouble shooting computers (and more likely to work digital modes) than someone over 60. I can see a mutually beneficial relationship forming here. Bottom line is don't worry about age. Have fun and enjoy this wonderful hobby!

CHAPTER 3: Blueprint for a DX Station – Hardware

Many hams struggle with building a station, especially their very first one. Some have a hard time even getting started. Yes, it can be a daunting task. The more you read, the more questions you may have. Even those who built a station which they had considered great at the time, may find that they did things they should perhaps not have. I sure did! More than once…I have made some mistakes and made numerous adjustments over time. In fact, I would argue that it is a never-ending process. And that is perfectly OK!

This will be a big topic in this book as hardware is rather important for DX success. Often when hardware and software are covered in other ham radio books, they are merged into a smaller chapter. I never liked this format. Therefore, I am going to break these down into two somewhat in-depth chapters, minus the math and things you do not need to worry about. I know many questions are hardware related, so it will be time well spent. This section is by no means intended to be a substitute for books covering these topics in more detail. However, it should provide you with enough information to get started and perhaps start making initial decisions. If nothing else, start thinking about things differently. Making educated decisions in these areas in my opinion is half the battle and is often overlooked. I know this from personal experience. I wish someone had told me.

You may be constrained by costs. Don't worry about this or let it stop you. It takes time to build up a station. Years if not more for many. I started with a small, entry level HF rig, a vertical and a power supply. And guess what, I loved it. You can work DX just fine with simple, inexpensive gear. Yes, it will be harder. Yes, it may take you longer. And no, you may not be able to "*work them all*" but you will still get enjoyment out of it.

Let me touch on something I always tell folks. Value and cost are different. Something that costs a lot, might not actually add value to your shack or work better. There is a lot of "*creative marketing*" out there to get you to bite. Beware of buzz words and exaggerated claims just as you would be outside of the ham radio world. You will likely pay "*extra*" for very little and sometimes frankly, nothing. Talk to folks you trust and are successful in the area of interest, such as DXers. Reviews are not always all that accurate and at times are even fake and/or paid. Do your homework. As you have guessed, this is where we start getting a little more technical. So put on your thinking cap and here we go!

Your Operating Position

No, you did not misread the above. This is the first item on my list. And yes, this is hardware. This is so very often overlooked, and I think it **one of the most important things to consider**. Afterall, it all starts here. With you! The station should be built around you, not the other way around. I recommend you give this some thought before you begin building the rest of your station.

Your desk and chair are both very important. Both need to fit the function, not the look, unless you enjoy back pain. Though you can have both. **The desk must be something that can support the hardware on it and leave you enough room to operate.** A typical desk is 29" (74 cm) high in the United States but 3" (7.5 cm) lower for typing. This, however, is not always the case. I am over 6ft (2m) tall, and I find the 26" (66 cm) typing height to be too low for me. Figure out what is best for you by visiting a store or a friend if not sure. Bring a tape measurer. You will be stuck with whatever you buy so spend some time planning and trying out things. You also need to take notes and be able to use a keyboard and mouse comfortably without causing strain. Basically, you do not want to be tense when operating and want to minimize your head and hand movement as well within reason.

Photo provided by **Anthony Dowler, K6BV**. Tony uses a corner setup with 3 monitors. The Amplifier is easy to reach on the left, Radio dead center and logging/digital PC on the right. Clearly, some thought was put into this, and this works very well for him.

It is also good if your station position is located so you can easily access the back panels if needed. In Tony's case above, he can easily pull-out components and this works fine too. This is important for troubleshooting, which likely you will need to face, eventually.

The chair is something you will use a lot, unless you are one of these standing desk folks. In which case, more power to you. So, it is important that your chair is not something left over from your last dining room set. Spend a few dollars here and think about your back a little. I would opt for **a proper, comfortable office chair**. For some folks a gaming chair also works. I use one and love it, but everyone is different.

The location of your operating position is also important. This means someplace where you have sufficient heat (don't count on your amplifier for this) or cooling. This you may really need to consider that on a hot summer day if you run your amplifier a lot, you will turn into a puddle.

As a side note to operating; I have observed some individuals living by the radio, literally. Which I somewhat get, I am on frequently as well. However, I make sure I have a balanced life and don't spend every waking moment on it outside of work. Those who treat it like a full-time job, I have noticed, tend to lose interest for a while, or forever retire from DXing. Sometimes after years of being super active even. At which point they may or may not return. Perhaps it's burn out, who knows…but as with everything, moderation is key!

Make sure there is enough power available to run your gear. Though this can be easily fixed in many cases. This is <u>not</u> to be confused with enough outlets. This is also possibly important, but I mean power specifically. If your shack grows you may need to pull in extra outlets from the main circuit breaker. You may also possibly need a higher voltage one (like here in the US where we use lower voltage) to run an amplifier if you go that route. You likely will if you get serious about DXing.

It is Important not to connect power strips into other power strips unless you like to roast marshmallows in your shack over a fire. You are asking for trouble here. Also, do not use poor quality power strips. The $5 special is likely not so special when you damage your gear, or the power strip melts. I have seen this happen, though luckily not in my shack. I have even witnessed this at work in a computer lab, so it is not just a myth! Look for the UL (**U**nderwriters **L**aboratories) certification.

To deviate a bit, I would like to mention a couple of important things which do relate to the above, though very give any thought to. These would be **getting up and moving around every 20 minutes** or so as well as not getting in the **habit of snacking at your station**. One way to deal with making sure you don't forget to move is by using something like Strech Break found at stretchbreak.com. Many smart watches also give you a reminder. Sitting too long is a silent killer. **Be careful with your morning cup of joe**. Many keyboards, laptops and ham radio gear have been destroyed by liquids. I personally know a ham who even lost an amplifier due to a cup of French roast. That became one expensive cup at the end!

Next, something many forget. **Security!** Your ham gear is likely not cheap and over time you may invest more and more. If you have a corner in the garage, for example allocated as your shack, make sure you do not leave the garage door open and unattended. I know of a ham personally who lost gear from the garage shack on a hot San Jose day. He was just trying to keep cool. Keep honest people honest.

Photo provided by **Dr. Antonis Papatsaras, AA6PP.** Corner stations are very popular with hams. Simple and yet very elegant layout for his Elecraft K4D as well as the FlexRadio 6600M.

A desk light or some overhead light is a must. This does not need to be an industrial flood light. I have seen some overdo it where planes might mistake it for a landing strip. You do not want to be fumbling around on an underlit desk. You want to be able to read your radio dials. It is also helpful if your radios dials are illuminated if applicable. Regardless, plan for it when figuring out your layout. LED strips work fine too. I also use these myself. But make sure its power source does not generate RFI (**R**adio **F**requency **I**nterference). More on this later.

I am fortunate enough to have an office to use and have ham gear in. Many use part of their bedroom or guestroom. These are perfectly fine too and for sure more climate controlled if you can fit things in without causing a hazard. Basement locations are also great, in fact, in your favor for grounding and cable runs. My office is on the 1st floor. This does create some challenges, but it can all be tackled. For non-US readers, this would be the second floor as much of the rest of the world calls the US ground floor the first floor.

Lastly, noise. I do not mean RFI but AF (**A**udio **F**requency). **Sound!** If you share an area, try to pick a location where you can comfortably listen to SSB or CW. Away from music, TVs, dogs barking and so on. Even with good headphones this can be an issue as none will perfectly insulate you from all noise sources. Clearly, this is not as important if you are primarily a digital operator, as you are not listening to audio likely.

The sound of FT8 can drive anyone nuts within a few hours, though I ran across a few operators who do have the sound on at a low level. I do not.

I know hams who turn their shacks in to man caves and technically there is nothing wrong with this if not everything is running at once. You certainly do not want to be running the TV while you are calling a DX station on SSB due to background noise or trying to listen to a weak CW signal. When on the air, even if only FT8, you really should be focusing on that anyway.

A special note for W6, KL7 and JA hams. I am sure you know by now; we do get earthquakes. Sometimes nasty ones. **Secure your gear**. Don't count on the coax to hold your radio up. Though I do have to say, during the 1989 Loma Prieta earthquake near San Francisco my TV was saved by the cord, literally. I came home to find it a few feet away from the ground hanging by the coax and power cord. But you may not get this lucky. Secure your gear to prevent it from falling and getting damaged. Simple and often forgotten precautions go a long way!

Bottom line, give it some thought ahead of time as it is not that easy to re-run coax and electrical (if you need to) once you set up a shack and realize it will not work for you. Everything can be changed, but good planning goes a long way and undoubtedly will cost you less.

The Most Important Piece of Equipment in your Shack…
Contributed by Bruce Bern, MD, K3NQ

Is your brain! Your ability to work rare and sought-after DX stations is dependent more on your situational awareness, your timing, and your cunning than it is on your amp. I'm always in awe of the consummate DXer's ability to determine where the DX is listening, when to call, and where to place the transmitted signal on the waterfall. These hams also seem to have the ability to extract the last bit of performance out of their rig to pull out the weak ones. Your mental acuity is tied intimately to your health: "Healthy Mind, Healthy Body." Elevations in blood sugar, blood pressure, and other measured "performance specs" of the body can exact far-reaching and long-term damage on clarity of thought and stamina during a long DX contest weekend. If you plan to volunteer for that DXpedition of a lifetime, being in optimal health will give you the best chance of being a valued member of the team.

Although health professionals spend a lifetime learning how to help you stay fit, a "quick start guide" necessarily includes some important points. First of all, if you smoke, don't! Quit before it's too late. Even if you don't develop cancer, your vascular system and lungs will become permanently damaged, which could lead to an early stroke or heart attack. If you lose your stamina, the wet landing at Bouvet will likely prove impossible for you.

Second, keep a close eye on your health stats. Discuss your blood pressure, blood sugar, cholesterol levels, and weight carefully with your doctor. It's important to understand that, like IMD, there are bad specs, mediocre specs, and good specs. Don't allow your doctor, or yourself, to tolerate *"borderline"* numbers. Healthy *"specs"* pertaining to these four factors will extend your life span. One easy modification you could make if your "numbers" are high is to avoid snacking while you operate FT8; if you need an energy boost, try some brain-nourishing nuts. Monitoring these simple markers will help you enjoy a healthy life. This way, you can remain independent rather than finding yourself fighting with a nursing home director who frowns on antennas and DX contests. To quote Anderson Cooper's mom, "You can never be too rich or too thin."

Finally, if you want to be able to remain active long into your "golden years," get up and move! All humans require at least 40 minutes of exercise daily. Mix it up so you remain excited about whatever you choose to do…walking, swimming, running, bicycling are all great. Watch your intake of carbohydrates such as bread, rice, potatoes, pasta, and cereal, which drive up blood sugar and facilitate weight gain. Drag yourself out of the shack every half-hour or so… take a quick jog around the block, and say hi to the XYL (or OM.)

By following these simple guidelines, you will keep your most important shack accessory- yourself- healthy and invincible in the next pileup!

HF Rigs aka Your Radio

There is certainly a large selection of radios out there to pick from and for many, the selection can be overwhelming. If you ask the membership of most clubs what their favorite radio (or even brand of radio) is, you will get a wide range of answers, generally followed with an unsolicited speech as to why. Sometimes even with a convincing argument about why you need to get the one they have. I am not saying that they are necessarily wrong or right. The bottom line is everyone has slightly different needs. Not to mention budgets and in some cases even space is a factor.

I used to be a brand loyalist and at times still am with certain things. But as with everything, computers (Windows vs MacIntosh), televisions, cars, sometimes it is best to take a step back and re-evaluate your next upgrade or even the initial purchase. The point is, just because you always buy the same brand, and this brand has something "*new*" does not mean it is the right fit anymore or you even need to upgrade. Does not mean it was the right fit to begin with either. Nor that you always need the latest. **Figure out your genuine needs, work style and how it integrates with the rest of your gear**.

If you are new to HF, some might be recommending a "*beginners' radio*". This does not mean it is a basic radio or a bad radio to use in the future as you learn more. It likely means an easy-to-use radio. Many seasoned hams use a radio that is beginner friendly. So please do not confuse "*beginner*" with not so good. With some minor and notable exceptions, this is far from the truth.

Most transceivers nowadays include an antenna tuner and all the general bells and whistles which were not as common even 15-20 years ago. Certainly, things have become more computerized, even if not always SDR (**S**oftware **D**efined **R**adio).

Chances are you will find all the basics you need to DX in most any radio manufactured within the past decade. **The differences may come down to very slight changes in performance, ergonomic layout, ease of use and how well it integrates with other components in your Shack.** The last is not so much an issue anymore. You can pretty much mix and match components to your heart's desire. There is no reason you cannot use an Elecraft K3S with a FlexRadio PowerGenius XL. Or a FlexRadio 6400 with the Elecraft KPA1500 or an Acom 2000a amplifier.

Regarding built-in antenna tuners, there is something to keep in mind. While they work and give you some basic functionality with the radio and the antennas system, if you decide to use an amplifier, you will not be using it. Rather, you will need to use an additional external antenna tuner with a power rating capable of handling what your amplifier can output and perhaps a little buffer for safety.

For example, if you are running a 1500W amplifier then the antenna tuner should be rated at 1800W or so since amplifiers can sometimes output (peak) above the legal limit. This Is not to say you should break the law, but you could accidently output more than intended and you do not want to end up damaging your gear.

The reception in most modern radio gear, as I had mentioned, will be very similar and except for the weakest of DX or the most crowded contest situation, you will find a very similar performance. Selectivity and IMD (**I**nter**m**odulation **D**istortion) will start to become important if you do need to deal with this, however, more on these later.

Dual watch feature is a nice one to have on a radio, but for a DXer is a must. Sometimes this is called dual receiver as well and refers to the radio's ability to receive two frequencies at the same time or even two bands at the same time. Most SDR radios, due to how they work, will offer dual band reception out of the box. Nice to have, though you will only operate one band at a time. This does give you the ability to monitor more, however.

You can get by with a radio that can quickly switch frequencies though when you key down, nevertheless this does have drawbacks. This feature is good to have due to split operating. The 2nd receiver can be setup for the transmit frequency while keeping the primary receiver on the DX transmit frequency.

Lucky for us, radio manufacturers know this. I have yet to come across a radio from the past few decades which does not do this at least to some degree. But, if you are looking for an older bargain radio, be sure to check if this split feature is **present and easy to use**. A very few older rigs involve several steps to go into split mode. Avoid these! Should be just a simple step. Most digital splits are generally done differently, even if they are in theory split. More details on digital split operating later as this is a completely different animal, at least for some modes.

Another serious consideration is to **get something you can learn to use**. I am shocked at how many hams do not understand their rigs. I often help folks with <u>their</u> radios which I have never owned. They never did the homework or even cracked open the manual. There are many videos out there as well if one is not a reader. Use what you paid for to the fullest, or at least somewhat. You will get more joy out of it.

You want to get something you are comfortable with. There are some models of radios which have such poor menu designs and counterintuitive buttons that frankly I would love to have a serious chat with the designers. I am not sure what they were thinking and no wonder some hams can't figure them out. I will refrain from mentioning any brands or models here. A quick search on YouTube will yield tutorials on most rigs that have been produced in the last 20 years.

Another issue is poorly written manuals. This may or may not be due to translation issues, although often I feel it is. Even bigger issue in the VHF/UHF world. I have even seen misspelled menu items on radios and in one case even a button! Makes you wonder what else they missed when quality control is so low. I will not mention brands here either, but these are usually on the lower end of the price range. Though that should not be an excuse.

The must haves for a radio are having a USB connection which allows for computer control of the radio. Sadly, some leading companies are still just adding an internal serial to USB converter, requiring the installation of half-baked drivers at times. I think this is done so they can say there is a USB port, but it causes a lot of headaches for many hams. Especially those not too experienced with trouble shooting 3rd party drivers. Some big names do this still. Hopefully they will wake up and address this soon. See the USB section for more on this.

You also want to have a good DSP (**D**igital **S**ignal **P**rocessing) on your radio. This will help you clean up signals you are trying to receive. I will go over how to use this to the fullest later in this book.

This day and age a panadapter display is common, usually with a waterfall display below it. You at minimum want something capable of doing the above to at least some degree. The waterfall displays signal history not only shows where the signals are now but where they were several seconds ago or longer depending on how you set up the display speed. Having a visual helps a lot! On SDR radios, waterfall and panadapter displays are generally much better than that of a conventional receiver, if they even exist.

Having a panadapter (especially with waterfall capability) will not only make your life easier but will help you locate DX more easily. It will assist in understanding pileup situations, confirm if and where the DX is split, help you avoid QRM-ing someone else and the list goes on. You can even have your spots displayed on a waterfall with some models and some radio software, which is very nice. As a side note a panadapter is sometimes referred to as a bandscope. So, if you see this term, basically we are talking about the same thing.

The point of the section is to have you think about branching out a little. Try new and "*scary*" things, perhaps outside of your comfort zone. Talk to other users and figure out what is best for you not just based on pretty pictures in magazines or what the guy who owns it says. As I said before, chances are it is right for him. Might not be right for you.

I have been asked a few times about buying American made radios by newer hams. Currently there are two companies who are based in the US. **FlexRadio** in Texas and **Elecraft** in California. The other "*big ones*" are **Icom, Yaesu and Kenwood**, all based out of Japan. The "*big five*" as we call them, all make excellent radios. At least in the beginning and for serious DXing, **I would recommend staying with well-established American or Japanese brands**. There are some super units from both sides of the ocean. The most expensive rig is possibly, and very likely, overkill. This is easy to evaluate. Some would say that if you find yourself only using a few buttons and settings on your rig, you have likely over-purchased. I tend to "*mostly*" agree with this statement. You do not have to buy the top-of-the-line unit in 99% of the cases.

If I had to add a 6th radio manufacturer it would be **Apache Labs** in Australia. They have some amazing radios with some great specifications. I would be a little cautious with anything beyond the above mentioned as far as HF when it comes to more serious DXing. But that is just my opinion, so take it for what it is.

Support might be an issue in some parts of the world and accompanying software has also been known to be a little "*sketchy*" at times for multiple reasons. From what I have seen working in the IT world, it is not getting any better either.

As for other American owned companies, there is no shortage in ham radio. Many antenna manufacturers are based in the US for example. Perhaps the most notable company which is based in the US is MFJ. MFJ is based in Mississippi and was founded by Martin F. Jue. His initials make up the abbreviation if you ever wondered. If you can think of it, there is a good chance they will make it. They also own and sell under other brand names, most notably Ameritron, Hy-Gain, Cushcraft, Mirage and Vectronics.

One last thing. For someone who is newly interested in DX, I would say the minimum power is going to be 100W for the transceiver. This will allow you to drive an amplifier in the future with 50-60W out to full power if needed. If you get a QRP rig, or something with 20W out, you will not be able to take full advantage of your future amplifier. Though if you think QRP is in your future, go for it!

Specifications (Over) Simplified

This is creates glazed over eyes for many hams. Could be for two reasons. Not understanding it, or not caring. Sometimes both. Many hams just want to get a good radio and work DX. I get that! And sometimes you try to understand these concepts and you get hit by math formulas and pages of charts. Enter deer in the headlights. I get that too. Though I do read those… Let's try and change this by giving you only what you really need to know. The good news is the differences in these specs are not "*really*" an issue anymore due to great receiver designs or have very little perceivable difference in reputable receivers. I am not saying they are all the same, but **most receivers per my recommendation above will do fine in the categories below.**

Signal to Noise ratio is one of the key performance indicators to look for. You may see this written as S/N or SNR, same thing. This is what indicates the difference between the background noise and the incoming signal. Pretty much any radio in recent history will do OK here. You will notice very insignificant differences in graspable performance from one to the other with reputable brands, for the most part.

Sensitivity is basically how well your receiver will pick up a weak signal. This is important for DX but, again, most radios made in the past 10-15 years will do very well in this category. The above goes hand in hand with signal-to-noise ratio. Receivers from major, reputable manufacturers will do well here and any one of them will work fine. You will only see very minor performance increases, though to be fair, it could mean the difference in some extreme cases of being able to hear a touch better. Just better enough to possibly make that QSO, though expect to pay several thousand more for these units in some cases. Side note, with a poor antenna system, this will not matter much though. Can't make signals appear if they never get picked up to begin with.

Dynamic range is perhaps one of the most important things to DXers and contesters alike. Basically, this is what makes the difference in being able to receive weaker signals among the stronger ones. Measured in dB and better than 100 is great. You can make do with something around 90db just fine. That's it! You can play around with preamp and attenuator to assist you in some situations to compensate for this.

Selectivity is basically in reference to being able to reject signals or pass signals you want to hear. This is mainly applicable with nearby strong stations. This is done in a few ways but essentially accomplished via filtering, such as IF filters and bandwidth adjustment. You may see references to **IMD** (**I**nter**m**odulation **D**istortion), usually caused by mixers when two frequencies are combined. LO (**L**ocal **O**scillators) have also improved dramatically in the past 15 to 20 years. This is part of the reason why I say most modern rigs will do great! Additionally, also why I recommend something no more than this age, preferably with a built-in sound card and USB and/or ethernet port.

Harmonics may come up in literature. This refers to signals created outside the intended frequency, basically a multiple of a given frequency. These are going to occur at predictable locations depending on what the two frequencies mixed are and what cause them to be generated. I know I am starting to sound like a parrot, but the above is not so much an issue in most modern radios, certainly not the newer SDR types.

In modern HF rigs, frequency drift is also not really an issue anymore and they tend to be pretty dead-on frequency. Though can be improved with a GPSDO if needed. I will cover these later. This is <u>not</u> true for some VHF and UHF rigs. Which is why if you are driving a transverter, this becomes more important.

If you want to go for the ultimate, check out the tests performed by Sherwood Engineering. Bob Sherwood, NC0B has this great site at sherweng.com/table.html. I have spent hours (if not days) here and learned a lot about what is important to worry about and what is not so much. I will not go into detail here as this book would be 500 pages longer in no time. I really encourage you to read up on this a bit more. You can pretty much throw a dart on the top half of this list, and you will be happy. Though I recommend printing it first! Paper is cheaper than monitors.

The ARRL also does a lot of rig testing, and these can be found in ARRL's QST magazine on a regular basis. The same is true for other international publications such as RadCom from RSGB (**R**adio **S**ociety of **G**reat **B**ritain) and so on. I get both personally. Online reviews can also give some insight into the good and the bad, so do a little research. But as I had mentioned, use some caution online when it comes to reviews.

Boat Anchors, SDRs and in Between

"*Boat anchor*" is an affectionate name given to older radios, usually to those with tubes. I always think of these as collector radios, for nostalgic reasons or as a second "*fun*" radio to experiment with. I would not recommend these for everyday use, certainly not for DXing or competitive contesting. Some older rigs may not even have an accurate frequency display and frequency can drift over time. This can be an issue, especially during digital operating and near band edges.

Most current radios will fall into two categories. Superheterodyne and direct sampling. Common direct sampling receivers used by many DXers would include FlexRadio 6000 series, Elecraft K4 series, Apache ANAN series and some newer Icoms. Superheterodyne radios liked by DXers would include the Elecraft K3S, as well as many Kenwood, Yaesu and Icom units. All make excellent radios.

For years I was hesitant to go to an SDR (**S**oftware **D**efined **R**adio) and I was even more hesitant to go with a fully computer-controlled radio like the FlexRadio. Then I got to thinking. I am using Ham Radio Deluxe for my rig control program and most radio knobs were collecting dust. Literally! Do I really "*need*" knobs? After looking at specs and reading non-brand loyalist reviews, I pulled the trigger and never looked back. I would have a hard time not using an SDR radio now for HF. I soon added transverters and upgraded my VHF/UHF transceiver to SDR as well. Even my Icom IC-9700 which I use for satellite work is 95% computer controlled. Though, going "*buttonless*" is not for everyone. I understand that.

I am going to be honest here and some may not like to hear this. If you are **buying brand new, go SDR.** End of story. That is how strongly I feel about this. I know that radios from the last century, especially those with glowing tubes (valves if you are in the UK) are neat, and they really are. I think they have maybe a place as a second, or third rig in the shack like I had mentioned earlier. I think of them as a bridge to the 20th century, fostering appreciation for ham radio history. Though, should not be your primary rig in 2022 especially if you are going to be doing digital work or really want to work those super weak stations.

Some non-SDR radio units, such as the superheterodyne receivers can perform well too. In fact, some from major manufacturers are quite amazing and they just keep getting better and better. However, the advantages you get from using an SDR radio will make you convert in a heartbeat if you just give them a chance.

SDR radios are different. There is no hardware mixer. There is no intermodulation distortion. No external Equalizer, it can all be done via software. Though many non-SDR radios are starting to include this now as well. If they do, use them! SDR has fewer cables to worry about, which is a huge plus for many hams who have shacks resembling a bird's nest with all the wires in the back of their units. I can relate…

With SDRs you get more than one VFO, without having to have a second radio tuner. The panadapters are second to none. With the right skills, this will help pull out things before anyone else as well as help you find a quiet spot to transmit when calling DX in split. I can even monitor several bands at once. I do this every day with my unit.

You are also likely to notice bad operators with a good waterfall display, which drives me crazy at times. Therefore, this could be a potential negative. On the waterfall one can often see overmodulated signals, both digital and SSB with serious splatter. Some even manage to have nasty CW signals. If you are close enough to another ham, you will probably be able to see their harmonics as well when they are transmitting. This is somewhat normal with most radios, nevertheless interesting. Many amplifiers do a poor job filtering these out.

It is OK to have more than one transceiver of course per above. **You may want to have a backup radio; in fact, I would highly recommend it.** Though you likely want to have a second rig that can handle digital as well. It is where the action is and likely will be.

Photo provided by **Dr. Antonis Papatsaras, AA6PP**. Picture above is an Elecraft K3S with the optional P3 panadapter. This provides an excellent visual guide for those hunting DX. The above makes for a very nice matching set for the avid DX hunter.

One last thought on the topic. See through the slick marketing. There are a lot of buzz words being thrown around, just as in IT. This is especially true with less expensive units, mainly sourced from the far East. There are no digital resistors, inductors and capacitors. I have read this once in a radio manual and box. I kid you not! My BS detector was in bright red.

Just like when they are trying to sell you an "*HDTV Antenna*". No such thing! Marketing; that's all! An antenna is cut for a specific frequency (or range) does not care what it is receiving. It can be an older legacy NTSC signal or a 1080 signal from a local HD station. You can even connect it to your scanner. Does not matter. Older style rabbit ear antennas will work just as well as the new overpriced "*HD rabbit ears*". They are made of the same metals, and even look the same. They want your money. **Not only is knowledge power, but it can also save you money.**

Most modern radios will be a combination of the two, analog and digital. No matter how modern or what the box or market says. Cut through the noise and buzz words. Look at specs, read reviews and ask around. **You will not see this kind of marketing from reputable manufacturers.**

The moment you take your new radio out of the box, it depreciates. If you try to sell an inexpensive off-brand radio to "*trade-up*" to one produced by the "*big five*" you are going to take a financial bath. Buy the transceiver you need a year from now. If you can't afford it, buy used. Just as with cars, they are already depreciated.

Buying Used

I already touched on this a bit. It is a good way to get started but do your homework and buy from someone you know or can reach in case there is an issue. Of course, this is not always realistic. As with anything sadly nowadays, there are some scams out there. This is not to scare you as chances are you are not going to have issues, just be alert and use common sense as you would with anything else. See and try out the product first if you can. Buy from a ham if you can.

I would be a bit cautious with buying anything more than 15-20 years old and for sure anything using tubes when it comes to radios at least. For Amplifiers, the use of tubes is perfectly fine, though there are some potential pitfalls. More on this later. Having a modern rig is rather important for those who plan on operating digital modes and requires at least some degree of frequency stability.

Many older units do not have built-in sound cards, USB connectors for computer interface and as I had mentioned some older units are notorious for having frequency drift even if they are solid state. This is not just while warming up but in general sadly. Ironically there are still new units that have this issue though not on HF as much. A specific VHF/UHF unit recently released by a major manufacturer comes to mind. Drift becomes more of an issue on higher frequencies.

I have seen signals visibly drifting on the waterfall display from these older units. I know this because when I see it, I often jump on QRZ.com to see if hardware is listed. Sure enough, usually an older unit. Since these units were designed before some of the current digital modes, it was not a huge deal then when drifting occurred. Times have changed though.

If you are buying an older radio for other uses or perhaps as a second unit to just rag chew on, go for it. But don't expect it to compete fully in all digital modes if you end up with one of these radios. You will have a harder time competing for a QSO, especially via very digital narrow modes.

Keep in mind the possibility or rather likelihood of repairs as well. While easier in some ways to repair older gear due to lack of surface mount technology for example, might be more complicated due to the possibility of having to replace discontinued parts. And I am not just talking about tubes (valves) but also specific transistors and ICs (**I**ntegrated **C**ircuits) which may no longer be manufactured. Obsolete LCD (**L**iquid **C**rystal **D**isplays) can also be an issue, I personally encountered this. Had to decommission a radio as I could not find a replacement display, even used. It is hard to find reliable local repair for most radios and most manufacturer's facilities are notoriously overpriced.

Also, **some older units have toxic components**. Make sure you know what you are getting into. And if you must, protect yourself and others around you. This includes children and pets who may get into things accidentally. Personally, I just stay away.

A Second Rig

You may have noticed that I have mentioned second and third rigs several times already. I am going to re-enforce the idea that it might not be a bad idea....scratch that...it is almost a must to keep a second unit around as a backup. Or at least have access to one if needed. This is especially true if there is a rare DXpedition coming, and you really need to work them. You have heard of Murphy's law, right?

The unit should preferably be not too old, but something you can easily put inline in case you experience problems with your primary unit. This is not an issue for most seasoned hams I know as there is a tendency to hoard gear. But something to keep in mind in case there is a filler needed, or God forbid even an ATNO (**A**ll **T**ime **N**ew **O**ne) that surfaces as your main radio is in the shop for repairs or all of a sudden starts experiencing issues. For a serious DXer this could ruin your year. There might not be another activation for years to come or even decades!

The Bare Minimums you will Need

Before I get into the rest of the hardware, let's look at a sample shack layout. I am doing this now, so it makes more sense when I am touching on individual components later.

For a basic "*barefoot*" (No amplifier) setup you will at **minimum** need:

- 13.8VDC power source / power supply
- HF Radio (Ideally with an antenna tuner)
- Basic Antenna (Dipole, vertical, etc.)
- Coax Cable, RG-8X or better. Preferably better.

For a Legal Limit (1500W in the US) setup you will at **minimum** need:

- <u>All of the above</u> PLUS
- 220V power source for the amplifier (Can be lower if 500W or less)
- HF power amplifier

- Legal limit antenna tuner
- Extra coax cable to connect the above. Coax will need to be able to handle added power. LMR-400 coax or equivalent recommended. Ladder line will also work. More on this later.

The antenna tuner will be needed since it goes <u>after</u> the amplifier. Therefore, if you run legal limit, the internal antenna tuner on your radio will be pretty much of no use. If you are only going to run 5-600W you may be able to get away with no 220V outlet. If you are in the EU, you are all set as you guys don't mess around with lower voltages. Lucky you!

Radio Accessories

I am going to be switching gears for just a little bit. Yes, I am putting computers under radio accessories first. I know…there is complete anarchy in this book! But trust me on this one. Many issues hams experience leads back to computers, specifically to the things I am going to be touching on. Currently, **almost everything you do will involve computers** and this includes the world of ham radio. From working digital, electronic logging, callsign lookup, DX clusters, transceiver control, rotator control, radio programming, and the list goes on. I think you get the picture.

Computers – MacIntosh vs. Windows vs. Linux

I am going to save you the suspense regarding using Windows versus the Macintosh debate, at least. **Pick the operating system you are comfortable with; you can make it work just fine**. The days of Windows only software is behind us now, so use what you like. Yes, there are more Windows software options, but do you really need 50 software logging titles and 30 propagation prediction titles which mostly pull data from the same source?

As far as the debate about safety and stability is concerned, I will put that to an end as well. Neither is perfect. Nothing is. I do IT for a living and worked on both equally as much. You can get a virus on the MacIntosh. Some will tell you no, I have seen it. Hardware failures can occur on both as well obviously, sometimes rather catastrophically. Some find it much harder to fix hardware on a MacIntosh due to design and build. Same is true for some windows-based machines, especially some laptops. Of course, we all know about bad updates. Happens on all platforms occasionally. Again, it comes down to your preferences, but would not base my pick on the above misconceptions.

Linux, if that is something you are comfortable with, is just fine for ham radio use applications as well. Sometimes even better than a Windows or a Mac to be honest but that is for another book. Ubuntu seems to be a very popular distro for hams users. Check out the software section of this book for more detail as I do cover some Linux software as well. I run all three platforms by the way though my primary machine is a Windows 10 desktop.

Geriatric Computers

I would not recommend using a "*leftover*" or spare computer, as the likelihood of issues is high even if you reinstall the OS (**O**perating **S**ystem) and upgrade them a bit. I heard countless stories of folks in the era of Windows 10/11, complaining about how their 8-year-old bargain laptop which they paid a whopping $300 with Windows 7 is acting up. No kidding people! And yes, I know that many hams are perceived to be cheap. That is a stereotype I have heard many times. Though I must say, I tend to see the two extremes. Nothing wrong with cutting some corners but your computer and antennas are the two where I would not do so with.

This brings me to this important and ironic point. **You are likely going to spend thousands on ham radio gear. Can you spend a small fraction of this on a decent computer?** Something to consider. You may have to spend significant time troubleshooting, assuming that is something you are comfortable with, and time is money. So just spend a little money upfront and get more enjoyment out of the hobby. Save yourselves headaches, possibly lost logs, software crashing in the middle of a rare QSO or having to ask others for help with random issues. As I had mentioned, I work in IT for my day job and do computer consulting on the side. I make a lot of money fixing stuff for "*cheap people*" who buy something for a few hundred less and end up giving me more in the long run fixing it, not to mention down time. Hey, more radio money for me!

One last word on selecting a computer. Look for **metal cases** or at least a well shielded case if you are getting a desktop. **This will reduce the chances of RF escaping the case as well as decrease the chances of RF getting into the case.** If the case is well shielded, the rest of the RF can likely be controlled with ferrites, bonding, and other methods. More on these later. As for laptops, this is obviously not an option but as far as RFI, they can be all over the place. Might be good to ask "*those in the know*" what they had found to be RF quiet. Ham radio forums and DX clubs are always a good start.

Ethernet vs. Wireless

Ethernet is for more than just the internet. Yes, your ethernet cables are used more now-a-days then just to surf the web. You can also use it to interconnect your computers, IoT (**I**nternet **o**f **T**hings) smart devices, NAS (**N**etwork **A**ttached **S**torage) or servers, and so on. They can be used to interconnect radio systems, amplifiers, as in the case of the FlexRadio systems components. Pretty much endless! These are not going away. Wireless is popular and convenient, but not always ideal.

If you are going to be using ethernet to connect your gear, I do recommend using a good quality switch with metal housing. Not the cheapest thing you find on sale or used, though does not have to be expensive. There is a lot of junk out there and you will be frustrated when connections drop, or the switch overheats. I have seen this more than once in the field. Having worked for over 3 decades in IT, I can tell you 90% of the problems come from bad cables and bad switches. Use good quality and current standards for both.

If you must run cables under the house or outside and have a known rodent issue, be extra careful. Try to route them carefully as rodents seem to love to eat plastic. This is also true for coax too by the way. This will cause headaches of course, and not for the rodents. One does not have to live in the country to experience this, I have seen a bundle of cables some rats have feasted on in downtown San Francisco. Nobody appears to be safe!

For ham radio, I would go a step further than your basic cable and say always use shielded cables and get a switch in a metal enclosure as I had mentioned, especially if you are going to be running some power (use an amplifier) or get serious about weak signal work. Will reduce chances of RFI issues arising.

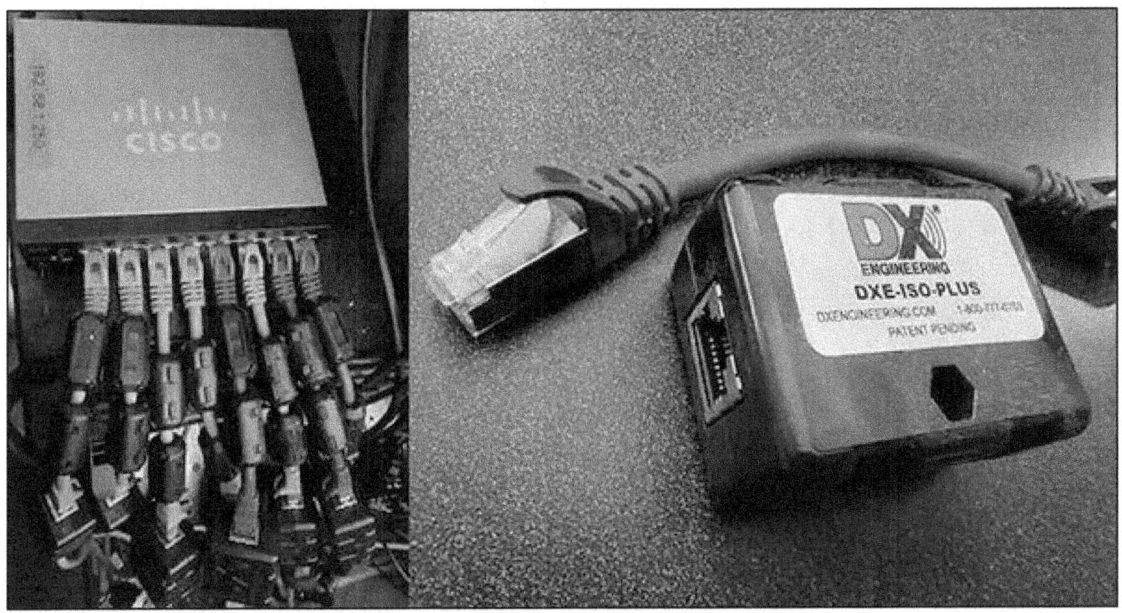

An example of an ethernet filter made by DX Engineering is pictured here. I own so many of these I feel like I should have bought shares in the company. I use these on both ends of the ethernet cables (per manual) and I go a step further and even snap on some extra ferrites for good measure wherever I can. Especially near my ethernet switch (left).

Paranoid? Yes! And since I also do low band DX and VHF/UHF DX I feel these may add a little extra. Is it just in my head? Maybe. But it helps me sleep at night knowing I did everything and then some. The Cisco smart switch I use has a metal case and I also added ferrites near the wall wart (DC adapter) as well as where the DC power enters the unit for good measure. I will cover ferrites in more detail later in the book.

This switch even has a fan, so not worried about heat. If you are a heavy ethernet user, I recommend getting one with a fan. It will last longer and less possible issues. I would stay away from switches in a plastic enclosure even if you are not a ham.

A lot of hams I run across have IT issues, so I will touch on this a bit here. Much of it is networking related. It seems to be one of the "*big three*" I see. There is a place for wireless, and it is great, usually fast, and convenient….and <u>way overused</u>. If you want to run wireless for your laptop and a few things, great! You should for sure. But keep in mind that if everything in your home is wireless, you absolutely will have issues with speed due to packet collisions.

For those not in IT, your devices use packets of data to exchange information. When they try and send these all at the same time, nobody wins. This is kind of like a DX pileup. This might be caused by frequency/channel interference, and you may not even realize it. You can resolve some of this by using both Wi-Fi bands (2.4GHz and 5GHz) but keep in mind it is not just you on the planet. If you live somewhere with neighbors, and 99.9% chance you do, they are also using wireless. Likely, like you, for convenience and everywhere. You will certainly share these frequencies in many cases and additionally these frequencies often overlap. If you have 6 houses line of sight, statistically speaking you will likely be packet colliding with at least one of their Wi-Fi systems. Likely more than one.

Keep in mind that everything from your smart thermostat, security cameras (and sometimes even alarm sensors), weather reporting devices, smart speakers, Ring doorbells, Roku, smart TV, tablets, laptops, Kindles, smartphones, gaming stations and so on are all fighting for the same bandwidth. Don't even get me started about Bluetooth!

So how do you address this? If you can use wired ethernet if it is not too big of hassle for you. Just plug it in. Easy! Turn off Wi-Fi on the device in question and know that you are getting much better speed, less drops (if any), better security, and now less traffic on the already overcrowded wireless network. Your other wireless only devices will thank you too! There is some confusion about this by many; **You can seamlessly mix wired and wireless on the same home network.**

Another thing to check is if you have a wireless router, make sure it is set to select a channel for you automatically based on traffic. Many routers, especially those from an ISP (**I**nternet **S**ervice **P**rovider) ship set to a specific channel. Many times, this is 1 or 6 (middle) out of a dozen or so. Imagine if we all tried to talk at the same time on the same frequency on a radio repeater. You get the point.

So why I am harping about wireless so much? Have I forgotten the book title? No. I follow many forums, ham radio, and IT and this topic comes up a lot. Save yourself some headaches and take the extra steps, you will enjoy the hobby even more if you do not have to troubleshoot IT gear. You can spend that time on the air instead, chasing DX.

I have helped hams and clients in the past who complained about issues with their internet and were quick to blame their providers. 50% of the time it was a wireless issue, maybe 5% ethernet and the rest was usually RFI (at least for the hams) which can be resolved with ferrites and a few other adjustments. And yes, the RFI issues at times did relate to both wireless issues and hardwired/ethernet. I will cover RFI later in more detail. Only a small fraction of the time was it the ISPs fault.

The 9pin Nightmare

Yes, I said it! I know some will say, *"but I love my serial ports!"* I am indifferent personally, and I will not stop anyone from using them. I do find that these seem to cause serious frustration for some hams, only second to audio and internet issues. To be clear here, I am referring to the physical cabling here.

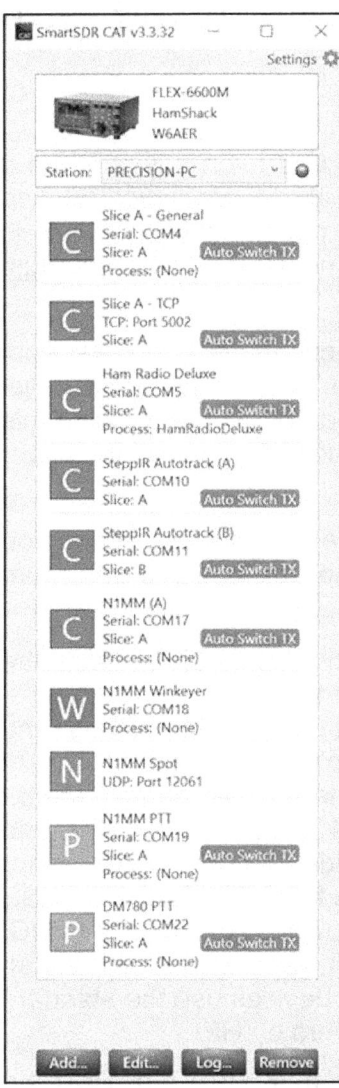

Serial ports or RS-232 are slowly departing the shack, following in the footsteps on the even more clunky parallel port. If you are over 40 you likely recall those. If not, you are lucky and did not miss anything. Thank God for being replaced with USB and other modern alternatives. Some may disagree with this, but I am still surprised to see these DB9 connectors on ham gear. I know they are there mainly for legacy reasons but this day and age, it becomes an issue finding PCs with serial cables and on the Macintosh, it is non-existent and has been that way for a very long time. Time to move on? They were generally very slow, often confusing because of null modems cables being thrown in the already confusing equation. Null modem is basically where the receive and transmit lines cross, at least to keep it simple here. Of course, half the time they (or you) did not bother to mark them as such.

And certainly, there are gender issues, but not what you think. Male and female connectors is what I am referring to and they never seemed to be the ones you wanted. And let's not even talk about 9 pin to 25 pin adapters, some of which were <u>also</u> crossover adapters. So, you constantly need a box full of adapters.

Wouldn't you rather just plug in a USB cable and be done with it? If you have the option of using a USB cable vs serial cable, clearly USB is the way to go.

I do, however, in the interest of fairness, must note that there are times when these connectors or similar-looking connectors are used. They may not be serial ports. Sometimes these have a different purpose or in rare cases cannot be changed to a newer standard. So be very careful not to assume all DB9 connectors are always serial connectors and as you can see there is a place and time for them. Just not where they can easily be upgraded to USB.

Now, there are also *"virtual"* serial ports, and I am perfectly OK with these. In fact, I do not see these leaving for a while. Sometimes these are used internally and at times you may even need to create a virtual share to control a device. Perhaps to allow two different pieces of software access to a piece of hardware or other software. Com0com is a very good option for a null-modem emulator software. You can find it at com0com.sourceforge.net.

If you are using FlexRadio's SmartSDR, they did a nice job with this as seen in this image. Ports can be virtual or physical and are tied to "*slices*". I edited this down a bit for simplicity as I have about twice as many ports, but as one can see your imagination is the only limiting factor. In this example, my SteppIR antenna controllers are physical connections, COM 10 and 11. They move in sync to my operating frequency. All others are virtual. I can create new virtual ports in seconds and edit others on the fly if needed. These also never seem to be affected by windows updates. Can't say the same for the physical ports!

USB 101 for Ham Radio

It is rather hard to miss all the different connector shapes and even colors when it comes to USB. In addition, there are many specifications, these are indicated with numbers you may associate with them. These are not directly related. USB is unavoidable in ham radio. Understanding USB to at least some basic degree is a must for hams.

USB 1.0 and 1.1 are mostly obsolete, though you may still encounter it in lower cost items. These started the revolution at the end of the 1990s from serial and parallel cables to where we are today. You may still run across keyboards and mice using this specification but that is about it. USB 1.1 is painfully slow compared to newer specifications. Besides the above, avoid them!

USB 2.0 is the most common (as of 2022) and will work fine for almost anything you throw at it. USB 2.0 is up to 40 times faster than USB 1.0 and clocks in at a maximum of 480Mbps. Though this is rarely seen.

USB 1.1 and 2.0 are usually associated with "A" and "B" style connectors. "A" is the computer end and the "B" is the peripheral end, such as a transceiver for example. Likely type used if your radio is USB controlled. This might be used to establish the CAT control and possibly carry audio as well from the radio's internal soundcard. At times some radios just have a serial to USB converter internally to "*modernize them*". Which in turn might require additional drivers. Type-A and B both have "*mini*" and "*micro*" variants, although these are not as common outside of phones with minor exceptions. Some SDR dongles, such as the **Airspy** and the **HackRF** use micro–USB type-A for example. USB 2.0 can supply up to 500mA of power to devices at 5VDC. Basically, up to 2.5W therefore not requiring additional wall warts (power adapters). This equals less wires and less potential RFI. The **SDRplay** devices use the standard type-B interface. Some hams use these as secondary receivers on HF.

The newer USB 3.0 increased the speed to 8 times faster than USB 2.0. USB 3.0 clocks in at a maximum 5Gbps. Again, rarely seen. It is closer to about 5 times the speed to USB 2.0 in most instances. Actual top speed depends on cable quality used and other factors, mainly peripheral hardware specifications. **You can easily identify the type-A SS (SuperSpeed) connectors by their distinctive blue color inside the connector.** These also feature 5 extra pins in the same form factor.

The matching type-B SS, which is slightly larger than the type-B used with USB 2.0 specification is somewhat rare outside of USB 3.0 external hard drives. On smaller external USB 3.0 hard drives you may encounter the mini-B SS. These will include the extra 5 pins on the side. Basically, **SuperSpeed aka SS connector, uses 9 pins vs the older 4 pin connectors.** There is some backwards compatibility to 2.0, but at a slower speed. USB 3.x can supply up to 1.5A via the cable over type-A and B and initially 3A over Type-C. This has been increasing though via specification updates.

USB-C also has a connector/plug shape that is unique. It now features 24 pins, 12 per side. Although, they are two sets of the same 12. Has a reversible connector, meaning you cannot connect it upside down by accident. In fact, there is no upside down, which is rather nice. **I think they finally got it right!**

With USB-C, enter USB 3.1 specification which is where you see these connectors used. However, you can still use the type-A SS and type-B SS connectors as well with certain specifications but only up to 10Gbps.

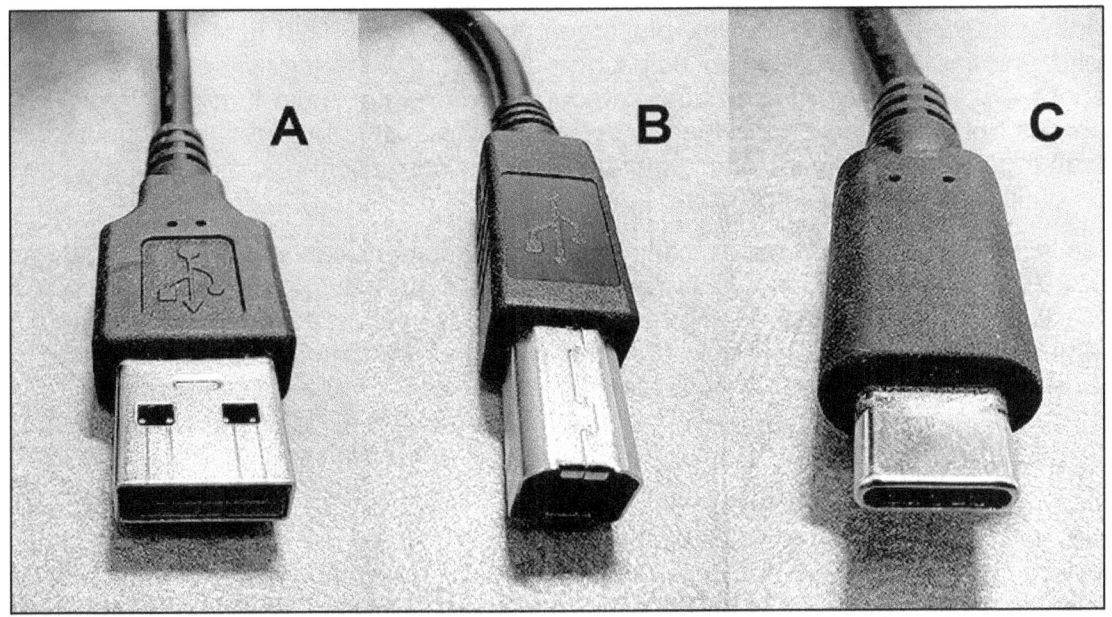

Above are the connectors you are most likely going to encounter in the world of ham radio. Type-A and type-B have a blue variant to indicate standard. Type-C is where everything seems to be heading. You will likely see more and more of these connectors.

There is also a USB-C DP standard which is for connecting monitors via the DisplayPort. Will also carry audio, much like some HDMI cables. **This is a great way to connect a monitor to a laptop in your ham shack**. Also, less RF noise than VGA cables according to some sources.

USB 4.0 is here, but you are likely not going to find anything yet using it. This standard uses the type-C connector and now claims a theoretical speed of 40Gps. There is already version 2 announced in late 2022 at 80Gbps.

If you are adding a USB extension card to your PC, be careful. These are also not all the same. There is a reason for price differences! Ideally if you can get one with a separate channel as they are called, meaning there is an extra processing unit onboard to take some of the load off your motherboard, get that one. I had issues with some SDR devices when using them via a USB hub. Sometimes the amount of data is just too much, so a separate card is needed. I would also recommend going with at least a 3.x version and maybe one which also provides a USB-C connector to future proof it. **You will start seeing USB-C more and more in ham gear!**

Monitors and their Placement

I have already talked about the importance of your desk and chair. You will spend a lot of time here looking for and hopefully working lots of DX. Monitor placement is a reoccurring issue with some hams. I think it is very important to address it. I have seen two extreme examples of what some hams do. One is not having enough monitor real estate (very common) and the other is a setup where the ham shack looks like a stock market day trading station. When using 6-8 large monitors I suspect it is mainly just for show and not much for function. But perhaps some of these are dual use desktops.

Another major issue with monitor placement is where you must turn your head in a weird uncomfortable way when adjusting your radio or generally operating. If you cannot use your station for an hour without neck strain or must turn your head 180 degrees to log a contact, you need to seriously reconsider your layout. Focus should be on ergonomics and ease of use, not the "*wow factor*". Your neck will thank me later, though likely not your chiropractor.

Ideally, you need at least one monitor which you can comfortably read with at least an HD (1920x1080) resolution. 4K is ideal if possible and I would recommend at least a 27" (68cm) monitor. Reason being real estate for what you need to see. Meaning your decodes in case of FT8, logbook and other things you need to keep your eyes on. If you find that your monitor is hard to read at this resolution, frankly, just get a larger one as you will struggle with getting all the content to show. They are not as expensive as they used to be even a few years ago. Also, two is better than one!

When running 4K there are some things to consider. You do want at least a semi-decent video card to go along with it. Likely this is not an issue if it is a newer Windows PC and on the MacIntosh this is seamless. The iMac does this by default, even 5K is available. You can also attach a 2nd monitor to a Mac as well. There is even a way to use an old iMac as a second monitor if you have one lying around.

Another thing to keep in mind is that some older applications you may need to adjust the DPI setting on if it does not scale (example font is too small) to what works for you. This might be the case for some programs written before 4K was around. Not a big deal, easy to do. Usually in the properties sub-menu.

Your monitor(s) can be oriented vertically, versus the standard horizontal. This is often done by desktop publishers and magazine/newspaper editors. I did this for a while and found it much easier to read DX clusters. Give it a try and see if having one monitor oriented vertically works for you. Very easy to adjust in your display settings and all modern operating systems support this feature.

Pictured above is my 2020 setup. I run two 4K 27 Inch monitors with 150% scaling. Meaning I get the resolution of 4K but the applications and the text are too tiny to read without this adjustment. This leaves me plenty of real estate to monitor SmartSDR, 2 instances of WSJT-X and 2 JTAlert programs at once with room for the cluster on the side as well as other applications as needed. Functional and just enough for the moment at least.

Radio Audio Interface

Most modern radios will include this internally, therefore you will not need to worry about an external unit to connect in case you want to work digital modes. However, if you do need one, there are many options out there for you. The good news is that once your hardware is set up for one digital mode, you are set up for almost all of them. You will likely only need to worry about the software which we will cover later.

The **SignaLink line by Tigertronics** is perhaps the most recognized brand for this purpose. I also feel it is the best all-around, especially for the money, if you need to add a sound card capability to your radio. These will run you around $140. Again, you will only need this if your radio does not have the built-in capability. You can read up more on these at tigertronics.com.

Normally, these are just literally jumper wires. In some cases, like mine, I had to use it with a Yaesu FT-847 and this required a capacitor and a resistor for the interface as seen above. The SignaLink unit is extremely versatile, and you can likely make work with any setup. You can also buy pre-made cables and jumper modules for this unit.

The above illustrates the SignaLink front interface and the image on the right is where you internally adjust your jumpers as needed for your specific configuration. They also sell preset plug-in modules.

Some hams are using the time tested, **TimeWave Navigator** as I did for a long time as well. This is a "*fancier*" interface than many others and these are widely popular. The price is close to $400. Their website is at timewave.com. Some of you may recall these as AEA, Advanced Electronics Applications.

Rigexpert also carries a very similar interface, read more about this at rigexpert.com. **MFJ** also has a solution in the MFJ-1205 series which works very well I am told, though I have personally no experience with it. These are around $130 and will get the job done. We can find MFJ at mfjenterprises.com.

West Mountain Radio has a variety of great products as well. Their RigBlaster lineup is outstanding if your radio does not already have a built-in sound card. Check out their lineup at westmountainradio.com.

If you must run an audio cable (vs. audio card built into the radio), it is slightly more likely to pick up RFI. I would recommend using a shielded cable for the audio, as well as for the USB cable if not already doing so. Additionally, add ferrites on both ends of the cables or buy USB and Audio cables with ferrites already on them. These are widely available on Amazon and are cheap. Then add extra ferrites.

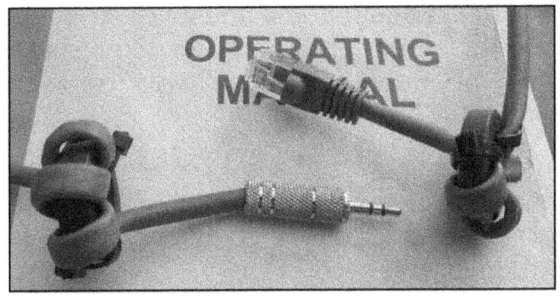

There are two schools of thought on cables. They both have a point, and neither is right or wrong in my opinion. One school is buying cables as short as possible to reduce chances of RF pickup. This makes sense, sure.

The other (the one I follow) is buy a slightly longer cable then needed, wind the ferrite with the cable on both ends to shorten it to what you need and what you end up with is the same length of cable with the added benefit of a ferrite or two…or three. Each time you loop the cable thought the ferrite increases how well it performs which is why you see this sometimes. More on this in the ferrite section, please stand by. In fact, these guys will keep coming up throughout this book for a good reason.

For ground loops, the best thing is to find the cause and/or use an isolator transformer, such as what you would use for any other audio issue. But frankly, I would recommend this only as a last resort. Best to find and resolve the source of the problem if possible. If all else fails, something like this will do the trick. Basically, these have a 1:1 transformer inside to provide isolation. Some do both, solve the issue and use one of these for good measure. That is perfectly fine too.

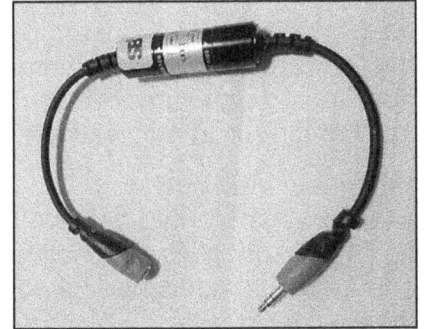

Headphones, Microphones and Speakers

Which combination is better? Depends on your style but you will likely draw the same conclusion as me at some point if you get more serious about chasing DX. I personally started my DXing career as an "*armchair DXer*". Sat back in my chair with a speaker and a boom mike. It felt comfortable, sounded good. Or so I thought. I then, just to mix things up, started to use headphones. Next thing I know, the speaker never came on. Then I **switched to a headset and never looked back**. A headset is basically a headphone with a microphone.

I no longer have to worry about outside or inside noises. Also disturbing my spouse or even neighbors at odd hours is no longer a factor. I do find I hear better, and I love the ability to move my head freely. Frankly, it also takes up a lot less space, which is an added benefit. But this is just one person's story, you may find this is not for you. Though I would be rather surprised.

Headphones good for music might not be the best for your HF station. Yamaha CM500 is a very popular headset as is the Koss SB-45. Both are priced at well under $100, have great audio and are super comfortable. You may also want to look at headsets used in the communications industry. I have heard of folk using military surplus headsets as well as those units retired by the airline industry. I never tried either personally but makes sense to me as they were designed for speech after all.

Some better-known brands all have their own advantages. Some sound a hair better to me, both for listening and especially microphone performance purposes. Some have more comfort, and this is also very important to me. Be sure to try them on first if you can.

Be mindful of your audio levels. Many hams already have some level of hearing loss, and you do not want to make things worse with excessive volume. Note that a receiver equalizer can be set to compensate for hearing loss if you have an audiology test in hand. There are online hearing tests which will help guide you in the right direction.

One thing I will note is the phasing available on some Heil headsets. This is something I love and find very useful. Basically, it is a switch on the headset which allows me to flip the phase of the incoming audio signal on one side. It sort of shifts the signal in your head 3 dimensionally if that makes any sense. Makes it easier to dig out weak SSB at times and is especially useful for weak CW. This is something you can also do yourself on other units if you have the know-how.

I still have a set of speakers connected to my main transceiver, though I do not recall the last time I powered them on. Some manufacturers sell nice looking matching speakers to complement their radios. I owned quite a few in the past. I ended up selling most of them and reinvesting that money into better antennas.

You may also find that you also want to use different microphones for various situations. This might be for different radios and tasks or even different modes. Such as DX versus ragchew (chatting) or SSB versus FM. You may have to use more than one as some radios need a completely different element. Some Icom radios come to mind.

Some settings can be adjusted via the transmit equalizer (not the receive, that is for listening) in your radio. Most modern radios have an equalizer, though it can be hidden in a confusing menu system and at times be very limited. Regardless, use them!

Microphones are one area I would not skimp if you are planning to use phone mode a lot for DX. Some great microphones, as well as headsets with microphones, can be obtained from **Heil** at heilsound.com and **RadioSport** found at arlancommunications.com. The **Yamaha** CM-500 and **Koss** SB-45 can be obtained from many online sources. I use multiple brands interchangeably. All units mentioned here are well made, though will differ in price. More expensive does not always mean better. As I always say, what works for me, may not work for you. Try things!

I currently like to use my RadioSport on my FlexRadio and my Heil on my Icom 9700 for satellite work. I also have a Sennheiser headset I use when I do not need a microphone, like when doing CW. I do realize that Sennheiser is more designed for music, but the comfort is second to none and with a receiver equalizer, it is just fine. The more variety the better I suppose.

For weak CW, I would not even consider using a speaker anymore. I do adjust the receive equalizer though to focus on 250Hz to 1Khz. For SSB I drop the lows a bit, under 300Hz and cut the highs above 4Khz. I also completely drop everything on the higher end, sometimes even everything below 2800Hz. This helps with white noise.

Works for me, though I do adjust these occasionally depending on noise levels, conditions and what I am doing.

To test the Microphone on the air, you have some options. Find a local ham who sounds good on the air and ask them to listen to you. The sound good part is important and here I mean a DXer preferably. You are not looking for high fidelity, you are looking for a punch but <u>with</u> clarity. Newer radios also offer a "monitor" function. When adjusting your microphones, you can set your transmit power to zero and then press the PTT and hear your own voice as it would be transmitted. Furthermore, you could adjust the compression setting (more on this later) until you are happy with what you hear. This will be accurate enough for most users.

You can also visit one of the many online SDRs and use headphones on your computer, transmit on a clear spot at just enough power to hear yourself. As in, leave the amp off. Find something closer to you if you are only testing microphone(s). Some good online SDRs can be found at websdr.org which has a list of various online SDRs around the world. This is where I usually look first. The site globaltuners.com is a similar concept to the above. Also offers a premium membership. kiwisdr.com/public is a list of folks sharing their KiwiSDR receivers. And lastly, airspy.com/directory is a list of users with Airspy SDR devices, which also happen to be one of my favorites.

The above is the station I used to use quite a few years back before switching to microphone headsets. Here I used a Heil Goldline Pro boom microphone, which I regularly got complemented on the sound quality. Got my first 100 phone entities using this microphone. I used the boom microphone with headphones sometimes still despite the speaker you see under the Yaesu FT-736R VHF/UHF rig on the right.

This is also a great way to compare not just microphones but also settings, such as microphone gain, equalization, and compression levels. There is really no excuse for having bad audio on the air.

Never ask "*how do I sound?*" when a DX is trying to work a pileup. I have heard this more than once. Not only is this going to get you on the naughty list, but they likely just want you to move along and tell you it is fine. So, you have learned nothing new and now everyone's time was wasted.

When I figure out how far to be from the microphone, I use my "two finger rule". This works for both the boom mikes and the headset mikes. Same as pouring 2 fingers of Scotch, 2 fingers side by side or about an inch to an inch and a half distance from the element when speaking. Then I adjust the microphone gain accordingly. When I am about to operate, it is very easy to check proper distance by just simply using my fingers if needed.

Especially with the boom microphone, you must be careful not to be too far. I sometimes hear folks who sound "*distant*" almost like they are transmitting from a large church or hall. Generally, this is due to being too far away and overcompensating with the microphone gain. Add VOX to this and you get the instant recipe for a serious audio mess.

A **foot pedal** for keying the microphone is nice to have if you find yourself working on SSB (**S**ingle **S**ide **B**and) often and calling DX stations. I would argue it is an absolute must if you are a contester as well. This really frees up your hands to do other things, such as logging or fine tuning the station. I think this is a much better alternative than using VOX at least when calling DX. There are also hand operated PTT (**P**ush **T**o **T**alk) switches to be used with a microphone and some have even made their own by simply using a momentary push button switch. I own and use both interchangeably. I generally do not use VOX though many serious contesters do. If you are on the air for 24-36 hours in a contest, certainly no replacement for VOX. More on this later.

Antenna Tuners & SWR

SWR is sometimes called the enemy. In some ways this is true and in other ways, if you know how to mitigate it, not so much. We can debate this but here are the facts. **Antenna tuners "*trick*" your radio into thinking the antennas is properly tuned.** They will allow for your amplifier or radio to operate at maximum power. But, if the antenna is mismatched, there will still be losses in the feedline. There is no free lunch and tuners can't fix a bad antenna. Sometimes antenna tuners are referred to as an "*ATU*".

There is a good chance you may need to use an antenna tuner at some point. There will be some losses when you do, even if in bypass mode, though minimal. This is in part due to the internal components. It's just simple physics. Therefore, if this is an issue for you, try to use a resonant antenna when you can. But I do understand this is not always doable.

Do keep in mind that power loss from high SWR is real in way of reflected power. That is literally what it is. Reflected power coming back to your transmitter and it will be dissipated as heat. This in part explains warm coax cables.

I like to operate at no more than 1.5:1 SWR and even this will cause a 4% or 0.18dB loss. Once you get to 2:1 you are talking about 11% or 0.5dB meaning that if you are running 1kW, over 100W is being reflected back to your transmitter. While this sounds like a lot, and it is a lot, not the end of the world. But there is a caveat.

Keeping in mind that 100W is more than an average ham runs. If you have very high SWR and running legal limit, this much power (or a whole lot more) can be coming back to your amplifier, warming up the coax, the balun (which can crack or worse), and causing other issues in the shack. Therefore, I underline the importance of not buying the lowest power rated components you can get away with to save a buck if you are serious about DX and going to the legal limit.

One way to remedy the above SWR situation is with antenna tuners. There are many types out there, not to mention makes. **Automatic versus manual** and **in-shack versus externally mounted** are the main categories which come to mind. Even within the automatic tuner types, there is **roller induction and relay switched**. Both have advantages and disadvantages. It all comes down to your needs, speed preference

(automatic roller induction is slower) and what exactly you are trying to accomplish. If you have an internal antenna tuner, this will likely be a relay switched automatic type.

If you have an internal antenna tuner in your rig as well as an external one, do not use them at the same time. You will likely need to obtain a higher power rated tuner if you add an amplifier. At this point, your internal radio tuner should not be engaged at all. It should be set to bypass mode.

Over the years, I have owned most major brand antenna tuners, **Palstar, LDG, MFJ**, and the **FlexRadio** TunerGenius XL (pictured here). There are many other great brands, certainly not a monopoly by the long shot. I have also owned both external and in-shack, automatic and manual, roller induction and relay switched varieties as well. Therefore, I can honestly speak from experience. As far as failure, both types of antenna tuners can fail. Relay switched tuners can have relays sticking or altogether failing though not that common with high power versions. I suspect some of this is due to too much power being applied and/or extreme SWR situations.

As they say in real estate: Location, Location, Location, and there is truth to this also with antenna tuners, tuners, tuners. Sorry, could not resist. Externally mounted antenna tuners generally will work better for most. Although this is not always practical. This is because they are the closest to the actual antenna. The problems with these are the lack of selection, power handling, and without going into details, sometimes also lack

of quality. If you are looking to match an antenna which is more of a challenge, you will likely have better luck with roller induction. These use a large variable capacitor(s) and an inductor to find the match. I also find they are much better suited for the lower bands.

The **Palstar HF Auto** is the one I would recommend in this category, pictured here. This unit is motorized and will find a match for you. Not the cheapest, but well-built, looks nice and has great support. Be ready to give up some space though, it is not as small as some of the units from other brands, or the relay variety but worth the space in your shack. Palstar can be found at palstar.com.

For roller induction you may need to do some maintenance if you notice matching issues. Usually, the manufacturer can recommend or even supply you with "*magic grease*" as one ham friends of mine likes to call it. This is a mixture of conductive and lubricating compounds and makes the tuner perform like new when re-applied. So, before you toss your tuner, or start to curse it out, investigate this first.

If your tuning speed is important, and your antennas are low SWR already, then relay switching is maybe the one for you. Some claim to be fast and maybe they are fast switching, but it may take a long time to find the right match the first time. So be aware of this. Once saved to memory, usually you are good. Though antenna tuners are speed demons, like the TunerGenius XL. This is the fastest relay switching unit I have ever owned or seen. This is also available in the SO2R variety, meaning single operator two users. Relay switched are the most common variety of antenna tuners out there. They basically switch inductors and capacitors in and out in combination until they find the right match. Then this stays in memory, theoretically.

I recommend automatic tuners as the time it takes to adjust with the manual tuner may result in DX being gone by the time you match. Though some seem to do it faster than I could ever, so take that with a grain of salt. If you do use a manual tuner, it is a good idea to write down settings, or mark the setting on the unit itself. This will speed you up a lot. Do double check occasionally as setting can change, even just based on weather! Rain, fog, snow are all factors.

As I had mentioned, one way to overcome SWR issues is to use resonant antennas when possible. This could be as simple as properly adjusted dipoles. That is one of the main reasons I started using SteppIR antennas. No traps, always resonant as the element changes not only based on the band but also on frequency therefore very low SWR on all bands and all frequencies. Monobanders and log-periodic will do similarly well but not always an option due to space, as in my case. On the lower bands, it is impossible to have an antenna with limited space which will resonate perfectly for the whole band. The 160m band seems to be the biggest issue of course. Even with my current setup, this is the only one band I still need to use an antenna tuner on.

Never adjust your antenna tuner on top of the DX. Just as you would not tune up your tube amp on top of the DX...I hope. You do not want to QRM those trying to listen and make a QSO with the DX. Move over a bit to a clear spot. You may think you are running low power, for example, and perhaps so, but keep in mind that even 5W can wipe out a weak signal to stations even just a few states away from you.

Power & SWR Meters

While modern radios, antenna tuners and amplifiers have built in power and SWR meters I still like to have a separate one to be able to glance at quickly and double check for issues. The ones I prefer are the **WaveNode** brand power meters, found at wavenodedevelop.com for two reasons. They have very large visual displays with colorful LEDs where I can easily tell if something is wrong even out of the corner of my eyes. Don't have to look for needles moving with my bad eyes. Secondly, they have a

variety of sensors I can use, all VHF/UHF bands and even for the 630/2200m bands. This is great as they do double duty for my non-HF gear as well. Additionally, both the build quality and the support on these is outstanding. I use two of these on my FlexRadio one for each side on the SO2R setup. Pictured here are the Wavenode WN-2d units.

Some of the other power & SWR meter units are analog using cross needles, others are digital, and some are even a combination of the two. Other units which you may want to consider also are **MFJ/Ameritron, Daiwa** and **Palstar**. If you are looking for the "*top of the line*" in the above with some extra features, **TelePost, Inc.** has the answer at telepostinc.com mainly their flagship LP-500 and LP-700 units. These do a lot more than just power and SWR and have a very attractive TFT (**T**hin-**F**ilm **T**ransistor) display.

Amplifiers

The question often comes up: Do you need an amplifier to work DX? Technically, the answer is no. A transceiver with 100W out will do just fine for many hams, in fact this is what most hams run on HF. Many have gotten DXCC using QRP (5W or less) and a select few have even worked them all using such low power, which is rather impressive. However, there is a stipulation. You will be miserable on 80m and 160m without one unless you primarily stick to FT8 or similar digital modes. Noise levels are high here and chances are you will not be using the most efficient antenna either unless you live on a ranch.

Can you get your DXCC award (first 100 entities) with just 100W? Absolutely! I got mine with just 100W and a multi-band vertical. In fact, many stations I worked using just 25W or less. Was it easy? Looking back, would have been much easier with an amplifier and a directional antenna, but it was not a big deal. I had fun doing it. It was especially challenging after a certain point from the US West Coast as I had mentioned earlier.

Using an amplifier is not a substitute for a poor antenna system! I had people debate this with me. Seasoned hams, mind you. All I can say is I am doubling down on this statement. Yes, you can throw more power at a poor antenna system, and someone may hear you better. OK, perhaps so…Sure! But you will still not hear better on it. You may call CQ and other hams might be trying to respond, you may not hear them. Even worse is when someone sees a DX spot and jumps in without hearing the station at all. They call and call thinking that maybe somehow this "*magically*" works without giving it a second thought. Newsflash…nope! It's science, not magic. Focus on your antenna system first, then investigate an amplifier.

The Ameritron ALS-1300 is a solid state (no tubes) amplifier which provides 1200W power from the 160-10m bands. The latter model, ALS-1306 also includes the 6m band. These are rock solid amplifiers!

Before using an amplifier, make sure you fix any RFI issues you may encounter when transmitting at lower power. As any issues you already have with RFI will also be amplified. No pun intended. This is even more so if you live in an apartment or a condominium where you share walls with neighbors. There is a good chance if you have issues, so will your neighbors. This can range from speakers buzzing to touch switched lights coming on and more series things like garage doors opening. Many consumer electronics devices are built with minimal protection from RFI so keep that in mind. Certain overseas manufacturers will remove a $0.002 filter capacitor to make an extra buck at the expense of quality. The extra zero above is not accidental, that was less than a cent and this is an actual example from last year involving a certain phone charger.

Less likely to be an issue nowadays is telephone line interference as most have moved to cell phones, but you just never know. Also, since we moved away from analog TV and cable TV companies have greatly improved their gear and have gone digital, those issues are also less likely to come up. You want to keep happy neighbors!

I have already touched on this briefly, but if you want to run more than 5-600W and live in the US where we are on lower voltage than many other countries, you will need a 220VAC source. You can run many legal limit amps on 110VAC but they will limit your output as they usually do a voltage auto-sense. There are many amplifiers out there in the 500W range that will get you more power than most folks use out there. You might be surprised to find just how many folks only run barefoot (no amplifier). Also, a large chunk of the world has much lower power limits than US hams. They compensate with better antennas and feedlines in most cases.

Some amplifiers in the US will ship without the capability of operating on 10m and 12m but rather have an add-on kit. Though I see this less and less. I specifically recall having to deal with this when I purchased an Ameritron amplifier about a decade ago. This is due to some amps being used for CB radio. CB radio has a 5W AM and 12W SSB limit in the United States. Certainly not 1500W! Once you prove you are a licensed ham, should be no issue getting this part. Installation is rather easy.

You may see amplifier classes referenced. These are rated by letters starting with an "A" and you will likely not be using anything lower than a "C". A class "C" is more efficient, but a class "A" will be the most linear. Class AB is a good compromise. There is a good chance you will not have to worry about this if you buy a reputable brand as they did all the engineering for you, but good to understand it.

Solid-state vs Glowing Tubes

Solid-state amps seem to be the new standard and they are getting cheaper and have many features which cannot be found in tube amplifiers. A perfect example of this, though on the higher end, is the PowerGenius XL from **FlexRadio Systems**. **Elecraft** also has outstanding solid-state units, one at the 500W level and another at the 1500W legal limit level. **Ameritron** (an MFJ company) has very well-made solid-state units worth checking out as mentioned earlier. They have been making them for a long time with various output levels. All three of these companies are based in the United States, though the PowerGenius XL has origins in Montenegro **(4O)**. **Acom** from Bulgaria **(LZ)** has also entered the solid-state market, though they are mostly recognized for their excellent tube amplifiers.

Tube amps are time-tested and predate me. There are strong feelings for tube amps by many and they are surely time tested. Do keep in mind they need expensive and harder to come by tubes replaced when the time comes. They also need to be matched if you need a set. Acom amplifiers, mentioned earlier, have a loyal following. Their 1000 series units are still used and loved by many, and the 2000 series even have an automatic tuning model. I have been personally eying this one. Ameritron also makes tube amps in addition to their solid-state variety and have never heard any serious complaints about any of these either.

Tube amps also need to be tuned up and warmed up, around 3 minutes. Some like to joke that by the time this is completed, the DX might be gone. I know folks this has happened to though actually, so maybe not so much a joke. This is especially true for bands with short openings such 6m sporadic E which can come and go in minutes as well as 160m openings that can be as short as 15-20 minutes at sunrise and sunset. Also referred to as "*grayline*" and more on this later. Bottom line, be ready! Or if you are not, just call barefoot until your amp warms up. You never know! Solid State amplifiers are generally ready to operate in just seconds.

Tube amps are more forgiving with SWR issues and generally less expensive upfront, though not always. They also do run cooler in most cases or at least do not have to depend on loud fans to cool them as much to put it more accurately. In other words, they are quieter! As an important side note, **if you are using a tube amp because of high SWR since it is forgiving, I would seriously address the high SWR issues instead**.

If you blow out a transistor in a solid-state unit, it may also cost you a pretty penny, though the gap is closing between these and tubes. But if you operate it properly, and there are safety circuits in place, chances are exceptionally low. An interesting fact is that many legal limit solid state amps are made up of 2 or more smaller amplifiers. For example, one can design an amp for 600W and double it. Now you have a 1200W amp. Several manufacturers have done just that. Just looks at many Ameritron amplifiers and their power ratings, this becomes clear fast.

As a friendly reminder, there are power limits on both the 60m band 100W PEP using a ½ dipole (formerly 50W, raised by 3dB recently) and 30m band 200W. Therefore, be sure to watch your power when operating with an amplifier on these two bands.

Those on the 630m and 2200m band have a 1W ERP (or EIRP to be more precise) limit. This is generally not an issue though due to the ridiculously high losses for most of these severely compromised antennas. You are lucky to get a 1% efficiency in most cases. Amplifiers for these bands will likely have to be homebrew.

As I pointed out in the antenna tuner section, never tune-up over the DX or over anyone calling the DX as you may bust their QSO. **Avoid getting on the lid list; Easy to get on it, very hard to get off it!**

Do not set the power out at the transceiver level to anything near 100W when using an amplifier. Start very low and see how it reacts. 100W is way too much for an amplifier, any amplifier. Most amplifiers will give you a full 1.5kW out with around 50W or less.

Final thoughts on Amplifiers. If you are getting a used amplifier, some of the older models do not cover the WARC bands. Meaning they will not work on 12m, 17m and 30m (which has a power limit anyway). Furthermore, quite a few amps do not cover the 6m band. So, if you are interested in the magic band, make sure to check.

Power Supplies, Protection & More

This is an often-overlooked part of your station. Not because people forget them, obviously not, as the radio will not come on. But rather because they give it very little thought. For ham radio you will be running 13.8VDC or thereabout. There are some minor exceptions, but those devices usually come with their own wall warts (DC Adapters) or built-in power supplies.

Let me get a myth out of the way first. **Both DC (Direct Current) and AC (Alternating Current) can kill you** under the right circumstance. We are made mostly of water; therefore, your body is a good conductor. Use caution when working on power supplies and dealing with any kind of electrical source.

There are two main types of power supplies, switching and linear. Some would argue the two types are "*good and bad*". But that is no longer true. Both types are "*generally*" good as of late. This was not always the case so be aware of older (and off brand) switching products. Some of these can have serious RFI issues, even if supposedly designed for ham radio use. The other kind of noise can also be an issue with some, especially the switching kind. I am talking about fan noise. If you are one who easily gets annoyed when a loud cooling fan kicks in, this is especially something to pay attention to. It's hard to hear a DX station when a fan is drowning it out, sometimes even when wearing headphones. I have returned power supplies due to fan noise issues before.

Older linear units can have leaky or dried out capacitors as well as other potential issues. This just happens with heat and age. Be careful when buying used ones, especially older models unless you know how to work on them. I had very good luck with **Astron** power supplies, both linear and switching, new and old. Though there are many other good brands out there. They provide excellent customer service, which is in short supply at times with some companies. Get this, they actually answer the phone! Pictured is a 70 Amp Linear Power supply that will take pretty much anything you throw at it.

In most cases you will only need about 25-30 amps unless you are running multiple units at once or own power-hungry peripherals. When you look at a power supply rating, the number used is often the maximum when it comes to the current rating. Marketing, remember? Therefore, if you purchase a 50 Amp power supply, likely around 30 Amp continuous is a more realistic number. If your radio requires 15-20 Amps, a 30 Amp unit is what you likely want unless you want the fan blasting and possibly experience a thermal warning. Or worse, damaging it!

Figuring out your needs is simple. Just add up the power requirements in your shack for gear you might be using at the same time, add about 20-40% overhead and purchase accordingly. There is also nothing wrong with having more than one power supply, I currently have four rack-mounted Astron's in use and a separate one for my Icom 9700 which is built into the rackmount. I never really hear fans as everything is chugging along with very little load.

As far as math is concerned, Volts x Amps is equal to Watts. Therefore, 13.8V x 30A is just over 400W at peak usage. Assuming perfect efficiency, which it will not be you can figure out the rest. Using 450W for the rounded-up guestimate (We are very scientific here) this will require about 4 Amps from your household AC outlet. Most are wired for 15 to 20 Amps, so you are good to go! Do your AC outlet load calculations accordingly.

I find that it is never a bad idea to take an extra precaution before feeding my gear from any power supply. Guess what? Ferrites! Why take a chance? You just never really know how well something is designed. Frankly, for me it is easier to assume that is poorly designed and quickly add ferrites than to do a full HF band sweep.

If you find that your power supply is running hotter, or perhaps louder than you would like it to, there are many modifications available on the internet which address this. Sometimes this is as simple as replacing one component or adjusting an internal variable resistor.

Also make sure you use adequately sized wire. I believe in overkill here as I use 8-10 AWG wire and sometimes even thicker depending on what it is. Generally, I go one size larger than recommend for good measure. Keep in mind, wires have loss, so the less resistance, the less loss. When there is resistance, heat is generated. When there is heat, this creates even more resistance. It is as simple as that! Sadly, this is also how many if not most electrical fires are started in houses.

Keep the length reasonably short and make sure everything is properly insulated. Also **stick to the color code standards, Red is always positive, and Black is always negative.** I know some may get a good deal on yellow or green wire (great for dipoles) but is it worth accidentally frying your station over saving a few cents if you mix them up? Speaking of frying, **always use a fuse!** Never bypass, remove, or use larger fuses than you should be.

Some power supplies can generate serious heat, despite of type or size. Since heat rises, I recommend keeping these lower, closer to the ground, but not on the ground if there is even a chance of water damage. They are also more likely to suck in dust if right on the ground. Always remember, heat rises! Make sure there is enough airflow around them and make sure there is nothing blocking the vent holes. Electronics in general have vent holes to allow for air flow. While some find these decorative or even stylish, they serve a function. Be sure to never block these. Mainly the back or sides in the case of power supplies as this is generally where the vent holes are. Since most ham equipment generates heat, the same advice applies to your radios. Don't stack them on top of each other. Use shelves to allow for ventilation.

While on the topic of power supplies, **Anderson Powerpole** connectors (pictured) seem to come up a lot. You can find an installation guide on my website if you want to find out more about them.

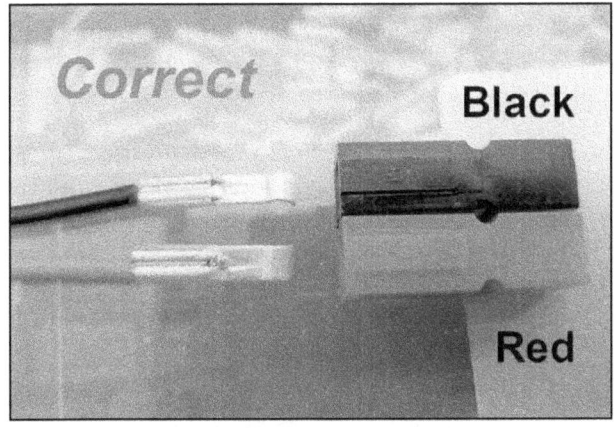

These are pretty much the unofficial standard when it comes to DC power connections in ham radio. While not perfect, they do the job well and you can help you to avoid possible polarity issues and provide protection from shorts. The plastic enclosures house stainless steel spring contacts which are rather durable and long lasting.

As for power strips, never daisy chain them and don't cut corners here. Yes, you can get them for a few dollars, but would you really trust these to protect your thousands of dollars' worth of gear? Or even your computer? I always opt for ones with metal enclosures and heavy-duty industrial designs. RFI shielding, possibly better ground and heat tolerance are just a few reasons as to why. A good surge protector at minimum is a must. I use many Tripp Lite products, and I am especially a big fan of the Isobar series. More information at tripplite.eaton.com.

Power surges are not caused by lighting, as folks sometimes falsely assume, but rather by power fluctuations. They are very likely to occur on your household AC (**A**lternating **C**urrent) line, but many times they are nothing to worry about. They can be weather related, mainly heat for example, as Air Conditioners (A/C, not to be confused with the other AC) usage goes up and down for example. You can help with voltage fluctuations in your house by using a voltage regulator or AC spike filter such as the one here. I run all my ham radio gear though these. Many corporate server rooms or places with expensive scientific gear require these.

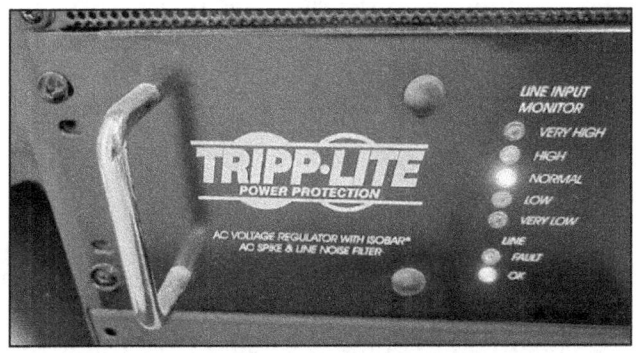

Some UPS (**U**ninterruptable **P**ower **S**upply) units are also an option, but some models do generate RFI which can cause issues with HF reception. If you have ever taken one of these apart, you will know that they already contain at least 2-3 ferrites in each unit. I regularly harvest these from broken units and add them to my UPS units, generally on the outside leads. Never overload a UPS unit, always follow manufacturer's instructions.

Antenna Analyzers

Unless you inherit a station with antennas already set up and intend to make no changes for a couple of years, you will need one or at least have access to an antenna analyzer. At some point though things will likely go wrong, and you will need to troubleshoot. I had weather related issues, both wind and rain (ice is even worse) as well as critters acquiring a taste for coax shielding.

There are a lot of units out there and for the most part they all do a pretty good job. Some are better than others as far as accuracy, features and especially build quality. You do get what you pay for here. I am somewhat partial to the **RigExpert** line now that I have tried a few others. They can be found at rigexpert.com. These are made in Ukraine **(UT)** and have great software to go along with built in memory and regular firmware updates to make the product even better. They can even be connected to a computer via a USB port.

The most popular and best-selling antenna analyzers, however, are from **MFJ** at mfjenterprises.com. They have one of the largest lines of antenna tuners out there without a doubt. I have also owned several of these over the years. You cannot go wrong with either of the above. There are other brands out there too, so ask around and check them out if interested.

Dummy Loads

No, this is not a reference to "*Lid Operators*". You may not have a regular need for a **dummy load** but believe me when you need one you will be glad you have one. A dummy load is basically something that replaces the antenna temporarily in certain instances, such as testing. You can think of them as essentially is large resistive 50Ω load. The larger in general, the more it can power handle. Since heat is generated when they are used, size and capacity does matter. If you need it to tune your legal limit amp, 2KW rating or better is what you want. Some can be oil based, such as those resembling a paint can. In fact, I am pretty sure they are paint cans.

They can also be air cooled and can even be as simple as an oversized resistor or a set of smaller resistors in parallel. If it gives you a 50Ω match and it can handle the load, you are in business.

There are many good options out there. Do pay attention to frequency ranges as some do not do well on the VHF/UHF spectrum. Some even have a high SWR on 6m if they are only rated to 30MHz. Usually, this range is indicated by the manufacturer though sometimes the specs are better than advertised. Does not happen often, does it?

Also pay attention to power rating. A little overkill here is not a bad thing. The load will stay cooler longer if needed. If you need to test 500W try and get something that is rated about 50% higher, for a good margin of safety. Double it if you need to key down for several minutes.

The above illustrates the spectrum of complexity and cost when it comes to dummy loads. On the left is my trusty 2KW rated Palstar DL2K which also features a cooling fan and an RF power meter which even has a light that can be turned on. This unit can be used with no extra power if the fan and light for the meter are not needed. These units are on the costlier side but are well worth it. I cannot say enough good things about them.

On the right is my MFJ 1.5KW dummy load. No frills but works dependably and takes up a lot less space. The cost is very reasonable too at around $100 and note that it can be used on VHF and some UHF frequencies as well up to 650MHz. This is basically a massive resistor in a metal box. Literally! I have opened it and checked. Can be connected to a coax selector and stashed under a desk so when needed, you are ready to roll. Dummy loads are not just used for testing, but some also use it to tune their tube amps to avoid QRM. Something to consider!

Rack Mounting

If you work in IT (**I**nformation **T**echnology) you are likely already familiar with the world of servers. These are generally rack mounted not only to save space but to create some order and even to create more efficient air flow. This is only an option and by no means a requirement for hams. An added benefit in the ham radio world might also be some added shielding. I like that! Not sure about the actual difference it makes, but I feel better knowing I did everything I could to improve my working conditions.

The rack component heights are measured in units called "U". Yes, I know, creative. I did not come up with it. 1U is equal to 1.75" (4.5cm). Width will always be 19" (48cm) wide and rack mount gear is always designed around this.

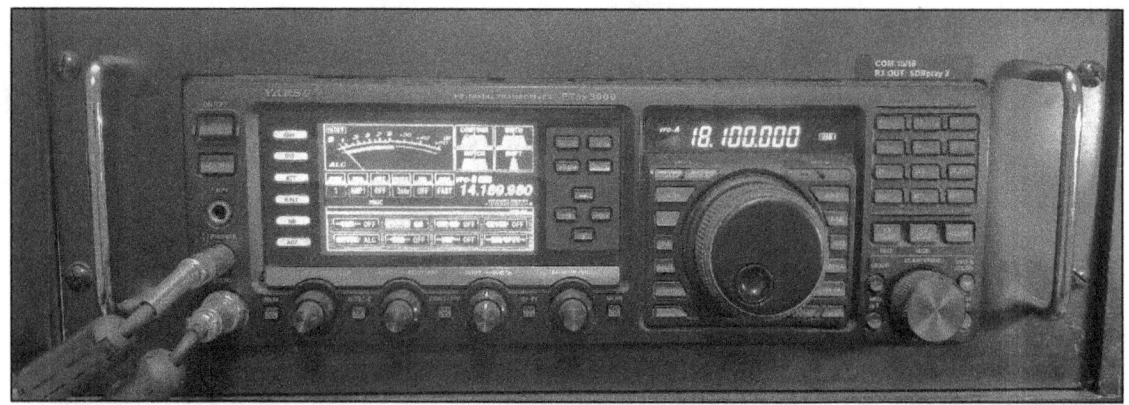

Most radios will require about 3-4 U of space in a rack configuration, such as this Yaesu FTDX-3000 above. Note the ferrites on the cords. I practice what I preach! The handles in my case are really just for looks and are optional.

Some of the places to shop for rack mounts and rack related items include **SNS Engineering** at snsengineering.com as well as **Novexcomm** at novexcomm.com . If you are a FlexRadio user, they have rack mounting hardware available for the 6000 series radios as well as their PowerGenius XL amplifier. I own both as well as several Novexcomm units and really helped to clean things up in my shack. I even had them custom cut a few pieces for me for 900MHz radios and scanners at one time. Pictured on the left. SNS Engineering also has a rackmount option for the new Elecraft K4 line of radios.

Samson Technologies racks are available in many sizes, the half-height version is what I use. This gives me 21U of space. Full height is considered a 42U (73.5") usable space plus top, bottom and wheels if applicable. You can visit them at samsontech.com. I like these due to their wheels as well as closed side design. But your needs may vary. Alternatively, you can take a look at SRO (Server Racks Online) for a wide selection to see what fits your needs. They are at server-rack-online.com.

Optional "Nice to Have" Hardware

Items such as **multimeters, meggers, coax crimpers** come to mind. Some would argue that you must have all the above. I think if you have access to them, you should be OK. Many clubs provide these as loaners to their members. Though, in the long run, I would add all the above to your toolbox.

I also like to keep around an **assortment of connectors, adapters and Anderson Powerpole** plugs. Often the latter can be often found in a kit form at ham fests or online. I highly recommend getting one as well as the crimping tool for it.

The Ham Clock I have already touched on in the hardware section when I covered Linux briefly. The paid alternative to this is the **Geochron**. You likely have seen these decorating the background in many ham shacks. There are two flavors of this device, a mechanical backlit device (older model) and the new 4K capable digital device which has a ton of amazing features. There is even a ham radio specific add-on for it, though it is subscription based. You can find more information on Geochron at geochron.com.

One Final Step

Now that you have got all this gear and you took all the precautions you can from mother nature's wrath, there is still one more thing to possibly do. It is optional; however, it may save you a lot of grief in case something does happen to your gear. Ham radio insurance!

Your homeowner's insurance or renters' insurance may cover some of these expenses but check with them to be certain. Again, I am not giving legal advice here so please do your own research. **ARRL offers an insurance plan specifically for hams**. Check their website for the most current information. You can find info on this at arrlinsurance.com.

Take photos of your equipment and record all serial numbers, save, or scan receipts. Do this every few years and save them in the cloud. If a fire or flood hits for example, your home computer or burnt paper will not be much use.

I already touched on earthquakes a bit. At least where I reside, my homeowner's policy does not include coverage for earthquakes. It is optional and a separate coverage. At the time I am writing this, it is around $1200 per year extra and I am sure this will only go up with time. This is something you may need to think about it if is worth it for you or not as they may not cover your ham radio gear if the ceiling comes down on your radio or your tower tips over. Your home may do well structurally, things inside your home may not so much. This is from personal experience, I lived in the San Francisco Bay Area in 1989 during our last big one.

Certain areas of the country, and I am sure worldwide as well, insurance may not cover certain acts of nature as I states above. Basically, they have exceptions due to higher probability of something "*unique*" to the area occurring in many cases. If you live near a volcano, get annual flooding, or live near an eroding beach are some things that come to mind. This is something worth investigating further depending on your risk tolerance.

CHAPTER 4: Metal in the Sky - Antennas

This is the most important part of your station. Read the last sentence again. This is what you should be focusing on likely before anything else. It does not necessarily come down to having to spend more, but it can help. Also does not come down to buying bigger, but that can also help. Much of it comes down to planning, picking the right antenna for you and of course placement. As with many things in ham radio, what works for me, may not work for you, and vice versa.

Be aware that like many things in ham radio, you will get some very opiniated folks if you ask them about this topic. I have had hams tell me what a great antenna they purchased and how I needed to buy it too, just to see them sell it or complain about it a year later. I personally like to buy an antenna, then a year later tell you if I think it is still great. To be fair, sometimes people do sell only because they want to go bigger or get lower SWR, like me. That is my main reason for the constant tweaking of my setup and in part manic curiosity. Just be aware that wars have started over antennas!

If you think you bought an antenna or two and you are set for life, I got news for you. Nope! For one, they do not last forever. Especially if you live in a harsh climate. And I do not just mean snow and wind. I live near the ocean and salt water is doing a number on my antennas. Wind does not help either.

You will also find that you can make slight changes, possibly get a better vertical, switch from a hexbeam (which is one of my favorites) to more elements or a SteppIR. Some find that what I thought would be ideal for them, maybe is not so much anymore or their needs change. You might change your antennas before your receiver. Of course, we are all different, but I certainly hear more antenna talk than any other topic when talking "*shack*" with other hams. This also reinforces their importance in the hobby.

Some hams, like myself, like to build homebrew antennas as well. These are ones not purchased but rather built and at times designed by hams. At the moment I have only 2 homebrew units, one for 30m and one for 160m. Both undergo constant adjustments to make them perform even better. Sort of becomes an addiction at some point, but you can be addicted to a lot worse things. So, I will take this, thank you!

Watch for Power Lines

Before I dive into the wonderful world of metal in the sky, I must touch on safety. Not to state the stupidly obvious but **never install anything anywhere near an electrical power line** if there is even a remote chance it can touch it. To state the even more obvious, do not use utility poles to hang antennas. Sounds obvious? Yes, but I know someone who did this on field day and even bragged about it. Yes, he lived to talk about it.

Additionally, power lines can also absorb some RF even at some distance. While this is unavoidable, if you have one near your home, just be aware. Best to install antennas as far away from high tension runs for yet another reason. Consider other parts of your yard if this applies to you and is doable of course.

Can you do it with Just one Antenna?

The short answer is yes, but it will be an adventure. I should know. When I got started and worked my first 100 entities with just a single multi-band vertical, all done with a Butternut HF9V. I am not going to pretend it was easy, but it is a good, robust antenna once you get it adjusted properly.

I know other folks who received a DXCC on a vertical and others have done it with a fan dipole. And no, fan dipoles will not keep you cool on a hot day. These are antennas which combine multiple dipoles or various lengths into a single unit fed by a common feedline. This allows a fan dipole, in theory, to support multiple bands. Though there are some interactions between elements at times, so it's not a perfect end all solution.

Sometimes simple antennas must work for folks due to circumstances, such as space, HOA rules and even income. On the other end of the spectrum, some folks are lucky enough to have enough property to build dream antenna farms as did Stax, KE5EE in Florida (Pictured). Here he has a stack of Yagi antennas at different heights mounted on a tower. This is all supported by some serious guy wiring. ***Photo courtesy Stan "Stax" Schwartz, KE5EE.***

Really the answer to the above comes down many other factors. Time you have to dedicate, money willing to spend on antennas and space you have to do all of this in. If you have all three going for you, you are in the 0.1% of hams for sure.

Location can be a factor as well. For example, if you live in Maine on the US Eastern coastline, you are a lot closer to being able to work certain entities, in fact a lot more entities. Europe, comes to mind. These are more of a challenge on 80m for example, for the US West coast. Therefore, your chances of getting an 80m DXCC more quickly with a less efficient antenna improve dramatically. And don't even get me started on 160m challenges from the West coast! Yikes!

On the flip side, as mentioned, East coast hams always complain about how hard it is for them to work parts of Asia. Especially on the lower bands. Depending on your goals, you may need to adjust and/or grow your antenna system. I know someone who was exclusively a 6m operator for decades and has more entities than I can ever hope for on the 6m band. Recently, he switched gears and is now focusing on the HF bands. Interest change, goals change and then antennas change.

A couple of last things which can be applied to all antennas we will cover here. When it comes to antennas in general, lower angles of radiation may require fewer hops. A single hop on the F layer (more on this later) is estimated to be over 2000 miles. Therefore, the lower the angles the more likely you are to make the DX contact...in general. Though there are always exceptions and depends on your target DX. However, you should always aim for the lowest angle possible when it comes to your antenna radiation pattern.

We want high radiation resistance and low ohmic resistance for an efficient antenna. Efficiency is radiation resistance divided by Ohmic resistance. Once you get more into learning more about antennas, this will make more sense if it does not already. Why is this important? To put it simply, not all antennas are created equal. In fact, without mentioning specifics, there is some serious junk out there!

Does size Really Matter?

Technically, yes, but not all that much. You do need to be at least somewhat resonant though for the frequency you are trying to operate at, and you will never be able to tune up a 20m dipole for the 160m band. In other words, in order to be efficient, you do want to be properly sized. Does not mean super-sized, just somewhat correctly at least.

Put up what you can and have fun! Yes, you will hear a select number of hams discussing/bragging on the air how many elements they have on a Yagi for each band and how much taller their tower is. Or how they got their license when you still had to know CW, and the young hams with their dipoles, FT8 digital, etc. Just ignore all that noise! It is just that...noise. Don't let any of that get to you.

Having more elements and antennas can and likely will help you, no question about that. It can also work against you at times when you go way overboard with elements. For one your beam width narrows, which again, can be good or bad, depending on what you are trying to accomplish. While you will hear better in a given direction, the exact direction will become more important. Therefore, you may miss some things on the sides of the beam. This can be disastrous, for example, during contesting as you are not sure where the next caller may come from. The angles of the incoming radio signal can also be different. Therefore, stacking like Stax, KE5EE (previous picture) did makes a lot of sense!

The general rule of thumb, the more elements, the narrower the bandwidth. Your dipole is a single element. It will have the widest bandwidth of all directional antennas. Of course, a vertical in comparison will have a 360-degree bandwidth as it sees in all directions. Pretty straight forward really once you start using both this becomes very clear.

I often switch back to a vertical to see where propagation is or if I am missing anything and then back to my beam to turn it to where the action is, for that exact reason. I recommend keeping at least one vertical if you can. Alternatively, turn your beam occasionally and see what you might be missing.

One place where small size could matter, is if you want to use more power. Generally, this only becomes an issue on the lower bands as the elements increase in size. A lot! So often the antenna designer must resort to using more traps, like the butternut vertical for example, to achieve additional band capability.

You may see antennas with different power ratings on different bands, so pay attention to this. For the 6m band for example, I had to move to a 5 element from a 3-element due to power limitations. Some 80 and 160m antennas will also have issues with legal limit and can only handle a few hundred Watts when compromised. Most are! In ads it might just be an asterisk indicating this with a tiny footnote. Therefore, it is important to check your specifications. For example, my Butternut HF9V could only handle 600W on the 160m band but 1.5kW everywhere else. The SteppIR BigIR also had a limitation on the 80m band up until they redesigned it.

What About Height and Location?

Yes, it matters to a point, but it is not a deal breaker. Just do what you can. We need to be realistic. I do not see anyone erecting a 100ft tower, which is half the maximum allowed in the US (200ft). Within the city limits of San Francisco, you are lucky to get up to 25ft, hence many turn to vertical antennas, dipoles and an occasional small Yagi. Even 55ft towers (which is perhaps the most common) are unlikely to happen for most of us.

The lower the band, you want some height above ground, when possible, with directional antennas. A famous DXer once said, for every 10ft (3m) you elevate your beam, a whole new level of DX opens up. If you can get an antenna, such as a dipole up about 30ft (10m), you are in great shape and can be very competitive! While, not ideal, I can work the world and have done so with this setup.

There is such a thing as going too high. Though, I doubt this is an issue for most. **Every band has an ideal recommended mounting height.** This can be a problem though for some. If you have a multiband antenna, like most do, it will never be ideal for all bands. Ideal heights for one band will not be the same for others. Therefore, assuming that you are one of the lucky few who can mount your antennas as high theoretically is ideal, all bands but one may have to be a compromise. Meaning all but one will be either too low or too high in ideal height.

If possible, you want your 6m antenna at 25-30ft (7.5-10m) for the best DX results. Higher is not always better here as it was pointed out to me by a couple of 6m gurus in the past. This is a lot more realistic height for your average ham than let's say the perfect height for a 20m band beam. This would be 125ft (38m), which let's be honest, for most of us is not happening. Again, just do the best you can, chances are it will work well enough to get lots of DX in the log.

As far as an antenna location, that can be a tough one. It was a nightmare for me, and I did not solve this one overnight. Trees are your friends…sometimes. And other times meet Mr. Chainsaw! Before the mail starts to roll in from environmental groups, that was a joke. Trees can be a blessing when they decide to grow strategically. Meaning where you need to hang antennas in your yard and two happen to be facing a favorable direction to hang a dipole on. However, you can get creative and use the side of your house or do what I did, pick a nice tall tree for one side and a push up mast for the other side to favor the direction you want. I have also done this with heavier duty PVC pipes as I like to minimize the amount of metal near other antennas, such as vertical given my small lot. When it comes to PVC, these are measured in "*schedules*", sometimes marked as "*SCH*" on the pipe followed by a number. The larger the number, the thicker the wall. It will also be heavier obviously as the number goes up, so something to keep an eye on. I generally just go with schedule 40 as this is very common. This will require some guying though. Anything rated lower, such as schedule 20, I would not use. Schedule 80 is a lot more expensive but is considered industrial grade. Get creative and use a map and compass!

A Quick Word on Loss and Gain

Before we look at antennas in greater detail, let's bring everyone up to speed. A "decibel" or dB is used to measure a variety of things, in this case, antenna gain. It is a logarithmic ratio measurement between two values, usually with a reference value compared to a 2nd value. **It is important to remember that dB is logarithmic.** The reference points used in antenna measurement are a bit odd to most hams. **dBi is used in reference to an Isotropic radiator and dBd in reference to a dipole antenna**.

If you have an antenna with +3dB gain, your 100W out will "*feel like*" 200W out. If you have a loss of -3dB, in your feedline for example (likely), your 100W out will only be 50W by the time it hits your antenna. I will go over this more when we get to transmission lines. Therefore, I always say, every dB counts! **Every 3dB gain represents a doubling of power. Every 3db loss halves the power.** 6dB is equivalent to 4 times; 10 dB to 10 times, and 20 dB to 100 times. Minus 6 dB is ¼ of the reference power. Gain is always a positive figure; losses are always negative.

When you see dBi, the "*gain*", or concentration of energy, is in reference to an isotropic radiator, which radiates equally well in all compass directions. An isotropic radiator can be thought of as an idealized vertical. Think of this three dimensionally. Hence, the name isotropic. A dipole radiates most of its power in two directions (at right angles to its length), therefore it has "*gain*" in comparison. There isn't really any more power, but the dipole has concentrated the available power.

When dBi is being referenced as 0, a dipole then would have 2.15dB gain in "*free space*". This would be the dBd reference antenna used by some manufacturers. "*In free space*" means that the figure does not consider location, soil (or water), height, obstructions and so on. So, when a "*shady*" antenna manufacturer tells you that their antenna has 5dBi gain on the 20m band and the competitor has only 4dBd, then hopefully you caught that the 4dBd is better. This is rather rare though, but I have seen it. I do wish that all would just use a single standard but until then, just pay attention to the letters. Obviously when they use dBi, the specs always sound better! Antennas front to back ratio will also be expressed in dB gain, but this is obviously only applicable when there is more than one element.

Antenna gain is just math. An antenna's energy is just focused. Energy that would be sent equally 360 degrees in a vertical, for example, is now radiating in two directions from a dipole and mainly from the front in case of a Yagi. The more focused, the more gain in a given direction, to oversimplify it. Therefore, when you see a dB gain increase in antennas, they do not generate more power or break the laws of physics. We will go over various antennas in more detail shortly.

Indoor vs. Outdoor Antennas

Do what you can, but **outdoor antennas will outperform indoor antennas**. The same antenna will always do better outdoors as it does not have to fight with walls and other obstacles. They can be used though and at times successfully. I do understand some may have HOA restrictions or have a landlord who is not OK with them putting up an outside antenna, don't let that stop you from DXing! Just be sure to practice RF safety and distances from your transmitting element. Read the section on RF exposure in this book, please!

People have debated with me about this topic, claiming there is no difference. I disagree. One way to confirm the above is if you have two antennas of the same type, let's say two 10m dipoles which is quite reasonable in size for an attic or even a larger bedroom ceiling, QRP of course. One can be mounted inside and one outside, about the same height or as close to as possible. Same length of coax to match losses, mounted the same direction. I do realize that there are other variables, but this is as close as we can get to the scientific method in most cases. Try to transmit and/or receive FT8 on both, one at the time of course using the same power and do a reception comparison for both your end and receiving end. PSKreporter at pskreporter.info is a good place to start. You will notice several dB differences in reports. This difference can vary a lot depending on the materials used in your home. If you have metal siding as I do, copper roof or solar panels on your roof as I also do, you will see the most signal degradation in both receive and transmit signals for your indoor unit. It is

important to also do this test within a short amount of time so ensure propagation does not dramatically change. This would affect your test results. This is also a good type of test if you are testing any two antennas against each other. I use this method all the time when I make antenna adjustments.

Again, if indoor antennas are all you can do, don't let that deter you. Just be sure to do it safely. Attics are a great option for those who have one and do not want or can have their antennas visible. You would be surprised how many hams have attic antennas. Just look around the pages on QRZ. I regularly encounter photos of folks with attic antenna installations, and I give them good signal reports.

Seen here is an example of how one can utilize a relatively small lot to work the world. Tony, K6BV has managed to tastefully install not only a large SteppIR antenna (right) but also a Gap Titan vertical with elevated radials/counterpoise (left) in addition to some other VHF antennas for local communications visible in the back. ***Photo provided by Anthony Dowler, K6BV.***

Resonant Dipoles - G5RV, Bazooka, Fan Dipoles

I will go over a few different, popular options here. I have bundled them as I feel they are the most closely related. Let me begin by saying **There is nothing wrong with dipoles**. Some, including a few DXers look down on them. They are one of the simplest and most reliable basic antennas out there. Practically all my 30m contacts (now 200 entities) were on a simple homemade dipole. I also used a dipole for the 60m band for a long time as well as a fan dipole and a trapped **Alpha Delta** dipole prior to that. If interested, they are at alphadeltaradio.com.

If you cannot install a beam, you may even want to use a set of 2 dipoles, if you have the room. Perhaps even with a coax switch mounted remotely so you can select different directions. Keep in mind that **your dipole radiates from the sides not the ends**. So, if your antennas wire side is facing the ocean on the West coast, you will be working more Asian and US stations. My recommendation is to orient it towards the most challenging direction for you, the rest might just fall into place. From the East coast I would pick around 30 degrees to shoot for Europe and the Middle East. This will also give you a nice path to Australia since dipoles are bi-directional.

Wire antennas such as these are great options for those with very limited spaces and even HOA restrictions as covered earlier. They can be easily hidden or disguised. If space is tight, you can even bend the ends slightly, though you may have to readjust the length a bit. I recommend getting dark green wire for outdoor use, that is what I use exclusively. Your local hardware store will carry this for sure as this the color is used for ground in AC electrical wiring. Try to keep wires at least 3ft (1m) away from metal or other branches if you can. Some would even recommend more. Mine are less!

An antenna trap refers to a way of shortening the antenna by using a wound coil, basically inductance and capacitance to oversimplify the concept. This reduces the space needed or even the need for multiple antennas in some cases. Making the wire electrically appear one length (longer) but physically being shorter in length. The drawback is rather minor, is a slight decrease in performance over non-trapped antennas. Though I must say, serious contesters and DXers often opt for trap free antennas for that reason. If you have limited space though, this is certainly an option. In fact, I would argue that trapped antennas are more common than not. Not having one might make the difference in not being able to operate on a given, usually lower, band. You can also build your own traps for dipoles rather easily. A short piece of light PVC can be used with wire wound around this. Many articles on the internet about these.

Many hams use a homemade dipole for 60m due to lack of commercially available resonant antennas. I think a dipole is one of the best ways to get on the **60m band**, alternatively many some use a vertical. Here is your chance to DIY.

To make your own dipoles is easy! Find some decent wire and cut two equal ends. Find the proper length for your dipole using formula or the chart below. 12-14 AWG wire is ideal, lower gauge (thicker) is also OK. Just don't go too thin. I would not go lower than 16 AWG. AWG stand for **A**merican **W**ire **G**auge and sometimes it is written as #16. This would be the same as 16 AWG. SWG is what is used in the UK and is not quite the same. I prefer solid copper wire as it holds up better against harsh weather in my area. Some like stranded wire as they claim it does not stretch as much and no metal fatigue, though I had these rip and even rust when I used non-copper. Stranded copperweld is a good option. Use what you prefer though and adjust for your climate. Attach a non-conductive end insulator and a good quality rope which will be used to attach to something like a branch or post. Non-conductive is important for both as the voltage on the end of the wire will be high when in transmitting.

$$\text{Length (ft)} = \frac{468}{\text{Frequency (MHz)}}$$

$$\text{Element (ft)} = \frac{\text{Length (ft)}}{2}$$

Band	FT8 (MHz)	Element (ft)	Total (ft)
6m	50.313	4.65	9.30
10m	28.074	8.34	16.67
12m	24.915	9.39	18.78
15m	21.074	11.10	22.21
17m	18.100	12.93	25.86
20m	14.074	16.63	33.25
30m	10.136	23.09	46.17
40m	7.074	33.08	66.16
60m	5.357	43.68	87.36
80m	3.567	65.60	131.20
160m	1.840	127.17	254.35

For reference I did the math for you for the FT8 frequencies of each band. Why FT8 for the center frequency? Two reasons. One, this is likely where you will be the most, so will not have to use an antenna tuner. Also, it is somewhat center to where you otherwise might be operating DX, CW is below and SSB is above this frequency in general. Except for 30, where there is no SSB. If needed, you can likely use an antenna tuner for the rest of the bands below the size you pick. So, if you build a 40m dipole, you can likely make it work on the 6-30m bands. To build an 80m and 160m dipole are going to be a little tougher due to total length. But for these you can set up an inverted V configuration. More on these later. I would also explore a vertical antenna for these bands first. Though if you have enough room for a full sized 160m dipole, I see a tower in your future!

The best end insulators will be ceramic or molded copolymer glass filled; nylon based basically with 20% glass. **Budwig** is a brand well known for quality. They can be found at budwigmoldedproducts.com. The second best is material is plastic, and is most widely found, likely due to cost. Most plastics will break down much quicker than the above and I would not use them with legal limit power. Some folks have been known to use electric fence end insulators; these will work fine too. There are also "*dog bone*" insulators, sometimes glazed ceramic or plastic. These are my preferences.

Solder the wires to a coaxial cable either directly or using a SO239 connector, also known as the UHF connector. This then can tie into your coax run via the PL259, UHF plug. Either method will require a good watertight seal. UHF connections (SO239 to

PL239) are not watertight. Below illustrates a few ways you may encounter these types of plugs and connectors.

I personally like to put a 1:1 balun right at the dipole before going to coax. This is not needed but rather a personal preference. I find it reduces noise and helps to match the antenna better. If you use a balun, always pay attention to the power rating.

The **G5RV**, designed by Louis Varney around the 1950's and named after his callsign, is perhaps one of the most popular antennas in some circles. Many hams at one time owned one or at least used one, perhaps during field day. It is easy and quick to install, relatively compact at around 100ft and if installed correctly works well. Meaning the right height and away from buildings and metal objects. There are a few versions around based on the original design.

I have also owned one of these variants as well. It had a set of symmetric dipole elements which connected to a 300Ω twin-lead (although I have seen 450Ω version) which then connected directly into a coax. Generally, tunes 20m band the best which is what it was originally designed for, though not always perfectly. You will likely still need an antenna tuner depending on what height you had installed it at. With an antenna tuner, you can make it work down to 80m but do not expect miracles due to the size. Also watch your wattage on the lower bands especially. There is a version of the G5RV HF doublet antenna by ZS6BKW which I am told is easier to match. This is something worth exploring.

I know folks who have modified these by joining the two sides by connecting them together to make it into a 160m antenna using a ground radial field as the other side, as in the ground end of the antenna. I suspect it performs similarly to an inverted "L" and A+ for creativity! There is also a **G5RV Jr.** for those with less space. With an antenna tuner, you can get pretty good results from these as well down to the 40m band.

Did you know you can make a dipole out of coax cable? A **Bazooka antenna** is like a coax stub with a wire on the end. These are very popular on the 40 and 80m bands since the design reduces the total length required. These also tend to have slightly higher bandwidth. I used one of these for many years as a second antenna for 40 and 80 meters with good results. Eventually a storm got it and is now in antenna heaven but if you see these around, they do work and work very well. They also tend to have a slightly higher bandwidth than a regular dipole antenna.

I have already touched on **fan dipoles** previously. Sometimes they are also referred to as **parallel dipoles**. I have owned two of them in the past, including the trapped variety as I had mentioned. You will need to get these up high to work well on the lower bands. The wires cannot touch or be too close so can be a bit tricky to install. If they include traps, this adds extra weight too so keep that in mind. One of these can easily get you resonant on a few bands.

Here is a fun project for the experimenter or those with limited space. You can make a **dipole with a Slinky**. Well, technically you will need two. The metal variety of course as some newer ones are now plastic. Obviously, the material needs to be conductive. Same idea as the method mentioned above, though the formula does change a little due to the helical nature of the coils. Look around online, many have successfully used this method to build even portable dipoles that can work down to the 80m band. Makes a nice portable antenna and in fact there are reports of them used by soldiers in the Vietnam war.

The last important consideration is that your dipole will and should move a bit freely. I know it may be a little weird seeing it flop in the wind, but if it is too tight it may tear or otherwise get damaged in the wind. So, use some common sense when mounting them.

The closer you are to the ground, the more omni directional the antenna will be. Therefore, whatever ½ λ is for the desired frequency, you want to be about that high up. Although, this has been up for debate amongst some, but I do not want to get too technical here.

Getting the antenna hung in trees can be tricky. I know folks who have used drones, bow and arrow, slingshot with fishing line with lead weights, like I did. Do whatever floats your boat but we careful with windows as well as people in your path. Also, check your local laws as in some places a bow and arrow and even slingshots are considered weapons. Yes, some places just love to overregulate.

There are also no fly zones for drones, like near airports, prisons, military bases, embassies, and some schools to name a few. Laws can also be very widely interpreted so as always, please do your homework.

You can check with an app on your phone for where it is OK to fly your drone. The app I use is B4UFLY. You also should register your drone. Though in the US an FAA unmanned aircraft license is not needed if not used for commercial purposes and under a certain weight. Check your current local laws as this is fairly new technology and things change. However, if you start a drone antenna inspection business, for sure you will need to do some paperwork.

The SWR of the antenna can change a lot when elevated vs closer to the ground, so don't be surprised if you need to re-adjust them. Using a pulley system can be very handy. I do this for all my dipoles and the photo example here is my own 30m dipole installation. Basically, PVC pipes, PVC end cap (keep the water out), pulley on the very top with something to hook it to the PVC cap to. When I need to inspect, repair, or adjust, it is as easy as lowering a flag. Even easier!

Non-resonant Dipoles - OFC & End Fed

I will be honest; I like balanced horizontal antennas such as above and therefore not a big fan of **OCF** (**O**ff **C**enter **F**ed) dipole antennas. Windom antennas are a type of these. I know plenty of folks who use them, and they are fine antennas, just not for me. These are basically antennas where one lead is of a different length than the other and is not fed in the center. I have tried them, and they work fine. They are relatively inexpensive but will require an antenna tuner. I know some love them as they fit into their yard better, cover many bands in some cases. But there is always a compromise. If this is all you can fit, and you are not running 1.5kW go for it. Units I have some across are rated for lower power. Though I understand there are a couple of models out there which do support legal limit. The priced does go up due to components required (like the balun) but if interested check at arraysolutions.com/antennas. These require a 4:1 Balun in most cases and it is included. The expected performance is about the same as a traditional dipole.

End fed half wave antennas are basically ½ λ long at the highest frequency it is to be used on. These will need a 49:1 balun and can be tuned up well for many bands. There are some issues when trying to run more power, mainly RFI and sometimes SWR issues depending on the model and design. Again, not all are the same, but I have heard of many RFI issues. If you only have room for this, again, go for it. Since part of the coax acts as a vertical, it will be a bit noisier than the traditional dipole. If you can opt for a balanced dipole, such as the ones in the previous section, I would opt for those instead. But that is just my opinion.

Verticals – Elevated Radials vs. Ground Radials

Technically a vertical antenna is just a sideways dipole if you really think about it. The radial system acts as the second half of the antenna. Though technically will behave slightly differently and the vertical resonant length will vary based on the efficiency of the ground radial system. As you add radials you may find that you must adjust things slightly. This is not a bad thing; this means efficiency is increasing. Technically verticals are the only truly non-directional antenna out there. All other antennas we are discussing in this book will be at least somewhat directional.

Verticals will suffer from ground return losses and far-field losses when compared to a horizontal antenna. Far-field losses are something you cannot do much about short of moving to a place with better soil or operating from the middle of ocean. Ground losses can be minimized when using radials and plenty of them.

Which one is better, elevated or ground radials? To begin with an elevated radial is technically called a counterpoise. These might be used interchangeably in some literature, so I thought I would point it out. I will also spare you the suspense, **there is no right answer here**.

At times this is a controversial topic and frankly depends on your situation, location, budget and space as well. The photo here shows my **Butternut** HF9V vertical with the 160m addon which I used a few years back. Once I got it adjusted just right, it was an outstanding vertical. Though initially took a little patience as there is an order which needs to be followed when adjusting it. This unit uses ground radials (120+) but the radials can also be elevated. And yes, it is on a hill and the radials are sloping. Not a problem!

I feel they are both good systems. I have used both methods. To oversimply this, the receiving seems to be a bit better on the ground radial version if at least 30 or more radials are used but the take-off angle is better with elevated radials according to folks who have researched this more than me. **Take off angle is very important; you want this to be low**. Generally, verticals will do well here. But you will need a good RF ground system for efficiency. This of course is not an issue for balanced antennas. **Verticals will have more noise.** That is just how it is, nothing you can do about it, though sometimes a 1:1 balun does help to remove RF from coax. More on this later.

Ground radials can get expensive if you run a lot of copper which I recommend over other metals. The same AWG as used in dipoles is perfect, I use #12-14. Wire has gone up in price over the years, so take that into consideration. Then there is also the time it takes to make them, put connectors on them and bury them if you do. You can also buy them premade, but you will grossly overpay, and I feel the quality can vary. I did not bury mine. Not much difference in my opinion but some studies have shown that buried radials work a bit better. I have not tested this, and my above ground radials seem to do the trick for me. Plus, mother nature will do the burial for you in a few years. I cannot even see mine anymore for the most part.

As a related side note, while the ocean's saltwater makes a great ground radial system, swimming pools do not, and this is due to safety. There are high voltages present in radials! I have heard stories of hams doing this, but let's just say it may not end well.

Two things to note here. If you bury your radials too deep, as in more than 4-5 inches (10-12cm) they will start to act like ground rods. I prefer to use insulated wire for ground radials. I also prefer this for elevated ones by the way. They will last longer for one, safety factor also increases. Voltage does get high while operating as I had mentioned.

Even if nobody is around the radials, I would still not go with a bare wire. I did try this bare wire test once and ordered some from a "*certain manufacturer*". Three years later they had almost completely disintegrated. To be fair, I am on the coast near salt water so you may not experience this as quickly. The insulated ones I made over 10 years ago are still running strong with minimal degradation.

I would recommend 8-12 ground radials minimum in different directions but do what you can. You can always add more, and I would. I also feel that when it comes to ground radials, length is not really that big of an issue if you have a lot of them. If you do not, go longer. I would recommend installing as many as you can within reason.

So, does it really matter if you use elevated radials or ground radials? Up to you really. I use ground radials as it is more practical for me and I sort of have the space. I know folks who use elevated radials/counterpoises, and this also seems to work fine for them, on a roof top for example. If you have a metal roof, like on an outdoor shed, this could work very well for a vertical placement as this will help you as well with both reception and transmit.

If a vertical antenna is being sold as not requiring radials or counterpoises, I would run the other way. I know some say they work fine. No, they do not. You can't beat the laws of physics. Even on your car vertical, the car is the counterpoise.

You will need to assemble verticals when purchased, though usually a lot easier than a Yagi in most cases. If you get a model with traps, this will add an extra layer of complexity as well as possible future failure points. However, if you do it right and with proper maintenance, you should be OK for many years to come. I will say, I have come across some poorly written manuals, and I am being kind here. Some even with mistakes, at which point I contacted the manufacturer. The response back was always appreciative, and I did notice some corrections made in future revisions. If you run into a similar situation, most antenna manufacturers do listen, therefore inform them of errors or lack of clarity if this applies.

I do like having a vertical even if I have a directional antenna for a given band. As I had mentioned, I can listen in all directions and size up the radio traffic and propagation then decide where to focus my attention.

My current setup consists of a **SteppIR** BigIR Mark IV vertical and a 160m vertical for the HF bands. You can find more about the SteppIR at steppir.com. I am using DX Engineering radial plates as well as a 1:1 balun from DX Engineering on both. There is an internal balun in the SteppIR which is just fine, but from personal experience I found this to perform even better. This is based this on my circumstances, so do some testing as I did.

I use a ground radial system on my SteppIR Big IR IV (pictured) as well as my K6MM 160m helically wound vertical, which I have built from a single trip to a local hardware store. Pictured on the left is the "brains" for the 6-40m bands. On the right is the much larger add-on to enable the 60 and 80m bands. This requires a second set of controller cables, which is not a big deal. As you can see, my radials are hardly noticeable.

You can find more information on my website about the 160m project as well as on the **John Miller, K6MM's** website. You can visit it at k6mm.com. He has also published an article on this in QST, in fact, made the cover with his "*A No Excuses 160 Meter Vertical Antenna*" article. He has pictures from at least two dozen others who have built this and submitted photos. Be sure to check it out! If you build the antenna, keep in mind that **160m antennas have a very narrow bandwidth** and depending on the number of radials, installation height, terrain, and size of capacitance hat, you may have to play around with adjusting the antenna wire length. These can get tricky so maybe not the best homebrew to start with. You will likely end up with 250-270ft of wire. I would try in 6" increments to get the correct resonance when building it. Start with more wire, cut as needed. Much harder to add wire. A **capacitance hat** is used in reference to the horizontal wires on top of a vertical. This can be a wire mesh or even a disc. It is used to reduce the resonant frequency and increase efficiency.

Both have a DX engineering radial plate and yes, **my 2 vertical antenna radial systems are connected!** Since I only transmit on one antenna at a time, this allows each vertical to take advantage of the other verticals ground radial system fully. Kind of like a 2 for 1 coupon! I often get asked about this and had no issues for close to a decade of having them connected. I hear very well for a city setting, at least even without the magnetic loop. I even hear surprisingly well on the 160m vertical. Some think this has to do with the sea of ground radials. Although if you are in a rural setting, I don't think I could compete for lower noise levels with you. That is ideally where you want to be for low band DX.

To those who say there is no room, I say if you have a yard, likely you can find a way as well. Above is my K6MM 160m antenna, with my modifications. On the left is my DX Engineering balun & radial plate with about 120 radials for 160m, very similar to my other vertical setup. I use this setup on all verticals I have ever owned and <u>highly</u> recommend them.

I also wanted to quickly mention **screwdriver antennas**. These are used in mobile installations and use the car/RV metal body as the ground. These can be used in home installations as well as portable use. You will however need to provide your own radial system, which can be done the same way as other verticals. Don't forget you will need to run cables for a screwdriver antennas controller. One good option for this is the MFJ-1924 which even has memory settings. There are many screwdriver antenna options out there, so do a little research if you want to go this route. Might be a great solution for an HOA situation as well, but if you can out up something better I would. These antennas are practical for some but very compromised. I do however know someone who used one within San Francisco city limits and has worked 200 DX entities with one.

Here is an interesting side note to the many 43ft vertical antennas on the market. The only band that may work on these without a tuner is the newer 60m band. This dimension was picked since the feed point impedance is such that it makes it easier

for antenna tuners to match. For most, these will work well. Though you will need at least a semi decent radial system. A large metal roof is always a good substitute if you have such a thing available, as I had mentioned earlier, though unlikely in the city. 43ft vertical will certainly outperform a 25ft vertical on 160m for example. So, in this instance, height does matter.

At one point the SteppIR Verticals with the 80m addon had issues with arching over when running higher power. This has been addressed now. I solved prior to it by adding a little extra spacing. This was accomplished by using plastic (do not use metal) spacers I happen to have left over from a flat screen TV mounting kit. These usually come with spacers and a variety of extra screws. A good example of why I never throw away any parts that can be useful at some point in the future. Point of the story, if you have arcing, distance is your friend.

I also replaced the metal plate they used at one time with the fiberglass one they provided which addresses the problem even further. My thoughts are better safe than sorry, so I had left the spacers in. Can't hurt! Distance is good. This antenna has been performing very well since even on the 80m band with substantially more power.

The Yagi and Other Multi-Element Antennas

There are many antennas in this category, and you will most likely encounter these the most if you visit a fellow DXer. These are rotatable, meaning you can turn them to give you the most favorable direction. They will have decent gain, even if only a few elements. Assembly can be a little difficult for some hams and needless to say, these do not come preassembled either and can get large quickly.

This group also includes **hexbeams** (sometimes called a hexagonal beam) and **spider beams**. Technically if you think about it, they just happened to be a folded Yagi. The old and proven technology in a new "*bent*" form. Hexbeams and Spider Beams are usually two element antennas per band, although there are some variants with less and some with more elements for some bands. These tend to be more homebrew though.

I have heard some argue against these being Yagi as technically they lack the 3 elements as it is sometimes "*defined*" for a Yagi. Many advertise them as a type of Yagi. I will vote, yes on this and say they are a Yagi since they are using the parasitic principle as defined in most textbooks, even if at times lack the full three elements.

Pictured here is my current main antenna, the **SteppIR** Urban Beam which is two elements on 6-20m and a single element on 30m and 40m. The reason they are single elements on these two bands is due to space. The 2nd element retracts completely on the lower bands, and the primary element actually occupies some of the space where

the 2nd element would be otherwise. This was designed as such to keep the antenna smaller. There is no way I could fit a 40m beam into my backyard if it was not for this. When in 30m and 40m mode, the Urban Beam is now acting like a rotatable dipole. No parasitic element is present as it is retracted, making it bi-directional.

I still feel that hexbeam antennas are some of the best value for the money. Before this antenna I used the KIO Technology hexbeam. This version was developed by Leo Shoemaker, K4KIO and I have to say it is built like a tank. This antenna survived many nasty storms and even an unintentional drop. Long story…Parts are easily and inexpensively replaceable if needed. Handles full legal limit with some overhead and covers the 6-20 meter bands with super low SWR though band options can be configured to your liking. The whole antenna is just over 10ft (3m) wide. You can read more about them at k4kio.com. MFJ also offers a hexbeam the MFJ-1846 and is very similar in design. You can find this at mfjenterprises.com/products/mfj-1846 . Though I do not have personal experience with this product, it does look like it would do well. On average you can expect over 5dBi gain on these antennas, but check the specs for the specific one you are interested in. It will vary from band to band. The front to back ratio is fairly decent on this unit.

Yagi antennas will have a combination of Reflector (back portion) Driven element (where the coax connects) and Director elements (front portion). Though when two elements either the reflector or director is dropped. If an antenna has more than three elements, generally speaking the director count will be the one increasing, so basically the elements in the front. Some longer antennas will "*grow*" in both directions usually somewhat proportionately. Rarely do you see more than five elements on an HF antenna, three seems to be the most common.

Yagi antennas are horizontally polarized, just as you see old school TV antennas (also Yagi) and this is fine for HF. Once you start to operate 6M and even higher frequencies on VHF and even more so in the UHF regions, polarity becomes important. By not matching polarity (antenna orientation) you will lose a lot of the signal strength, several dB in fact, above the HF frequencies.

For the 6m band **the generally accepted convention is that FM is vertically polarized, and SSB/CW/Digital (non-voice) is horizontal.** You may even see vertically rotated Yagi antennas here but that is for another book.

This "*digital*" reference above is not the be confused with things like DMR and C4FM (Yaesu System Fusion) which are digital voice modes. These are not used much on HF except with the rare exception, specifically D-Star (also on VHF/UHF) and FreeDV come to mind. A quick side note to prove a point. There is also something called circular polarization that you may run across. This is used on higher frequencies, even on satellite TV as an example. Here stations can coexist simultaneously on the same frequency using different linear polarization (vertical and horizontal) simultaneously due to the much higher frequency used. Therefore, using left hand and right-hand polarization plus vertical and horizontal, up to four signals can coexist on a single frequency. Pretty amazing! Why do I bring this up? Because this demonstrates the point

that if you are polarized "*wrong*" you may completely miss a station. Again, this is more for DXing above HF but it is an important concept to understand.

Some hams use **Moxon** antennas as well. These basically act as 2 element antennas via the use of a parasitic element, therefore giving directionality. They are rather compact and easy to build.

Horizontal Loops

The loops here are not to be confused with magnetic loops (covered later) which are usually or rather should be, mounted vertically. These loops are not magnetic but rather a full 1λ (wavelength) of wire and are mounted, per the title, horizontally. They do not need to be as high as dipoles and there are many designs on the internet for these, some of which are experimental in nature. Directionality and gain are also a bit better according to some articles. Directionality in this case can also work a bit against you, but it all depends on your situation. Some of these antenna designs even have the ability to switch directions via relays. This is one of the few types of antennas I have never used and is on my list of things to build and try myself. Many hams report very good luck with this antenna type. Due to how it is mounted, can be very space friendly.

Vertical Loops

Yes, this can also be done. These are oriented the same way as a magnetic loop would be. I think the most famous deign is from Gary Breed, K9AY and you can read about it at aytechnologies.com/TechData/HowToBuild.pdf. This loop does need a preamplifier and a grounding rod to work properly. This antenna will work best for the lower bands. This appears to be great for limited space as well and I have heard very good things about the performance. I have not built one, but this is also on my bucket list.

Low Band Antennas

This is not a type of antenna but rather are group of them which will work on the lower bands. I will cover some in more detail further down. These are generally what antennas **40m and below** are referred to, some would even say starting at 80/160m and below. Somewhat open definition, I will go with 40m as that is how I think of them. These can be verticals per above, wire antennas such as inverted L or inverted V. These look exactly as the letters suggest, more on these later. Slopers are also a practical option for hams in some situations as are horizontal loop antennas which we already covered. Loops can be a confusing term as there are magnetic loops as well as horizontal and even vertical loops as I have mentioned. For low band DX, you will mostly encounter the magnetic variety. Of course, there is always the beverage antenna if you have got the room.

Some folks have even shunt fed their towers for the lower bands, mainly 160m. I have personally not done this, nor do I know anyone who has, but this is an option worth exploring.

For a serious low band DX setup, many recommend having multiple antennas on these bands. A low noise receive-only antenna can provide a critical advantage in working low band DX. But of course, this is not realistic for most of us who do not live in a fantasy world or have acres of land. Though, there are some workarounds which I will go over. Just try and do your best with what you got to work with. It can be done!

I would focus on these antennas during low sunspot conditions and winters. So, if you are trying to decide what to put up, start with the ones which make the most sense given the solar conditions (where we are on a given solar cycle) and seasons when reading this.

Magnetic Loop Antennas

This is the smallest option on this list, but by no means should you judge them based on size. Sometimes hams just call these "*mag loops*" for short. Because you know, it takes so much more effort to say "*magnetic*". These antennas do come in two main

flavors, receive only and ones that can be used for both receiving and transmitting. When they say, <u>receive-only</u> they really do mean it. I have in the past accidentally transmitted into a receive-only loop antenna. The photo below is the result of this mistake. Burned a set of resistors and damaged a transistor nearby. You can actually see the holes created by too much power on some of the components. About 60W did this. The repair was not horribly expensive however the shipping was another story. How did this happen? I was using a Yaesu FTDX-3000 at the time. The 3rd antenna input was used for the receive only loop and therefore I set it to RX only. Then I performed a firmware update and forgot to set the 3rd antenna input back to RX only. Oops! So be very careful and perhaps have a checklist if you have a special configuration and need to reset and then readjust everything.

I have owned **AOR** and **DX Engineering** loops, both performed very well! I am still using my RF-PRO-1B regularly on the low bands. These will perform about the same as a dipole but have more directionality. They will also have a lot less noise. A small TV antenna rotator is more than sufficient for these as they are light and wind load is not a big factor. AOR can be visited at aorusa.com and of course DX Engineering at dxengineering.com. For many die-hard DXers, this is practically their home page. I know I have quite a wish list here!

Another option is the **W6LVP** loop, check it out at w6lvp.com. **MFJ** also makes a couple of loops, some of which also transmit, though generally do not handle much power, for sure not even close to the legal limit. I have heard good things about **PreciseRF** magnetic loops, and these can be found at preciserf.com and their HG3 QRO antenna can even support legal limit on most bands and with an optional second radiator loop down to 80m. **Chameleon** receiving loops are also very popular with some. I've never heard anything negative about these, which is a good sign. And believe me hams talk! These can be found at chameleonantenna.com.

Not to state the obvious, but it is not a good idea to sit too close to magnetic loops when transmitting, if supported that is. This is for obvious health reasons. Especially if you operate them indoors! If you can, put them outdoors.

These antennas are extremely directional and are great at getting a null as well as removing noise by rotating them away from a noise source. This is one of the reasons they are so very popular for receiving on the low bands, though usually as a second antenna. I feel for 160m work they are almost a must have within city limits. I often listen to 80m and 160 on my magnetic loops. I would recommend these for the lower bands, or at minimum a separate receive only antenna if you got the space.

Some folks do mount these horizontally and apparently this works for them. I have seen this on cruise ships (salt water likely helps) as well as apartment balconies. I would recommend sticking to vertical orientation on these but what the heck, try both.

Some if not all magnetic loops connect to your radios PTT circuitry usually via an RCA type cable. This is in case there is too much power heading towards them, they can switch off the preamplifier to prevent damage. Make sure you use this if available on your unit. Obviously, units which transmit will already do this for you.

Magnetic loops which also transmit can be very temperature sensitive. You may need to adjust your tuning from day to night and from season to season you may find the settings will be very different.

There are ways of using 2 loop antennas and even phasing the signal. This can really be good for removing local QRM and general noise cancelling as well as providing better fine-tuning capability with more precise directionality. Using the DX engineering DXE Pro line with a DXE-NCC-2 you can gain an extreme DX advantage. You will need rotators on both loops of course. I will talk about this more later in the book.

Beverage Antennas

These are <u>receive-only</u> antennas and primarily used on the lower bands, though they can be used somewhat higher and with the added benefit of requiring less space than their lower band counterparts. The name is the last name of the inventor, Harold Beverage. Sorry to disappoint but has nothing to do with liquids.

These are unusual antennas still as instead of needing to be up high, they are designed to be down low, as in eye level. They have very low noise and are extremely directional though will require a very long run, depending on intended usage frequency. You do not want to put any sharp bends in these installations or slope them much. Ideally you want them level and straight as much as possible.

DX Engineering as well as KD9SV both have very good systems available. All you need is to provide your own wire or 450Ω ladder line, depending on your configuration. All the above is available at dxengineering.com and if you are serious about low band DXing and have the room, I would highly recommend them. If I had more room, I would have these all along my fence lines!

For receive antennas, I would recommend a receiver guard. This would be a good example as to where they could come in hand. These protect your receiver front end in case of a strong nearby signal. Those with a small lot with many antennas, such as myself, come to mind. For example, if you are transmitting on one transceiver while monitoring the same band on another, these are a must in many cases. Bandpass filters are an option as well, assuming operation on different bands. Some radios already have these built in.

Inverted V Antennas

Sometimes these antennas are also called "*Vee*". We can look at these antennas as a dipole, in fact they are a dipole, though sometimes cut slightly differently in this configuration. They may need to be a touch longer when installed in a "V" formation versus a dipole configuration. An antenna analyzer is your friend here. The beamwidth will also be a little wider than a dipole.

These are to be installed with the peak or center at the highest point and the arms closer to the ground. Therefore, the resonant frequency may shift a hair. The top angle should be 90 degrees or more for the best performance. Think of them as an upside-down V with a support in the center, basically.

In an ideal world, you want non-conductive support in the middle even if using coax. A tree is good, even though it is slightly conductive but be careful with branches touching the wire as this can make your SWR go nutty. These are going to have less directionality than a dipole but work well, though not as well as if you mounted a dipole at 180 degrees. Ideally, you want them to be at least 30ft (10m) high, or even higher for lower bands. The ends can be as low as 8-10ft (2.5-3m) above ground. Some use ladder line to bring it down to a tuner, others have a balun at the peak and coax down to the shack. This is the way I did it with a 1:1 balun. These are one of the most popular choices for DXers for the lower bands due to space, only second to verticals.

Inverted L Antennas

Now that we have covered the "V", here is another letter for you! These are often used as a cheaper alternative to a vertical and behave somewhat similarly. Basically, this is a wire antenna where part of it is vertical (going up) and ends in a horizontal suspended wire after the 90-degree turn. The sharp bend is like in an "L" hence the name. The "L" can be tipped either way. Shorter height and more horizontal portion or higher up and shorter horizontal section. I have seen designs for both, seems the main factor is space and to a degree depends on the frequency it is designed for or directionality desired. My opinion, go as high as you can given these are likely to be used on the lower bands. It also will save you space. Just as a vertical antenna, it will need a radial system. These can be elevated radials (counterpoise) or ground radials. Though I feel most who use these opt for ground radials.

As with the Inverted V, try and use a non-conductive mast if you are using one or at the very least, offset it as much as you can. Thicker fiberglass or firm PVC can work well too or a tree. These are very popular antennas when space is at a premium and work very well. If you do not have the space for a full horizontally extended dipole, this works well due to lower space requirements. Just like the Inverted V, should be easy to install.

RF Exposure

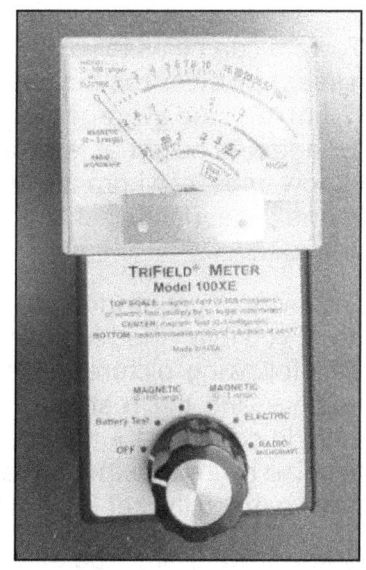

This is a good way to close out the antenna section by dealing with another safety topic. I am sometimes blown away but stories I hear about equipment misbehaving when radios transmit. Frankly, some of them I find rather entertaining. I used to have a PS3 which ejected DVDs when I transmitted on 80m band with any power. I addressed it right away with ferrites.

I also heard stories of garage doors opening (now, that is scary) and Christmas tree lights glowing in sync with CW. So why am I mentioning these stories? **RF is real and you need to be very careful with RF exposure.**

One way to check levels is by using something like a TriField Meter by **AlphaLab, Inc.** pictured on the left. This one is model 100XE and is very widely used. This model measures from 50MHz and up, which is where you really need to start paying attention to safe levels. Their digital TF2 version works down to 20MHz. These units measure magnetic, electric and radio frequency radiation, hence the "*tri*" in the name. More information can be found at trifield.com.

As a side note, **RF and RFI are often confused and are not the same thing, though related**. RF is simply Radio Frequency, RFI is Radio Frequency Interference. This is what causes issues for both reception and RF generated can interfere with other electronics therefore causing RFI. RF is needed for radio communications and is not a bad thing. Without it, no ham radio. Too much in the wrong place is bad and can turn into what we call RFI.

The lower the band/frequency the safer it appears to be, meaning you do not need to do an RF evaluation if you run less than 500W on the lower bands for example (40m and under). However, once you get up to 6m and 10m, this drops to 50W, meaning you really need to make sure that you are far enough (as is everyone else) from your antennas when transmitting with some power. I keep antennas between 75ft (23m) and 150ft (46m) from the house and all are elevated above roof level. Since my 6m antenna is the closet to the house, in part to prevent signal loss via coax, I try to not use serious power when pointed in the general direction of houses for that reason. Higher power levels as well as focused power can become an RF safety hazard as I had mentioned earlier. I encourage you to visit the ARRL RF exposure calculator at arrl.org/rf-exposure-calculator.

This also applies to those who have antennas on relatively short towers on masts mounted near your house. I have seen folks beaming right into their bedrooms. Really need to be careful, especially if you are experiencing some of the above issues already. Be safe folks!

Ham Radio Towers & Masts

This topic can take up a whole book. Most licensed hams do not have a tower or even a push up mast, and frankly most hams do not need one depending on their interests. But nearly all hams dream of one.

Towers can be as tall as 200ft (61m) in the US, assuming no airports (due to flight path issues) or other restrictions nearby. The shortest practical masts start at around 25ft (7.5m) and the average height seems to be about 55ft (17m) for both masts and towers for antenna height at least around the San Francisco Bay area. This is based on folks I know, most of whom are DXers. My push up masts are also at this height or just below in one case. Push up masts are a much more practical solution for many, not to mention cheaper.

One way to get around not having to push up a push up masts is by tilting a mast as I do. If you even tried to push one up once vertically, you are my hero. Not an easy task! The following picture is one of my setups. Tilts are available via many ham radio dealers and the rest are of the supplies are from my local hardware store. These include concrete for the post holding the tilt and pipes. This makes it a one-person job to raise or lower without the aid of cranks or motors. Though you can add one. I also have a much thicker push-up mast for my Urban Beam, that is motorized due to weight.

Some towers are free standing and others are guyed. Same goes for masts. If you require guying for any of the above (recommended) make sure it is not a tripping hazard.

There are also towers and masts which are motorized (like mine) or sometimes hand cranked. They can be lowered during storms easily as well as for maintenance. Some ever tilt over once lowered. Of course, the cost goes up, but it will save you a lot of time and stress if you need to make adjustments. Not to mention the cost if you must hire someone to climb your tower. There are some considerations if you have a crank-up tower. For one, you will need to use somewhat more flexible coax and hardline is out unless you switch to more flexible coax at the base for example. Chances are, if you have a tower, you have the room to tilt it once it is lowered. I do recommend doing this if your wallet agrees and you have the space. This may require 30-35ft (9-10m) clearance, possibly more with larger towers.

When purchasing antennas and towers, pay attention to wind load capacity. Also consider how it will be delivered and installed. This is not only for the tower, but the concrete needed for the base. Is the location easy to get to? In my case I had to move concrete up the hill side 60lbs (27kg) at the time and mix it there. I am not doing that again!

The permit process can vary a lot from place to place for a tower. Some make it next to impossible to erect anything, as is the case in many California communities. Other places could care less what you do. Always check your local jurisdiction to get details. Rules can also change with time. If you do need a permit, the cost and time involved will go up obviously. Regardless of permit or not, follow good engineering practice and recommended specifications for pouring concrete.

Most of the time masts will not require a permit, but again, check locally to make sure you are following the law and are doing this in accordance with the local code. I have mentioned this already, but it is worth repeating here. Never erect anything (this includes a vertical) if it is anywhere near a powerline. Rule of thumb, take twice the height and if the base is close then that to a powerline, it is not safe.

If you decide to erect a tower attached to your house, you still need to take precautions. Maybe even more so. It will need to be secured to your house with braces. This will add extra stability in harsh weather and in the case of California, even earthquakes.

An example is illustrated here in this photo on how it is properly performed. Ideally you want a tower 25-50ft (8-15m) away from structures, but I know this is not realistic for many hams.

Rohn is perhaps the most recognized name when it comes to ham radio towers, and they can be found at rohnnet.com. **US Tower** at ustower.com has everything a ham could want as well. **Tashjian Towers** at tashtowers.com has a variety of nice units as too, they also tend to be more visible at ham radio conference so pay them a visit if you see them. DX Engineering also sells towers as does Ham Radio Outlet.

You need to know a bit of chemistry when it comes to towers. Yes, you heard it right. **Do not connect copper directly to galvanized metal or aluminum**. You basically just created a battery if you do. Oxidation will occur and structural safety issues will arise. Use stainless steel as a transitional metal if you run into this.

For push up masts check out **3 Star Inc.** at 3starinc.com.They carry many brands and I have personally purchased several times from them. There are others out there, but these are likely the ones you will run across at conferences and are what other DXers are using. Though, always worth asking around.

One last thing. Remember to think outside the box as Stan, KE5EE said on his QRZ site. You can visit it at qrz.com/db/KE5EE. Your options are only limited by your imagination. The mast pictured here has a pair of hexbeam antennas. Stan also has a four hexbeam version now, which is likely the only one in existence. Why not do it if you have the room. Wonderful antennas and gain is now increased. *Photo courtesy of Stan "Stax" Schwartz, KE5EE.*

The height of the antenna's installation does matter a bit when it comes to towers. Different bands have different ideal installation heights, especially if you are looking for the right take off angles. If you have a tri-bander at one height, one band might be ideal (or close to) in height, the others may not be. Is that bad? No, of course not. It will work just fine. If you can get a tri-bander or a SteppIR up 100ft (30m), you are going to be in heaven regardless of ideal height or not. However, for 6m operating the 100ft elevation would be way too high. If you are using multiple antennas, also be aware of possible coupling. This is something you may need to contact the antenna manufacturer(s) about as they tend to deal with this question a lot and likely have some great suggestions for you. You can also always use an antenna modeling program, though that tends to be a little more advanced topic. Many excellent books are available on that subject matter.

Other Things to Consider

Some hams seem to forget damage control. They get the antenna up and they think they are done. By damage control I mean, good guying for wind, grounding for lightning, sealing connectors properly and just general all-around safety. I have seen all three severely neglected by some hams in the past. Not to worry, I will take the names to my grave.

Remember that all corners of a tower will need to be properly and separately grounded! If you cannot go deep, increase the number of grounding rods. Remember to space them accordingly. There is usually additional information available from your tower dealer as well as online if you ran into this issue.

Seriously take into consideration guying, even for masts. If you are in a windy area like me, you may need to double up on them. I added extras myself for good measure. **Guy wires, and especially ropes do stretch with time, even if the manufacturer claims they do not.** It is impossible for them not to stretch at least a little over time, so don't believe otherwise. If they insist that theirs is special and you believe them, you might as well get a can of "*DX lubricant*" for your antennas from them as well while at it.

But in all seriousness, it is a good idea to tighten these ropes up and at least to check on them before bad weather rolls in. **Mastrant** brand rope from the Czech Republic **(OK)** is my personal pick now after having some bad experience with supposedly good rope from my local hardware stores. They do have different grades so buy according to your needs. They are upfront about stretching and load, I always appreciated this from them. They are at mastrant.com but can also be found at some ham radio conferences.

Some like to use metal guy wires for push up masts but I think it is overkill. Plus, I like to limit the number of metal objects near my tightly packed antenna farm. If you have a tower, you want to use something more serious if it requires (or even recommends) guying. For my verticals, I use all rope support and have not had an issue with them since I switched to a reliable brand.

When buying any rope be sure to check the UV rating if possible. **Black rope does best** in the sun, lighter rope, as a rule of thumb, not so much. The same is generally true for coax, which is why they are usually black. Some ropes degrade faster from sunlight, specifically UV. Some even have issues with extended exposure to moisture. I have even seen mold even in my setup. **Dacron is also very good** and available online from multiple suppliers.

Distance from what you are trying to support is also rather important when attaching a guy wire. You do not want to be too close to the base as it will not provide as much support as if you spread out a little. Many companies tell you the ideal angles and/or distances. Also, there is nothing wrong with some redundancies as I had mentioned.

I use wire ties to secure things outside quite often. The same goes here, use the black ones as they put up with UV light much better when outdoors. White or translucent ones I found to dry and crack within a year in some cases. Save those for indoor use or give them to someone you do not like. Just kidding, do not do the latter. The darker the color, the better they tend to hold up.

Lastly, making things watertight is not an option. I **use high quality 3M electrical tape**. No, not all are created equal! I also use **Coax Seal** in conjunction with the above. Electrical tape first to keep the metal from getting sticky. Then, Coax Seal and one more wrap of electrical tape on top of that to keep things from sticking to it all.

This combination seems to do the trick for me and had no issues so far. Been doing this for many years, never had a leak. Always wrap during dry conditions so you do not accidentally seal in moisture. Water can not only cause rust but mess with your SWR as well as ruin coax and possibly other connected gear. This is no joke and not a good place to save a few cents.

Rotators / Rotors

This is what will be used to turn your beam. Both terms, rotator and rotors seem to be used interchangeably, even regionally. These come in all sizes and qualities…Yes, there is junk out there too. I have always been partial to the Yaesu line of rotators; however, I use the **Green Heron Engineering** controllers with them. I find that this integrates into my setup better, and to be honest, they are just pure works of art. The owner, W2FU, Jeff is also very responsive and listens to his customers. These controllers can also be computer controlled via a simple USB 2.0 cable. Well worth the extra money and you can easily rewire the controller boxes if you decide to change your rotator to a different one. Visit them at greenheronengineering.com.

MDS also makes very nice after-market controllers. These are also USB connected and therefore computer controllable. **MicroHam** has a new rotator Controller which works with pretty much anything out there called ARCO. You can read more about it at microham.com/contents/en-us/d50_ARCO.html.

Rotators are rated by capacity and wind load mainly. The price seems to correlate with what they can do or how long they will last in <u>most</u> cases. Some have brakes, meaning when in a given position, the wind is less likely to cause issues, as in move them. If you live in a high wind area, I will say this is a must have. Your TV antenna rotator will likely not do well for a beam, however it can support a hexbeam antenna if it is not heavily used, as in for contesting for example.

One word of advice on the bolts for rotators. **Bolts can and will come loose. A little Threadlocker goes a long way** and may save you a small fortune. Use it! I go with the blue version personally. Do not go too strong though as you will never get the bolts out. These are color coded based on strength or rather, locking ability. **Purple is the weakest, Blue is the medium strength, Red is high strength.** Often there are others, and this can vary with brands, but these are the ones you will run across most likely. **Loctite** brand is one I also had great luck with.

Sadly, I learned the above the hard way when a pair of bolts flew out of my rotator after a few years of use. Never found them either. If you need to replace your bolts then, always get stainless steel. Some, like Yaesu will be metric variety so be careful. Your local hardware store likely carries them, but quality does matter for outdoor use. At the very least get the galvanized variety, preferably stainless.

Depending on your setup, you may need to use a thrust bearing. Chances are you will though if you get serious. Examples pictured here in a tower installation as well as those by Kurt Andress, K7NV (SK 2022). These basically help to take the load off your rotator. Most towers will have an obvious place to install these but those with limited space using a push up mast, may not. I would still recommend using one of these. This can be accomplished by purchasing a set of **Barenco** brand brackets from the United Kingdom **(G)**. You can find them at barenco.co.uk. These allow the mounting of your rotator on one and a thrust bearing on the other using a separate, shorter mast. These are very well made and work amazingly well. I use two sets for both my Urban Beam

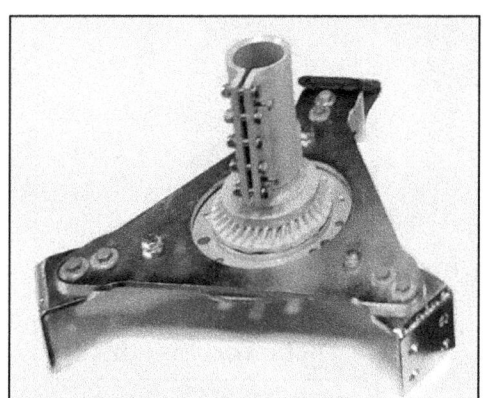

installation and on my 5 element 6m M2 antenna. Could not find anything similar in the US at the time, hopefully this will change. I am sure I am not the only one who needs these.

Lastly, it is recommended that you use a more flexible coax around the rotator so it does not experience fatigue as much as stiffer coax would from the constant rotation. Perhaps using a flexible 8ft-12ft section joined to a lower loss, but stiffer coax for the rest of the run is the way to go for most.

Since it relates, I want to throw a word in here about beam headings. North is going to be 0 degrees but also 360 degrees. Some rotators go to 450 degrees. Many Yaesu rotators do this for example. They are not breaking the laws of mathematics but rather they can go past a full circle to save you time when turning.

Your beam heading will change based on your QTH. So, if you see someone beaming from South Africa (**ZS**) to Hungary (**HA**) at a certain beam heading, mine in California (**W6**) would be different.

Transmission Lines

Also sometimes called feed lines. As I have already said, **every dB counts! This is especially true for DX.** The length of the transmission line plays a part as well of course. For HF this is not as big of an issue, but it does add up. Do not expect older or used coax to behave the same. 10-15 year old coax is pretty much junk if it has been sitting outside exposed to weather. The sun, specifically UV, breaks down the outside plastic and water will get in eventually. UV resistant is not UV proof. If you get snow and ice as well, it will be even more so. You likely will need to replace them! I am right about at that time where I also need to do so. Your indoor coax cables might be still OK after decades of use.

Loss in dB per 100 feet (30.5 meters)			
	1Mhz	30MHz	50MHz
AVA7-50, Heliax®	0.019	0.105	0.137
LMR-400®	0.2	0.7	0.9
LMR-400 Ultra®	0.2	0.8	1.1
RG-213	0.2	1	1.5
LMR-240®	0.24	1.3	1.7
RG-8X	0.5	2	2.1

Coax cable loss is usually measured in 100ft (30.5m) Increments. **Loss also goes up as frequency goes up.** The most loss is on 6m for an average DXer, the least is on the 160m band. At 50MHz (6m band) the RG8/X loss is 2.1dB, RG-213 will be 1.5dB and LMR-400 is at only 0.9dB of loss for 100ft. Big difference!

These numbers in my table do vary a little from various manufacturers but generally fairly accurate for general reference. The LM-400 or equivalent cables do have better shielding than the RG-213 so this is where the main difference come in when it comes to performance. **Try and use the best, lowest loss, coax (within reason) you can**, and this is even more important when the cable run is long. If you have a tall tower further away from your shack, do explore a hardline option. But if you are using a dipole with 100W 25ft from your operating position, your needs will be very different. **The minimum I would use is LMR400 coax or equivalent.** In fact, that is what I use everywhere on HF. I know some will disagree and think I am exaggerating, but dB loss is your enemy. Also keep in mind that

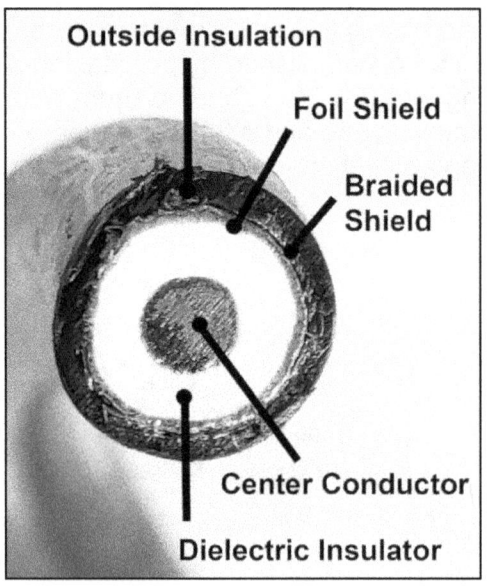

some coax does not handle legal limit power if you ever decided to get an amplifier. Therefore, might as well plan and be ready. You can get away with less expensive coax on 160m than on 6m. Some coax cable may handle legal limit on 160m, may not do so on 6m or particularly 2m if you run power there.

Photo courtesy of Stan "Stax" Schwartz, KE5EE. Above is the size comparison of the LMR-400 and the Andrew AVA7-50 hardline 1 5/8" coax. The LMR-400 fits inside the AVA7-50. I am sure it is obvious which has less feedline loss.

Especially important to calculate losses once you get to 6m as it really starts to hurt your signal. A long coax run I would define as being over 100ft. I am right around this currently. While you can use a preamp for 6m to compensate for this as I have, it will introduce some noise and switching/bypass adds a lot of extra complexity most hams do not want to deal with. However, you also do not want to be too close to your 6m antenna as RF can be an issue on this band. This is not only because it is bad for you when running power, as I have already covered, but also because RF can get into your shack gear and create a mess. I had the most issues with 6m and 40m in my shack even with all the precautions I took.

Avoid using couplers if you can. Run a cable directly to the antenna or your antenna switch and then direct to the radio. Couplers add loss, though not too much. But they also increase the likelihood of issues, like water getting in, etc. Even the best quality ones can suffer from this.

LMR600 is a lower loss coax cable and there are also a wide variety of hardline cables out there. These are much harder to deal with for the average ham and frankly there is very little extra to be gained from using them once you compare cost and the added complexity, at least for most hams on HF. If you are swimming in money, have the time, know-how and are very serious about DX, go for it. Just do not forget to invite me over to operate.

Another option is to use 450Ω ladder line. There is also a 300Ω version and you will have a lot lower losses than with most coax, but you also must be careful. These may require a balun to match your radios 50Ω input and need to be away from metal or anything that can create issues. Some antenna tuners do have connectors for direct connection to a ladder line if you can run it directly to the tuner. If you can manage to install these without coupling problems, this might be the way to go! **Ladder line loss is 0.5dB only at 100ft (30m)**, which is amazing. Although KE5EE's setup still beats it. Sometimes these are also called twin lead and window line at 300Ω and 450Ω respectively. Though as with many things in ham radio, the names are sometimes used interchangeably.

How is this done? In theory the radiation from the two wires cancels. Each lead is technically an antenna after all. This is if it is at least 3ft (1m) off the ground and at least that much away from metal objects or anything that conducts electricity. If you live in an area with true four seasons, check with local hams and likely there are other things to consider. Think snow & ice!

After a while, you may notice your SWR might go up and down. This could mean either an antenna failing, or the coax is failing. Or even both! Higher SWR is usually, but not always, means the antenna is failing. SWR is calculated based on the return trip, as in reflected power. In part because of this access to a quality antenna analyzer is a critical piece of equipment for any ham.

If the loss in the coax goes up due to damage, the SWR will appear to be miraculously down. Especially if the shielding is damaged, which is the most likely thing to occur. Think of old coax and rodents. In one case I know of, a lawn mower! Normally this low SWR would look great, but this is not good news. Power is not getting to the antenna like it used to. **Antennas SWR curve should look the same as long as the weather conditions are the same.** Seasonal changes are somewhat normal. Always keep your eye on changes. Documenting readings from your gear as well as an antenna analyzer is not a bad idea.

I do recommend the direct burial of coax if you can. Most coax can be buried but check with the manufacturer. If you have issues with critters chewing your coax above ground, this might be a solution. Some like to bury their coax in a protective sleeve. A common one is PVC pipe. Make sure you use large enough PVC to allow for easy removal or replacement, basically at least 1" diameter. Did you ever think there was so much PVC pipe consumed in the DX world?

Some hams even attach a string to the coax end before removal to make it easier to pull the replacement through and to make it easier to add extra cables, such as rotator cables, in the future. Be sure to caulk both ends to keep critters out once the cable(s) are in. Also drill small holes on the bottom of the PVC to allow water to escape in case it does find its way in. Of course, not to state the stupidly obvious, do this before you put the coax inside the PVC.

If you need to go from hardline to an LMR-400 coax for example, this is best done before entering the shack such as this photo from Stax, KE5EE. ***Photo courtesy of Stan "Stax" Schwartz, KE5EE.***

If you are unsure of coax quality or suspect possible damage, there are a few steps you can take to test. To test for shorts or opens you can do a basic test by measuring for resistance with an ohm meter. The reading should be zero or very low if measuring from center conductor to center conductor or shield to shield. If you get a reading between center conductor and the shield you got yourself a short. This happens sometimes, especially if the ends were put on wrong and some of shielding wire is touching the center conductor. Neither is good! Another test is to connect your center conductor only to the radio without screwing it on and compare to fully connected with no antenna or dummy load attached. If the noise decreases, you may also have a short.

You can also use a megger device. These are not that common in most shacks but maybe they should be for more serious DXers. **A megger is used to measure insulation resistance by sending a high voltage down the coax**, or whatever it might be measuring. Megohmmeter is the official name, so one can see why megger rolls of the tongue more easily. These cost around $100 on average and are a lot cheaper than tossing a good spool of coax. Or not tossing it and thinking it is your antenna!

Before installing your coax, inside or outside, label them or color code them. The easiest way to do color coding is by using color electrical tape. I use green for my ANT1 and red for ANT2 for example and have a system for all my VHF/UHF cables as well. Looks like a rainbow threw up, but I can always trace things! I just do a few turns about 1" below the connector. This makes it easier to trace coax later. Obviously if you are color blind adjust accordingly. Labeling is also easy, using a Dymo labeler for example which I use for my band specific labels. You may know which is which now, but a year or two down the road I doubt it. And I hate to say it but will not get better with age.

Remember that there is 75Ω coax out there, such as used by your local cable TV provider, though rarely used in ham radio. Theoretically you can match these as well but why bother. I do know some frugal hams who have used this type of cable and one is now off the air due to frustrations, the other is still active but seems to have a lot of issues. I think you get the point. There are times some antenna designs use 75Ω coax sections for matching purposes, but this is different than using them as your primary feed. Some receive only antennas may also use this type of coax and it is perfectly fine in those instances. Always follow the manufacturers recommendation.

One last important thing. Be careful with cheap, "*knockoff*" coax. Yes, you will find some deals online, but you may not be getting what you think you are getting. Anyone can print a number on the coax. I do not think I need to repeat where these are generally from. Given the cost per foot, not a big surprise. **Get your coax from a reputable dealer** such as Ham Radio Outlet or DX Engineering for example.

Baluns & Chokes

Baluns are defined as a passive coil which basically balances signals. This is greatly oversimplified but just go with it for now. Short for **BAL**anced to **UN**balanced. Baluns are used for transmission line matching purposes. An example would be two different types of feedlines.

Common types of Baluns are 1:1, 4:1 and 9:1 but there are others out there, some go double digits. Dipoles for example, would use a 1:1 balun as I had mentioned, and I use one with my 30m dedicated dipole. This really helps with feedline noise which we will cover shortly. Baluns come in a variety of power ratings from QRP (low power) to many kilowatts. **Always pay special attention to the ratios as well as the power rating.**

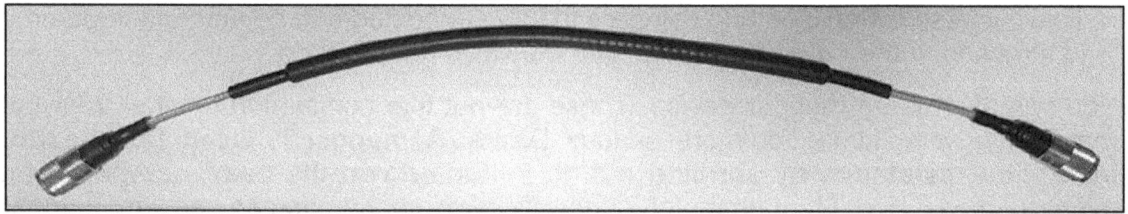

Choke baluns can be just a bunch of ferrites installed on the outside the coax to trap return current as pictured above. I use this liberally. Kind of fitting for San Francisco, isn't it? Never mind…moving along. A choke balun is <u>not</u> a transformer balun when it has just ferrites on the outside. In fact, technically some would say it is not even a balun but a choke. I would tend to agree with that. There are many who use these terms interchangeably in literature, therefore use caution. Also, it is important to make sure that the size and type of ferrites used is correct and fit snugly around the coax. If they are loose, it will not work as well, or perhaps even at all.

Coax 1:1 choke baluns can be made without ferrites. These are called air baluns and the ugly balun in the next section is an example of one. They will introduce a bit of loss though as you will likely add another 20-25ft (6-7m) of cable at the lower frequencies. This is still less than a dB in most cases, but the tradeoff can be huge in noise reduction. Not a big deal when it comes to HF frequencies but do keep track of your dB losses. They do add up. Basically, these are made by creating a loop of 5-9 turns then tying it all together with wire ties. Lower frequency usage may require a bit longer. There are many formulas online to reference if you want to learn more about this.

If you go from a 450Ω ladder line feed for example, back to 50Ω coax you would certainly need a balun. 450Ω to 50Ω would require a 9:1 balun. Divide the large number by the small number. Sounds a lot more complicated than it really is.

Never snap ferrites on a ladder line. Will not only not do anything positive but will degrade performance. In fact, you will likely never see a good SWR. I have seen a photo of someone online doing just this. Maybe it was a joke, but just in case, it is not a good idea nor does it many any sense. Off center fed dipoles often use a 4:1 balun for example and end fed antennas are usually 49:1. As you can imagine, this involves some serious winding due to the ratios. This likely explains the increased cost of these baluns as well.

There are two camps when it comes to baluns. Those who do and those who do not. Some love them, some even hate them. There are many reasons and explanations as to why people do not like them or like using them. Frankly, some I get, some I do not. At times it comes down to a single bad experience, such as using the wrong ratio and never getting an antenna to work right because of this.

I have run antennas both ways and I find using baluns to be very helpful. I found that my antennas to be easier to tune when needed as well as help with return current. **I use baluns abundantly and all my HF antennas have baluns. I am a big believer in using chokes especially.** General "*guidelines*" also state that anytime one goes from balanced to unbalanced you need one. But this is not always the case, in fact, many hams never use them. An example is the common dipole. Many connect these directly to coax without any baluns. Is that wrong? Nope. Your call!

If you are a "*user*" I would recommend getting something that will handle more power than you use. If you are running 100W, get one that can handle a couple of hundred Watts at least. If you run the US legal limit of 1.5kW then I generally opt for a 5kW balun. Overkill? No, not really. If your SWR is off accidentally or something goes wrong, you may even manage to blow a 5kW balun. I have seen blown ones, so it can be done! There are even 10kW versions out there for a reason.

Baluns can get and will get hot and can also fracture if they contain ferrites rings. This can be due to heat generated (as above) or being dropped. The higher the rating, the less likely a heat fracture will occur if you exceed specifications. There are many good books written on the topic and I would recommend picking up one of these if you get very serious about extending your antenna farm.

Ugly Baluns

These are some fighting words! Do not let the name fool you, they do work, and I don't find them ugly. These are generally used to fight common mode noise created by common mode currents. Insert inline and really helps with the issue, though unlikely to

completely remove it. With ferrites on your coax in addition to one of these, you should be able to control common mode noise well. I use them in conjunction.

When I looked around the internet to build one, the most common one I ran across is made using a 4" diameter PVC pipe (here we go again with the

PVC) using about 20ft of coax. Some folks use larger PVC, but does not make too much difference. In fact, that is what is pictured here and is used in my antenna farm, currently on my 160m antenna. Read about a great design at hamuniverse.com/balun.html and to see a video as well instructables.com/Air-Choke-Ugly-Balun-for-Ham-Radio.

You can also build an enclosed version which may be better for some scenarios. You can read about it at hamuniverse.com/kg5dcmencloseduglybalun.html. Oh, and guess what it is enclosed in….Say it with me. PVC!

Connecting it all Together

Getting Cables to your Shack is something that many do not think about initially. Then panic sets in at times since it will require drilling, and this may require some creative planning for many. Once the coax enters the shack you want to make sure it is watertight as well as easily accessible to fix issues in the future. And likely there will be issues, if nothing else at least an eventual replacement will be needed due to age.

Bringing coax into the house can be a little tricky depending on the complexity and your options. If you have a brick house or a concrete wall, might be a little harder for example depending on what other openings are nearby. Stucco and drywall are a breeze to drill. In my case, I also had to go through metal siding.

Make sure you do not drill into or run cables near electrical wiring, water pipes and especially gas pipes!

The solution I came up with can be purchased from most hardware stores. It is basically a watertight plastic electrical enclosure that I mounted and on the top portion I applied plenty of silicone. Not the bottom, in case water gets in, it does need to get out. It is tightly secured against the wall with large wood screws into studs. The back side of the unit has holes leading into the house.

This has not leaked so far, and it has been 10 years now. Might be due for maintenance come to think of it. The LMR-600 cable on the top leads to my discone scanner antenna. A discone by the way, is a vertically polarized cone VHF/UHF antenna. The box cover can be removed with 4 screws if I need to feed more cables. Although I am maxed out now as one can see. This method also makes it easy to do repairs, replacements, etc. with ease. On the bottom I drilled holes large enough to attach electrical conduit adapters. The inside has threaded nuts. Openings are about 1" (2.5cm) times 10 holes. Each hole can support two coax cables with ease, and I managed to run rotator and SteppIR control cables as well parallel to these. Should be plenty of space for other nutty hams like me with way too many antennas.

An alternative to this feed through is available from MFJ. The MFJ-4600 through MFJ-4605 feedthrough panels might be ideal for many who have certain types of windows or cavities they can use these for.

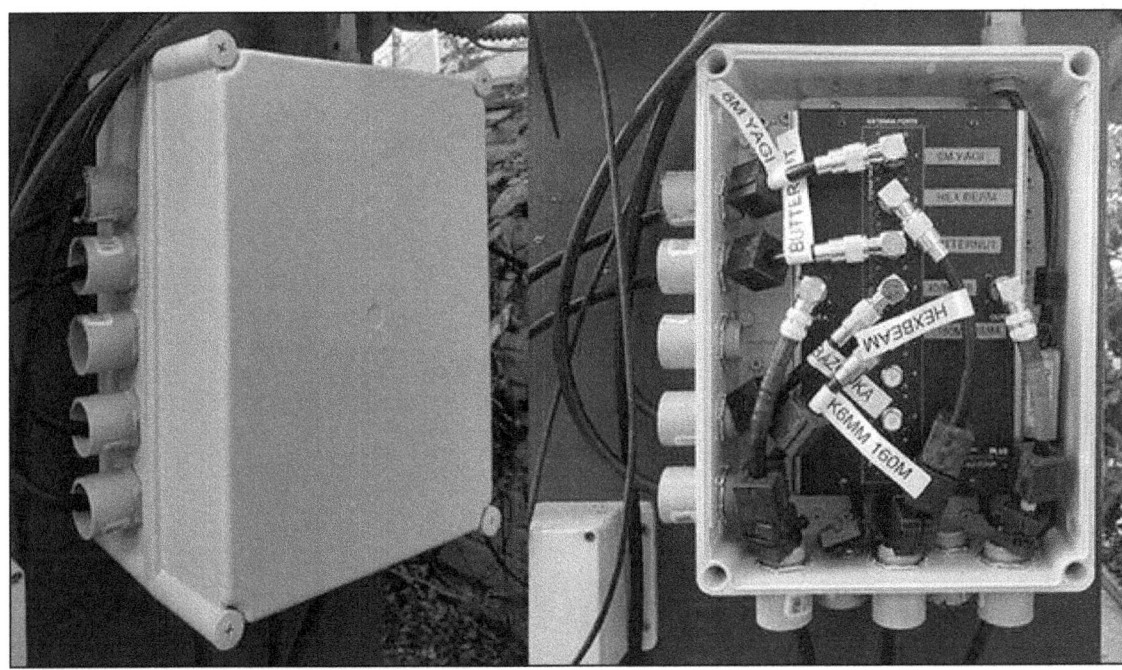

Before my cables get to this junction box, I also employ a FlexRadio/4O3A 8x2 coax switch (above) which is controlled and integrates with my FlexRadio. This unit uses ethernet and a separate 13.8VDC source to operate. It is not POE (**P**ower **O**ver **E**thernet). Note the labels and angle connectors to reduce tension and the overkill (or not) of ferrites. This photo was taken a few years back, my antenna farm is in constant metamorphosis. Ideally, I would like to use a larger box to eliminate the need for the right-angle connectors and move this box closer to the actual antennas.

The above configuration takes all my HF and 6m antennas and reduces them into just two cable runs for SO2R (**S**ingle **O**perator **2 R**adio) operations. This is accomplished via the 4O3A AntennaGenius per above. Basically, these go to my FlexRadio A and B inputs. This way I can operate on any band on any slice. Allows me to listen on different antennas on different band combinations at the same time.

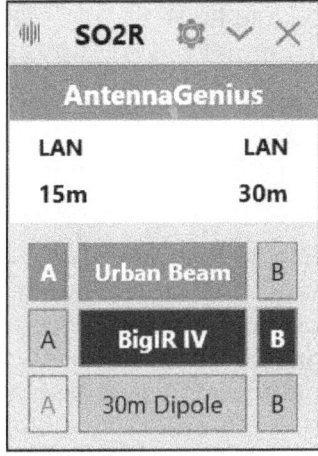

The software is very straight forward. Basically, the "A" side and "B" side refer to the SO2R setup in the FlexRadio in this case. It shows what antennas are available for the given selected band. Ones not compatible for the given band are hidden. Less confusion. I love it! There are times you will have multiple options for a given side, meaning if for example, you have multiple antennas for a band available, let's say 30m. In this case it would show my SteppIR Vertical as well as my 30m dipole, therefore I can toggle a directional (dipole) and an omnidirectional antenna (SteppIR Vertical) quickly to check where I need to focus my DXing. The picture shows the configuration screen for each antenna you define what bands it will work on. Ones with no specific bands specified will not show up as an option.

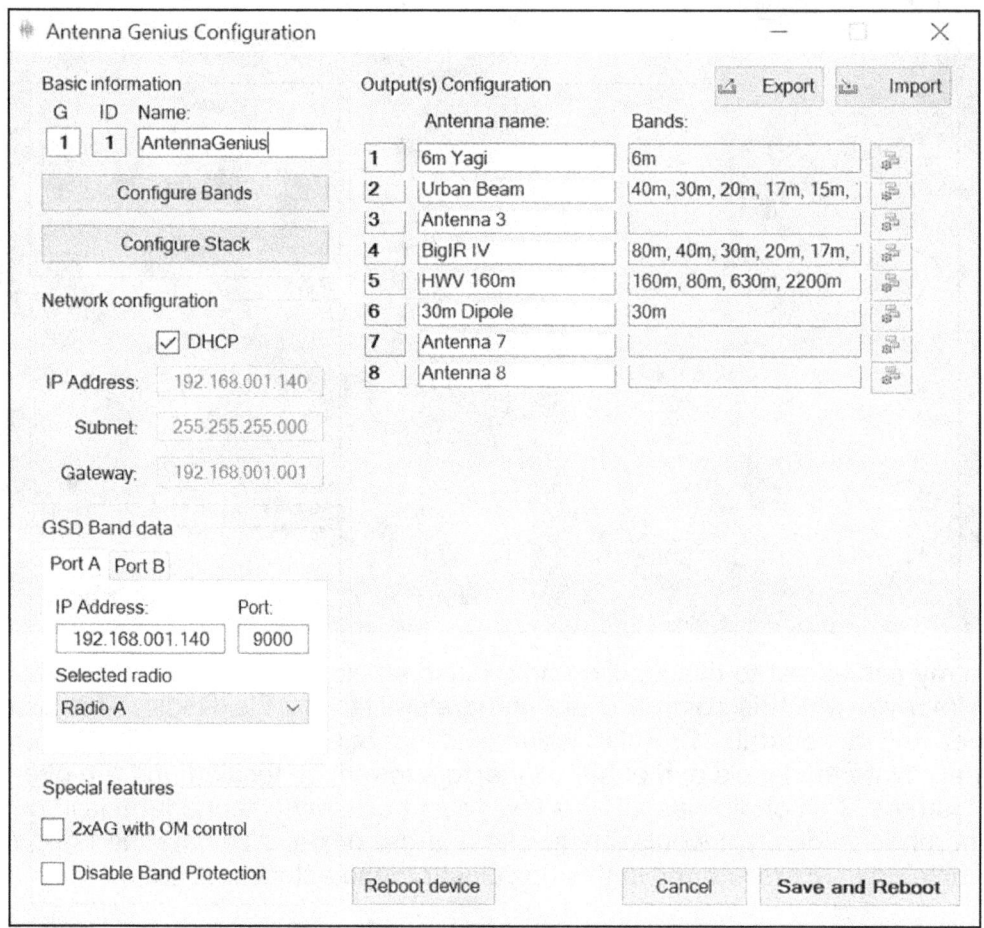

The left side of the setup screen is used to configure network settings as well as making connections to a remote device for auto tracking, such as a FlexRadio and the TunerGenius XL in this instance.

There are many antenna switches out there. If you look at ham radio catalogs, I am sure you noticed. Some hams use simple manual switches inside the shack, and these are just fine if from a reputable manufacturer. From personal observation, I would say most hams do it this way.

Others may use manual control boxes with an external unit to perform the actual coax switching, generally with a mast mounted unit outside. The advantage of a mast or outside mounted switch box is the reduction of the amount of coax needed, much as I did. Since good quality coax does not come cheap and if you want to minimize the number of holes used, antenna switches are a good way to go. Let's do some math. If you are switching 6 or more antennas and reduce 30ft or more of good quality coax run to the house, you just saved money by using an antenna switch outside.

Array Solutions at arraysolutions.com , **DX Engineering** at dxengineering.com and **MFJ** at mfjenterprises.com all offer great external antenna switching solutions. MFJ external switches are sold under both the MFJ and Ameritron branding. **FlexRadio Systems** also offers one via **4O3A** at 4o3a.com per above called the AntennaGenius which is what I use. These are available in 8x1 or the SO2R (**S**ingle **O**perator **2 R**adio) 8x2 format which is ideal for the FlexRadio 6000 series. FlexRadio Systems is located at flexradio.com. Green Heron Engineering also offers a mast mount antenna switching unit as well as an 8x2 SO2R remote coax switching unit. They even offer a 12x2 for those who need to invite me to over to operate. Given their great track record, I suspect these are also very well made.

Check them out at greenheronengineering.com.

Grounding & Bonding - Fighting Mother Nature

Lightning arrestors, bonding & grounding rods comes to mind first but of course it does not stop there. Far from it! Depending on where you live and where your shack is physically located, these may need to move to the top of your list as these can prevent damage to the gear in your shack!

Grounding comes in many flavors, and this is a rather important topic. There is a reason there are so many questions about this in FCC license exams. Depending on who you ask, there are at least three types of ground out there. Electrical/AC ground, Lightning ground and RF ground are the main umbrella categories.

The term "*bonding*" is used in reference to connecting your gear together. This is done to bring gear to the same electrical potential. Grounding is what you do when you run your gear to a grounding rod. These are often used interchangeably in literature therefore confusing many hams.

One type is not optional, even if you daringly skip lighting ground and RF. **Electrical ground is no joke and a must have. Never bypass or remove the 3rd grounding prong from your gear.** This could not only save your gear, but possibly your life.

Many sources online, including some videos, confuse RF ground and AC ground. Several use the terms interchangeably as well making folks even more confused. Early on in my ham radio career I often left more perplexed after watching videos and reading articles than before I started. There is a lot of misinformation out there. Not on purpose, but just because it is a complicated topic. Now, I do not claim to be an expert in the topic either, but these are certainly not the same. I think we can all agree on that.

There are direct lightning strikes (very bad) and indirect lightning strikes (still bad). Direct lightning is what most think of as a lightnings bolt hitting something, usually a conductive object and grounding to earth. Indirect lightning can couple into wiring, which includes antenna systems! It is important to protect against both.

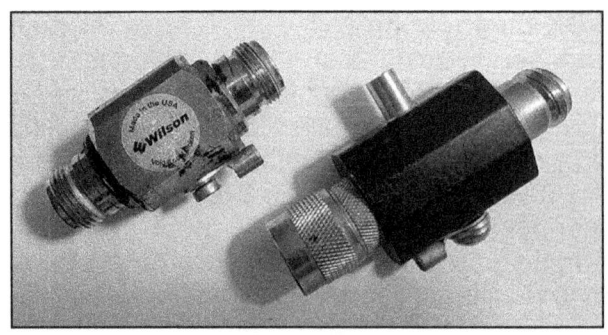This is where **Lighting arrestors** come into the picture, and they are not all the same. Some are a joke frankly or are no more than a barrel connector with a grounding screw. These are not lighting arrestors, though sometimes are labeled as such and I have seen these installed by hams thinking they were. These are likely just a coupler with a ground wire option. The dead giveaway is the cheap price and often also the source. I will leave it at that. The ones you want to use have gas filled arc tube inside which may need replacement if triggered. You want ones with the desired connector on each end for HF use. The ones pictured here are the "N" and "UHF" type. You also want to make sure you get the one rated for your station power. So, if you are running 1.5kW, do not get a 500W rated one. Most pass DC but some do not. If you are sending power over your coax to power equipment such as a preamplifier or a remote antenna switch, be careful which one you purchase. You can find some very nice ideas on how to accomplish this at kf7p.com.

Lightning can strike anywhere but is more likely in the Southern part of the US, and of course in many other climates worldwide. According to some numbers, there is lightning on some part of the planet at any given time, at times as many as 100 lighting strikes per second. Here in Northern California, we do not see too much of this but still precautions need to be taken. Lightning can strike up to 20mi (32km) away from the actual storm and these rogue strikes may even be more destructive than others. If you hear the storm approaching, then it is likely around 10 mi (16km) away, your gear might already be in danger. If you get a direct lighting strike, there are no precautions you can take which will save your gear. The only certain way to protect your equipment during an electrical storm is to completely disconnect it. This includes all power and antenna cables. The good news is that the chances of being stuck by lightning is very low, some say one in a million. The bad news is your chances of being struck are still higher than winning the lottery.

Never use existing electrical conduit or even worse, a gas pipe for lighting grounding. You will have a very short DXing career. Most newer houses use PVC pipes for water, so that is also out as a way to ground. This leaves you with using actual grounding rods. These will need to be installed before the coax enters the house.

When installing AC **electrical grounding rods**, follow your local electrical code. Most homes will already have this done but if yours does not, make sure you address it. In general, grounding rods are 8-10ft (2.5-3.3m) long (will be about 95% buried) and at least 5/8" (1.6cm) diameter copper clad though you sometimes find thinner ones. I have two of these at my house for AC. When I had my solar system installed, I had them add a second one for good measure near the combiner. I use the same practice as above for RF ground rods for my shack but shorter due to bed rock issues. I made up for this by using more of them.

I also have ground rods installed near the antennas (both plural) and even an extra one near the shack. My 160m vertical has 6 ground rods evenly spaced around it and my other HF Vertical has 4. Overkill? Maybe, but due to poor ground conductivity where I live, I want to protect my gear as much as possible as well as lower the noise. Besides, I got a great workout from hammering down all these rods into the ground. Who needs a gym membership when you are installing antennas and grounding! If you are using multiple ground rods, you do want some separation between them for them to be more effective. Mine are generally about 4ft apart, though does not need to be an exact science.

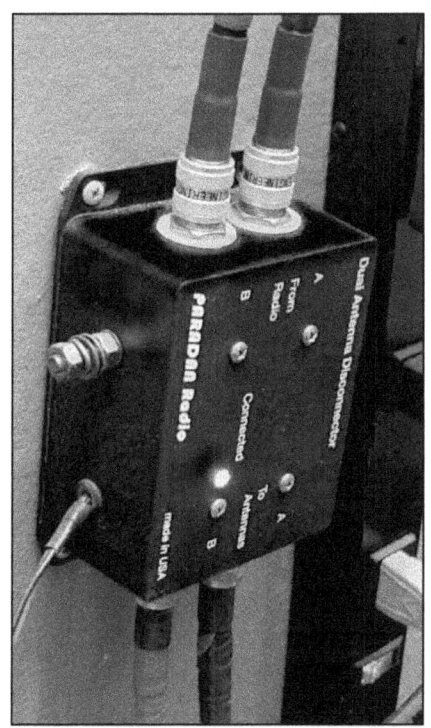

The typical things done by most hams when a storm is approaching include lowering the tower if possible and disconnecting radio gear. These are good steps! I automatically disconnect my radio gear using **Paradan** Antenna Disconnectors. Basically, these units keep the antenna disconnected until power is applied to the unit. I run this off the same power source as my primary HF rig. Therefore, when the radio is on, the antenna is inline, once off, no longer physically connected. Simple! I like simple. You can find more information on these at paradanradio.com. These are available in both single and dual configurations. This unit is certainly going to help as it does disconnect both the inner conductor and outer shields. It is, however, not to be used alone but rather in conjunction with other protection methods as discussed earlier.

Lastly there is **RF ground**, and this is a slightly different animal. This is used to reduce unwanted RFI from causing interference in your radio gear. This is caused by unwanted voltages and current flowing around and likely will never be 100% under control. RF Grounding is a controversial subject! Very polarizing in fact. Some believe RF ground is a myth. Again, this is not the same as an electrical ground or not believing that lightning is real. I will not get into details about it here, but I do recommend reading up on it. My personal option is that the truth is somewhere in the middle.

To perform RF ground bonding, use a ground bus or bar close to your equipment. Braided strap is recommended here. You also want your RF ground wire to be "*beefy*" regardless of type. Some guides recommend #6 or #4 AWG wire. This will not do as well as a braided strap due to skin effect. Some hams also use large strips of copper. The greater the surface area the better, electricity travels on the surface of the conductive part of the wire or strap. Braided strap will have the most surface area and will outperform pretty much anything else out there. This is pretty much the same concept when it comes to coaxial cables. Use braided, though not from old coax cables! Connect all straps to this central point and from here do your RF ground.

Keep in mind that you want to keep RF ground wires short, even if braided. If you do not, it may start to act like an antenna and then you have a whole new issue to troubleshoot. When I say short, I keep mine 4-5ft (1.5m) long at most. These need to go to a common central point. Do not daisy chain, and from here go directly to a proper earth/outside ground. If you can!

So, this is where it may get controversial again. Why connect all the gear even if not going to connect to an outside RF ground? To make sure that all are at the same voltage, in other words the same RF potential. When everything is at the same potential, you do not need to worry about rogue current causing issues in a piece of gear....as much at least.

If your device does not have a ground screw terminal, figure out a way to connect it to the central grounding point. Therefore, creating an electrical balance between all your equipment.

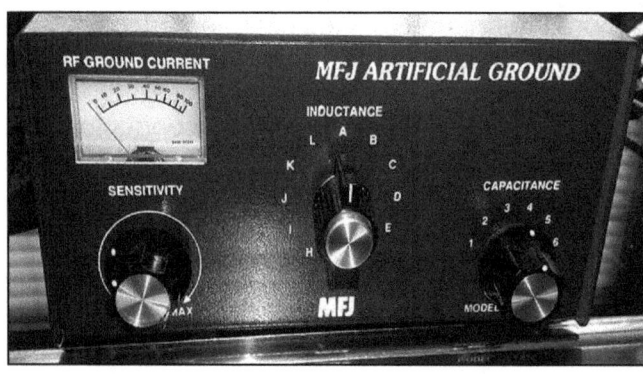

An MFJ-931 Artificial Ground is something you can purchase to assist in difficult situations when RF ground is needed but not an option. This system uses a piece of wire as a counterpoise followed by adjustments to balance everything. It looks like an antenna tuner in some ways, and it also uses capacitance and inductance to match.

Keeping Everyone Happy

There are a few things to consider here. Besides the HOA, other acronyms to keep happy are your XYL (Wife) or OM (Old Man, Husband). My wife has offered to hang laundry on my hexbeam, when I had one and Christmas lights on my verticals. If this comes up, you may have to explain potential SWR issues in simple terms.

Some folks get this weird idea that a garden is not intended for antennas. Yes, I know. Some folks are just plain weird. But seriously, they can coexist. If you can camouflage some of your antenna gear, do it. For example, my 160m home brew antenna (which I covered earlier) is painted dark green with green wire wrap. Yes, I still see it, but it is not as bad as it would be if it were bright white with red wire. Some of my antenna posts have some ivy growing on them. If it does not touch the active elements, you are good and may even hide them a bit.

Don't forget to be considerate of your neighbors as well. If you have a neighbor who does not want to look at your antennas even though it is your property just ask them nicely to move. Just kidding…do your best to keep the peace. Meaning, do not block views or at least do your best to minimize doing so. Make sure your antenna does not enter their property line or any public areas, such as the street. Check for this as well when the antenna turns as some Yagi antennas for example might not be the same length on both sides due to the need for balanced weight distribution. A little pre-measurement and math is rather important here before installation.

Speaking of neighbors, the FCC is generally on your side. Rules vary regionally on how much say they have, but most of the time, technically *"none"*. Of course, if you live in let's say Manhattan Beach, CA in a 4 million dollar home with an ocean view and you block your neighbors scenic view, I can certainly see issues arise. So, use common sense, be a good neighbor and work with them or find alternatives if necessary.

CHAPTER 5: Getting Binary - DX Software

Most of the ham radio software, as most software in general was written for Windows. There are historical reasons for this, but this appears to be changing or at least doing some serious catching up. MacIntosh users do not have as many options when it comes to ham radio software but for the most part you can find something you can use for a particular purpose on any platform. Most of these options are rather good.

Notably, FlexRadio Systems somewhat recently released a MacIntosh version of their SmartSDR software. WSJT-X is also available for Apple users as are many logging programs. Some only for the Mac. As I said before, use what you are familiar with, you can likely make it work just fine! I will go over the basics you need but if you are a MacIntosh user, you can always visit machamradio.com for the latest and greatest.

Something I would like to point out is that a lot of the ham radio specific Windows software out there is outdated and, in many cases, even completely obsolete now. This might be due to the original author being SK (**S**ilent **K**ey, passed on) or simply losing interest and moving on to other things. Also, it is worth noting that some software was written for older versions of Windows and was never updated to be used with current operating systems. This can be on occasion a real hassle with newer versions of Windows, sometimes a deal breaker even. Theoretically one can get the most programs to work in "*windows compatibility mode*" at times but frankly, it might be time to look for something current at that point.

Linux is something not to be forgotten, there are certainly many tools for this platform, in fact some are better in some ways than those available for the Windows or the MacIntosh markets. There is even a release of Linux prepacked with all the ham radio goodies for operators as I had mentioned earlier.

Don't forget that there is software out there for iPads, iPhones and Android devices as well. You may find that even repurposing an older tablet is perfect for these applications instead of using them as an oversized coaster or cutting board. As these come up in given areas, I will mention them. Your mileage may vary, so do your research.

Logging Software

Logging ham radio contacts is no longer required by many licensing authorities for general ham radio operating. This has been the case for quite a while now, but with minor exceptions such as 630m and 2200m bands as well as the 60m band under certain circumstances. As a DXer, you certainly want to keep track of your contacts for many reasons.

Logging is always done in UTC format. I will explain this a bit more later. **It should include the callsign, report, date and time, mode as well as the band.** If you are a 6m grid hunter or work Satellites, additional information as well such as the grid and the satellite. Some folks also log power used or if they exchange additional information.

Examples are 10-10 numbers and IOTA (**I**slands **o**n **t**he **A**ir) info, this information can be very helpful when applying for future awards. See the last chapter.

The issue of paper versus digital logging often comes up. According to some estimates, only 50% of all licensed US hams keep logs. Around 90% of hams on HF log their contacts and of those about 20% still have paper logs. Yes, one can also write a book with pen and paper. Would you? I sure didn't!

Much of what would take hours or even days to compile can be checked by computerized logging software in seconds! Most can also pull detailed user information from the internet based on callsign from sources such as QRZ. You can track accomplishment for things such as awards. Easily check for duplicates, errors, or search for missing entities. I would also argue that electronic logging helps you avoid mistakes and will save you a ton of time and aggravation in the long run.

Furthermore, you can also easily back up electronic logs in the cloud in addition to your local computer. If you want to keep a paper log for nostalgic reasons or perhaps out of habit, go for it. But I would really encourage you to have an electronic copy as well. If you ever decide to submit a contest log, this will likely be a requirement.

The ADIF (**A**mateur **D**ata **I**nterchange **F**ormat) is the one you will likely encounter most of the time. This is what you will be using more than likely for logging your DX contacts. There are multiple versions of ADIF, but frankly not something you need to worry about much. If you would like to read up on this more, visit adif.org for more information.

Some software can also save in XML (**Ex**tensible **M**arkup **L**anguage) format, and these tend to be program specific with extended fields and comments. XML is more of the standard in the IT world. It is easy to work with it generally and easy to convert to and from. There might be, in fact very likely, data in an XML file which will not transfer to an ADIF file. Going the other way around issues could also arise. Meaning, if you are trying to restore from an ADIF file to a logging program and you had additional comments at one time in the original XML file, these will be lost. If unsure, save in both. I backup in both formats and often.

Additionally, you may encounter the Cabrillo format, which is used more frequently in ham radio contesting. This format was developed by Tray Garlough, N5KO. You can find info on this at wwrof.org/cabrillo. Most any logging software can import and export to this format if needed. If you also contest or move between programs, pays to get familiar with it for sure.

There are some popular and rather good logging software titles out there which are DX Friendly. After all, we are talking DX here. I would personally recommend two of them. **Ham Radio Deluxe** which is a paid software title and is part of a suite which includes a logbook program, **HRD Logbook**. It is extremely easy to use and does everything you can think of. You can find HRD at hamradiodeluxe.com. The first year is $99.95 (as of 2022) and renewals will be $49.95 and there are often discounts during ham radio conferences and such.

By default, Ham Radio Deluxe does not use a SQL DataBase, though there are plans to move to SQL. The one it currently uses is fine, but once your log data grows, most computers start to slow down with it. You can run HRD with MS SQL or MariaDB though it involves an extra few steps at setup. This was one of my deciding factors for using it. You can find the MariaDB (SQL) Logbook install instruction for Ham Radio Deluxe via video at youtube.com/watch?v=C7f-RZOnmDs.

My second recommendation is from **DXLab** which also offers very good software called **DXKeeper** which is part of their suite. It is completely free, thanks to author, Dave Bernstein, AA6YQ. This is also part of a suite and includes many additional features and functions as well. You can read up on DXKeeper at dxlabsuite.com/dxkeeper.

Both suites are great, try both and see which one fits your style and workflow better. I have been known to go back and forth and my above order does not necessarily represent my order of preference. Both offer great support, stability, and performance. **It really comes down to your computing style and preference.**

Most logging software will include award tracking, label printing for QSL cards, DX clusters and some even have a grayline map as well as many other features. Others logging software worth mentioning are as follows. Some of these do have a contest concentration but keep in mind that there is lots of DX in contests. Many hams still, like me, may use multiple pieces of software but keep one master log.

LOGic 10 from Personal Database Applications, Inc. can be found at hosenose.com/logic at the writing of this book they are on version 10 which includes many great features. The cost is $129. This one is used by many hams as well.

Another logging program which is widely used is **N1MM** Logbook. It is used in contesting primarily and I often switch to this myself during contests. This appears to be still #1 for contest logging and is also my #1 pick for contests. Following the contest, I import my data and merge with my master log. You can get your copy at n1mmwp.hamdocs.com. Be sure to download the latest updates as well for this. The software is updated often and is well supported. The cost amazingly is free.

For the MacIntosh users **MacLoggerDX** is a great option and can be found at dogparksoftware.com/MacLoggerDX.html. The cost is $95.00 and is a great all-around option. **SkookumLogger** is another option for the MacIntosh, though the focus is contest logging. The cost if free, and you can find out more about it at k1gq.net/SkookumLogger or if you prefer **Aether** is another option for $38.99 with more information at aetherlog.com. Both are available from the Mac App Store.

RUMlogNG is available for both Windows and MacIntosh. Written by Tom, DL2RUM and more information can be found at dl2rum.de. **HamLog** by Pignology, LLC is designed for the iPad or your iPhone. This only costs $0.99 and is perfect if you are in the field or traveling. More info at pignology.net/hamlog.html.

Another option is from N3FJP called **AC Log**, sometimes called, Amateur Contact Log and can be found at n3fjp.com. At time of writing, this retails for $34.99 though the contest addons must be purchased separately. These are generally under $10 each. The complete version can be yours for $59.99.

Log4OM is a very clean, compact, and flexible piece of software. The cost is free. log4om.com. **Logger32** by K4CY was one of the most used loggers "*at one time*" I am told and has many extras in addition to just log keeping. Get it at logger32.net and the cost is also free. **SwissLog** is free as well and is considered software for the "*sophisticated ham*" (not my words). You can find more at swisslogforwindows.com. **DXtreme Station Log** can be found at dxtreme.com and the cost is $90. This one has a very loyal following.

DX4WIN is $90 per but upgrades from previous version are between $35-$45. Very feature packed and wide support. Check it out at dx4win.com. And finally, **WriteLog** clocks in at $30, pretty good but more contest oriented. For info writelog.com is the main site.

TQSL & LOTW

Many think of LOTW as a web-based service, and it is. But there is also a software component called TQSL which you likely want to install to use along side with using LOTW online. Frankly, there is no other way around it short of having every single card manually checked and submitted. While this is how it used to be done for many years prior to LOTW, it is not a recommended option anymore. Unless of course the station you need to submit for does not have an LOTW account. I will cover later how to deal with this.

TQSL is available for Windows, Mac and Linux. There are many tutorials online on installing and configuring TQSL, I urge you to take the time, get it setup and get familiar with it. You will thank me later. Here is an online guide to help you get started by Gary Hinson, ZL2IFB available at g4ifb.com/LoTW_New_User_Guide.pdf.

Many logging software titles out there integrate with LOTW, including my top two recommendations. Once you get it setup and configured, uploads are easy and even automatic in some cases, if you choose to set it up that way. You only need to upload new contacts, no need to reupload everything. Most logging software will track for you what you have already uploaded making it essentially seamless.

There is no cost to download to LOTW, to use the software, or any fee to upload. However, if or when you do apply for an award, there will be a minimal cost for processing and possibly mailing costs. I will cover this in the awards section.

Automation & Remote Access Software

Automation software comes in many flavors from rotator and amplifier software controls to antenna selectors. When you research hardware, I encourage you to check if there is software available to control it via a computer. Even if you think you do not need this now, it may come in very handy later. If a device has a USB port (or even serial port) there is a good chance either the manufacturer or someone has written software for it. If not, if you have the know-how, you can too!

Much of what runs your station now has a software component available. Sometimes it might not even be obvious, but a little digging at manufacturer websites can reveal a lot. This, of course, makes remotely accessing your computer possible for many more who cannot be physically in their shack. Furthermore, it is often a lot easier to configure hardware via software than via the traditional "*knob and button*" method.

There is nothing worse for a DXer than to miss an activation due to other commitments. This is when "*remote access software*" comes into the picture. This is not to be confused with "*remote operating*" which we have discussed earlier. Remote access in this instance refers to a ham controlling their own rig vs "remote operating" someone else's rig. I realize there is some overlap but allow me to clarify. There are times when, for example, we must work, travel or just simply cannot be at the controls. Via the marvel of the internet, this is now easily accomplished.

There are two major ways of doing this. Via software provided by some radio manufacturers or via remote controlling your desktop. When it comes to manufacturers, the best example is FlexRadio SmartLink which allows either the use of the Maestro (a hardware head unit) or their SmartSDR software on a remote computer installation. Both methods perform fantastically well, I have used both.

The second method works for more radios in general. If you can setup whatever you may need locally on your desktop, you can remote into it and have full control. Meaning radio CAT control is a must. Nice to have controls for the rotator, amplifier, antenna switching, tuner and so on. This method allows you to run WSJT-X via your local computer and access it from anywhere in the world. Basically, your computer in your shack does all the work, and you are using the remote session computer, essentially as a viewer. This is ideal for digital work, possibly CW with preset memories. The previous method, via a head unit, is more recommended for phone modes. Maestro also allows for a CW key connection therefore also CW friendly for remote operators.

Good options for remote software are Windows Remote Desktop (RDP is built into Windows), though has some limitations or Chrome Remote Desktop at remotedesktop.google.com. From the paid variety, TeamViewer at teamviewer.com or my favorite Splashtop which can be found at splashtop.com. You do get a few extras for paying, but frankly for most users, any of the above will do just fine.

There are others out there, however, I find that either they are a bit buggy, lack security, use high bandwidth, cost too much or bundle things most of us do not need to justify the extra cost. Sometimes more than one of the above. There is one specific company I will not mention by name here who used to charge around $30 per year, now ask for several hundred dollars and justify it with bundling cloud storage and other "*goodies*". To me these moves scream, "*we are hurting*".

Two things to test though when looking at options, especially other ones. Sometimes the waterfall does not display well via the remote session, or even at all. If you use this, clearly those are out. Also, reliable connections can be an issue with some locations with slower internet connections. This applies on both ends. If your home internet crawls, likely this option is out for you.

Do your research and see what fits your needs, budget, and hardware you will be using for remote access. Technology also changes fast, so check around before you renew. As I said before, always re-evaluate your needs when it is time to renew or repurchase. What was great yesterday, may not be today. A company which had great products and support might be lacking now. Good to keep an open mind.

Radio Control Software / CAT

No, you did not pick up a book on pets by accident. CAT stands for **C**omputer **A**ided **T**uning though sometimes it is defined as **C**omputer **A**ided **T**ransceiver. Same basic concept, no matter how defined or by whom. To put it simply, CAT makes your radio respond to your computer and vice versa. Some of the suites I had already mentioned above in the logging section have built in radio controls to do just that.

Some SDR radios also have seamlessly integrated software available. Manufacturers like Yaesu have their own software as well which usually resembles the actual unit on the screen, allowing full control and access to most features. Though I will be honest, I find these a bit hard to use with the mouse and I can only seem them being practical in remote sessions. Icom has the RS-BA1 IP remote control software which works remarkably well but does require a purchase. This has a slightly different interface than what you would encounter on the radio but can be figured out and mastered quickly. This is for sure one of the better ones I have seen.

You will encounter CAT connections either via serial, USB, firewire and ethernet connections. Older FlexRadios used firewire, but they no longer make these units. This is the only manufacturer I can think of which ever used them. Units made before USB, or those slow to update their hardware use serial technology still. USB and ethernet connections are significantly easier to configure. If you see a reference to CAT over TCP/IP, this is refereeing to CAT over the network using ethernet cables, though at times WiFi.

For serial connections, you will need to pay attention to Baud Rate, which is the speed basically. This needs to match on the computer as well as the unit you are trying to control. Higher is better, though may not always be compatible with some gear. On some older devices if the speed is set too high, they may occasionally not respond. This is due to errors, possibly due to poor connection, software issues, hardware issues or just not having been designed to reliably handle the speed. The solution is to just lower the baud rate on both the computer and the unit in question until it responds reliably and consistently.

Other things to check are data and stop bits as well as parity. Parity is error checking; the others are for data transmission format and control. Make sure these matches both the computer end and the radio and you should be in good shape. The only other thing to make sure is you are using the **correct COM port**. This can be checked on the device manager and the radio needs to be pointed to the correct one. This seems to be the number one issue for most hams when it comes to serial ports. Make a note to save yourself future hassles once you get it all working.

You may see the term CI-V in reference to Icom radios. This is Icom's version of CAT based on the RS-232 protocol. There might be some additional parameters you need to set when using it. CI-V is kind of like RS-232 on steroids, minus the bad side effects. Refer to your radio's user manual or the software's user manual which you are trying to configure as they likely will cover CI-V connections extensively.

CW Decoders & Skimmers

If you are a newer ham, or just never considered it, one way to get into CW is with the aid of a CW decoder. These are commonly used during some contests as a sanity check or even as a band skimmer to help identify and spot stations. Many DXers also use them for the same reasons, even if they are proficient in CW. As far as CW decoding software is concerned you have quite a few good choices.

CW Skimmer by VE3NEA is likely the most versatile option I have come across. It is also the most expensive on this list, but it is well worth it. There are a lot of things you can use this for besides just getting away with not learning code. $75.00 is the cost. Download it at dxatlas.com/cwskimmer.

MRP4066 is a nice option too. I love this one as well and use it often in conjunction with CW skimmer. At times I find this software decodes better than the above. The cost is €52.50. Not as versatile but does what it needs to and does it very well. One drawback is you will need to contact the author to re-activate it every time you reinstall your operating system or do a major upgrade. For me, this is often. Something to consider. You can find it at polar-electric.com/Morse/MRP40-EN/index.htm.

DM780 is part of the Ham Radio Deluxe suite. Does the job well in a low noise environment and will be OK for most basic CW users. If you have the suite already, give it a try. You might just find that this is enough for you.

And lastly on the Windows side, **CWTY Decoder/Generator** by WD6CNF is free and you can find it at hotamateurprograms.com. For the MacIntosh there is **Morse Decoder** by HotPaw Productions for $19.99 which appears to do a great job and appears to be the only iOS solution I have heard of or could find.

If you are using an SDR radio or one with a built-in audio card, the above will be a breeze to install and configure. Just point to your audio source and you are mostly up and running. Some radios may require an additional sound card to feed the audio to a computer so the software can do its magic. Additionally, some radios may even have their own decoders, however they greatly vary in quality from my experience.

There are also hardware decoders out there as well as some apps for tables and phones. I do not really recommend the latter, but you may find these are just fine for you. As far as hardware options, check out MFJ at mfjenterprises.com if that is your cup of tea.

Clock Sync / GPS Timekeeping

There are two main reasons why you need an accurate clock when DXing, though there are others. The obvious one is for keeping accurate logs, no matter what mode you are using. It is worth repeating, **always remember to log using UTC Time** which hams agreed to use to avoid issues with time zones. This is also sometimes referred to as **G**reenwich **M**ean **T**ime (GMT).

While to some I am stating the obvious, you would be surprised how many physical cards and eQSLs I get with the time being wrong, as in written in local time mainly. Another thing to make note of is that when you are entering the time in UTC, it might already be "*tomorrow*" in some places. In other words, here on the West Coast where we are about 8 hours back, from UTC. Around late afternoon, on the West coast, need to make sure we log tomorrow's date. Of course, if you are already using an automated or computerized logging program, this is usually taken care of for you. Yet another reason to ditch the paper. It would be horrible to find out that you logged a rare DX entity, just for it not to match up in ClubLog, denying you the QSL.

You want to be at least a within a minute or two when logging ideally. Some online logs such as LOTW, eQSL and ClubLog forgive up to +/- 5 minutes of error when matching QSOs. Sometimes perhaps even more, but don't count on that. Not worth your contact not matching up! If both parties are off in the opposite direction, that can be a problem.

The second reason for wanting **accurate time is a requirement with some digital modes**. The FT and JT digital modes for example, require synchronized time to achieve correct decodes. Here you do not want to be off more than a second. In fact, really should be a fraction of a second for the best results. This can be hard at times. If you see that the DT is consistently off by a lot in one direction when using the WSJT-X software for example, you know you are off. More about this in the digital section.

A good website for quickly checking if your clock is off is time.is. There are others but this is the most used. Don't count on your computers clock to be accurate enough. Frankly, this is rarely the case without some additional software assistance.

You can also use a GPS (**G**lobal **P**ositioning **S**ystem) to set your clock as I do. Some of you are now thinking...GPS? Isn't that for navigation? Yes, but it can have other great uses. GPS satellites contain multiple atomic clocks which keep very precise time. This is used to calculate position based on how long it takes for the signal to get from various GPS satellites to a given point. You generally need about 4 satellites locked in to get a good sync. This is done so it can compare ranges. To put it simply, the receiver uses four satellites to figure out longitude (North to South), latitude (East to West), altitude (elevation), and time. This last one is what we are mostly interested in.

There are many used GPS units available for cheap. Some of which used to cost 10x as much a few years back. Here is the image of the aftermarket GPS I use with my FlexRadio to get a fast lock with its optional GPSDO (**GPS D**isciplined **O**scillator). I split this signal received to set the computers clock via compatible software. I also use the same received signal for frequency calibration of my **Q5 Signal Transverter** as well as other devices. If interested in these read more about them at q5signal.com and **Down East Microwave** found at downeastmicrowave.com for additional gear, as well as a great 10MHz signal splitter.

There are also now a set of satellites from Russia called the GLONASS system. The screenshots from **NMEATime** show the reception of both the US GPS system as well as GLONASS on the right. The left screenshot indicates the GPS being perfectly locked. This is about as accurate as your time can get. Do you need to be this accurate? No, not really. But if you are in the field, this might be the best option and can't hurt at your QTH either.

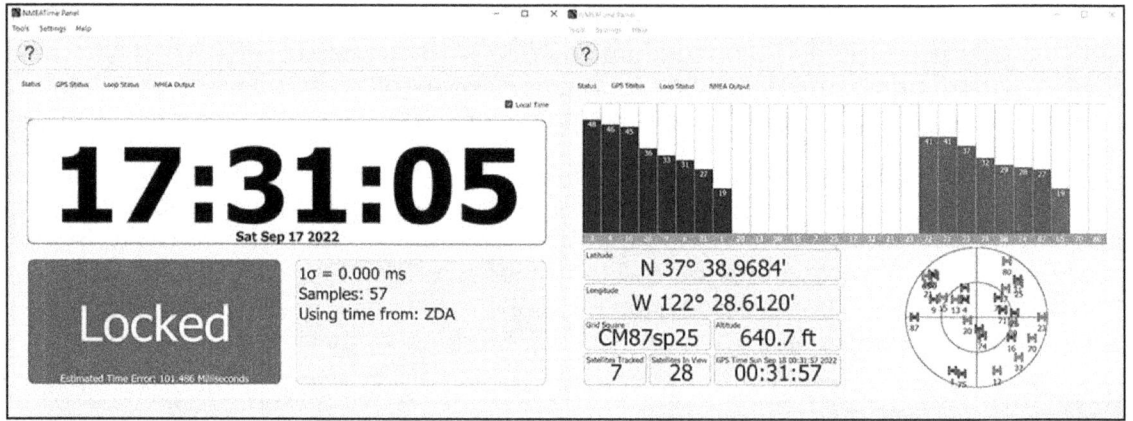

To keep my computer clock accurate, I use Mania Radio **BkTimeSync** by Mauro Capelli, IZ2BKT on my Windows PC but have used others. You can obtain this at maniaradio.it/en/bkttimesync.html. The image here is the actual settings I use. I like this piece of software since it can use GPS as well. Great for in the field use or as a backup in case the internet fails. For the time server I use the Google default time server, the Apple time server also works well here. This may work well for you if you are in the San Francisco Bay Area. If you want to shave off a few milliseconds, you can also use the direct IP of 216.239.35.0 for the Google NTP server.

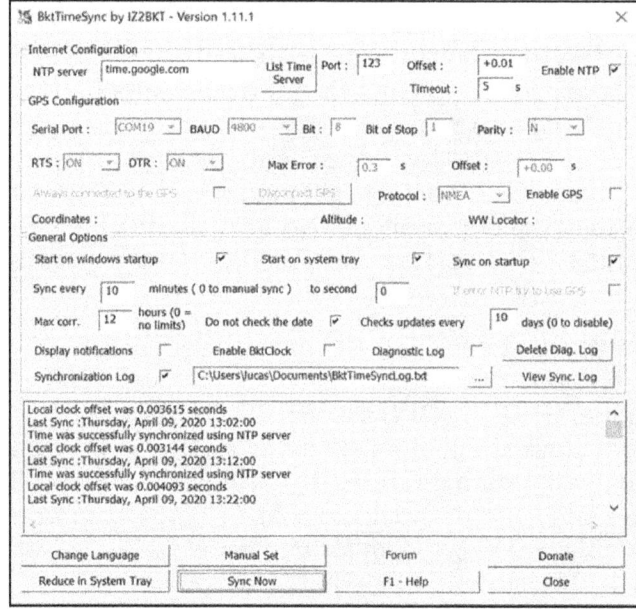

Alternatively, use something like an **NTP-Tool** at:

download.cnet.com/NTP-Tool/3000-2085_4-75718996.html

to see which **NTP** (**N**etwork **T**ime **P**rotocol) server is fastest in your area to keep your clock as accurate as possible. You want it to be a server as close as possible to your QTH. I also had good luck with **ISP** (**I**nternet **S**ervice **P**rovider) specific NTP servers as well. These you can easily look up online. Your computer may only adjust the clock once in a long while or not always to the best service. Both the frequency and the source are often adjustable on your computer, but usually not ideal. Software out there for the purpose of setting your clock out there is numerous. Clearly it is an issue.

Other common time adjustment software include **Atomic Clock Sync** which can be found at atomic-clock-sync.soft112.com. NetTime can be found at timesynctool.com and is simple as well as dependable. The author welcomes donations if you end up liking it and using it. Of course, the very popular **Dimension 4** by Thinking Man Software can be found at dimension-4.en.softonic.com.

There is also a piece of software called **JTSync** found at dxshell.com/jtsync.html which is simple, lightweight and can even adjust your clock based on other user's time offset which is a very cool feature. To my knowledge no other software does this. **Meinberg NTP Time Server Monitor** is rather popular with some hams. It can be found online at meinbergglobal.com/english/sw/ntp.htm. For the MacIntosh there is **NTP Clock Sync** found at macdownload.informer.com/ntp-clock-sync.

Another option is to use GPS to set Time via **NMEATime**, download is at visualgps.net. I used this before as well, but I feel it is overkill for most ham users though works very well. Requires a USB or serial GPS as does **BkTimeSync**.

DX Clusters

There are always going to be folks out there who think that using DX clusters is cheating and usually these are the same people who think that FT8 is the work of the devil. They are entitled to their opinion, though it seems much of the ham community will agree to disagree. I feel when DX clusters are used properly, they can be wonderful tools. Properly configured and with filters applied they are even more powerful implements. You may hear the term "*spotting networks*" and this is the same thing used interchangeably.

For those newer to ham radio, DX clusters are basically all over the world, generally connected over the internet (though there are private ones) and share spots as they are reported by other hams. A spot refers to a "*station*" basically. This can be DX or can frankly be any station in general. Perhaps one using a special mode, special callsign or running a special event even. **Something that is not DX to you, might be DX to some.** Just remember that! So, if you live in New York and get annoyed at a Delaware or Rhode Island station being spotted, trust me when I tell you many on the West coast (**W6**) and Australia (**VK**) may need that station and really appreciate the spot.

Generally, the DX station callsign is accompanied by the spotter callsign, frequency (or at least the band, though not as helpful), time spotted, mode and perhaps additional comments like the grid. More specific mode, for example FT4 versus just digital. It is also a good idea to mention that the station is split if applicable, for example "*Listening up 5*". If you say simplex, be sure it actually is, as some don't bother to double check, though they always should. This could create chaos for the DX and those trying to work it. Some clusters you may see the expected "*DX*" heading but also a "*DE*". DE is in reference to the spotter, meaning "*from*". The station who submitted the DX spot. **If you are the one submitting the spot and have not yet worked it, work it first!** You may just invite a hundred of your closest friends to compete with you. I usually spot things after I worked them. Is that selfish? No. I will share but will not make things harder on myself. Trust me, most if not all DXers do this.

Some of the original methods used for DX spotting before the internet took over are still in place. Specifically, I am referring to using 2m radio to alert other local DXers that there is something that might be worth working. Our local DX club, the Northern California DX Club, has a very active 2m repeater for example. W6TI is often active during morning and evening hours with spots. It is a great way to keep the repeater alive and sometimes even get to DX before it hits the internet. Lesson here is an older, less used method may just give you an advantage.

Packet radio clusters pretty much disappeared along with the dial up modem era. Although might be interesting for someone to experiment with an APRS (**A**utomatic **P**acket **R**eporting **S**ystem) like system at some point, perhaps at an alternate frequency. Just a thought!

Other modern technology has also been embraced by many hams in addition to the traditional DX cluster, such as Twitter, Slack, and email alerts. Though, these are not nearly as widely used and sometimes there is more garbage than usable information.

Speaking of which, sadly, some get this idea that DX clusters should be used like social media to send messages to the DX, other hams or even the world about things that do not belong here. Often rude comments are seen or requests to move to a certain band or mode. Trust me, most DX do not read this. And if they do, likely could care less. Use email or JTAlert if you really must notify them about propagation or your personal needs. If you _really_ must. Also, if you have something negative to say on the DX clusters, just keep it to yourself. Please! The same goes for political messages which seem to increase during elections, invasions, and other political events. Believe me, you are not going to change anyone's vote, but you might be ignored and/or maybe even blocked by some. **DX clusters should have only one purpose, to post spots and DX relevant information.**

If you miss out on a DXpedition or just do not get a QSO, there is always next time. And if there is not then so be it. Certainly, no need to bash the DXpedition on the DX cluster or on the air. These folks work hard on getting you a DX entity you may not otherwise have a chance to work. Not to mention the costs involved. Work on your skills, antennas and be more ready next time. It is after all, _only_ a hobby!

I use **VE3SUN DX monitor** software, and this has a super interface and many useful features. Also, **has a great "_lid filter_"**, as they call it, where you can add calls _and_ key words to a list kept locally on your computer. The above stated problems makes me appreciate this even more. My lid filter list has gotten huge over the years sadly. I even share this with friends who also use this so we can really clean up our displays. It is so nice to see a clean, usable cluster with no profanities, folks complaining about idiotic things and just a general lid fest at times. I highly recommend it! **Other software titles and perhaps even websites should implement this feature.** You can check out this software at ve3sun.com and if you like, you can buy it for $39.95 with future updates included.

There are also often calls on DX clusters which are posted incorrectly. This is <u>especially</u> true for CW. Mainly due to folks mishearing code, reporting from decoders without double checking, and so on. For SSB I often see spots that are off frequency, sometimes just due to poor tuning on the part of the poster. Some folks also use RIT/XIT and forget to turn it off set back to zero. This is common during contests, but easy to fix if you hear the station being slightly off.

Accidents do happen when one posts, but please double check the frequency and callsign if you post. Lots of folks will see these. While most will appreciate the post, even if not perfect, there are always a few who must comment and sometimes rudely. If this happens, don't fuel the fire. Send a correction or just forget it and move on.

Self-spotting is often frowned upon. I do not think it is such a sin, especially if you are in a wanted grid, unusual mode or something that may be of interest. For example, I would not spot myself from W6 land stating I am on 20m SSB. Likely <u>nobody</u> cares. But if I am on SSTV or PSK31 mode sure. I personally look for these at times myself.

Something I have noticed though is that some more popular spotting networks at times have "*information overload*" with no off switch. Data is nice, but there should be a way of filtering. Some do this, some do not. For example, if you live in Utah **(W7)** land. You likely do not need to see local spots from someone from Sweden **(SM)** or India **(VU)**. Likely we are not hearing the same things and for sure not the same strength even if you are. When a rare DX comes on, I do want to see <u>all spots</u> but, filters are good.

I usually set a radius on my local software to not even show me anything outside of 600 miles (960km) from my QTH. I do run multiple clusters and with different filters and settings as well as use different ones during contests of course only when allowed per contest rules if I am going to submit my log.

SpotCollector is an application which is included in the DXLab suite, mentioned earlier, and is truly outstanding as well. If you are already using the DXLab suite, you are pretty much set. Though many serious DXers, like me, use more than one spotting application.

DX Summit can be found online at dxsummit.fi and **DX Heat** calls dxheat.com/dxc home. No list is complete without dxwatch.com which is perhaps the most popular of the three I had mentioned so far. **DXscape** at dxscape.com is one some DXers I know like to use. If you are looking for a telnet directory try dxcluster.info, they also have some other great resources here.

As a sidenote if you are a **US county hunter**, check out ch.w6rk.com for a place to assist you. **For those interested in VHF** (including 6m) a great resource is **DXMAPS** which can be visited at dxmaps.com. Relating to the above Ping Jokey is a great scheduling site for VHF users at pingjockey.net/cgi-bin/pingtalk. We will cover scheduling a bit later. If you are on the West coast, mainly Northern California there is an extra resource. The Lodi Amateur Radio Club, N6SJV maintains a site for spots at n6sjv.org. There is also a nice **cluster website for IOTA** (**I**slands **O**n **T**he **A**ir) hunters at iota-world.org/iotamaps.

Award Tracker Software

These are basically applications to help you figure out if you qualify for certain awards. I will cover awards in more detail in the final chapter of this book. Many logbook applications have these built in, such as Ham Radio Deluxe. But these standalone ones might fit better for what you are after and might be award group specific. I use these in conjunction with the above.

Award Tracker by W5DJT is the all-in-one solution for most users! Easy to use, works with your electronic logging programs and ADIF files well. Even if your logging software already does this, this software likely does it better. It is my favorite and updates often. You can get yours at www2.w5djt.com/index.php/software.

FISTS Log Converter is used to apply for CW awards via the FISTS CW club. This club is devoted to the preservation of morse code. You can use your electronic logbook exported ADIF File to see what you might be able to apply for an award. Membership to the FISTS CW club is free for under 18 and over 80 and $10 in-between, as in, everyone else. You can obtain this software at fists.co.uk/index.htm.

Ultimate AAC Covers multiple clubs and countless awards. I have seen some added over the years. These clubs are mainly EU based but at least one is based in Brazil. Regardless, these awards are open to anyone worldwide. Some awards specialize in bands, modes, continents and so on. You can also apply for membership to the free clubs right from the startup screen. You can obtain the software at epc-mc.eu/index.php.

There are of course other award software titles out there, so look around based on your interests. Sometimes you might be surprised at what you find!

Optional "*Nice to Have*" Software

If you spend a lot of time looking at logs or text files in general, you are likely familiar with Notepad if you are a Windows user or TextEdit if you are on the MacIntosh. Well, maybe it is time to try a couple of more powerful alternates. This could apply to you if edit contest logs or want to search the WSJT-X internal logs.

Notepad++ is one I would recommend. It is free and is a great upgrade to the notepad which comes free with Windows. Much faster at loading larger files and has many additional useful features. You may encounter huge text files during your ham radio career, such as logs or config files, this will save you a lot of headaches. You can find it at notepad-plus-plus.org.

UltraEdit Studio or UEStudio as it is called is very popular with coders and if you work in IT, you likely have run across it. This takes Notepad++ to a new level. However, it is not free but well worth it if you already are a heavy text file user or have other needs for it, such as work related. Runs on multiple platforms. You can find more information at ultraedit.com.

Changing gears, a bit, **CATsync** can be found at catsyncsdr.wordpress.com and this software will allow your radio to run synchronized with an online SDR receiver. This is great for testing your audio settings and tweaking the microphone for that optimal DX sound. I know some ham who also use this for nets when there are dead spots for them, or propagation is poor. The full version costs €9.95 or about the same in US Dollars at the time of this publication.

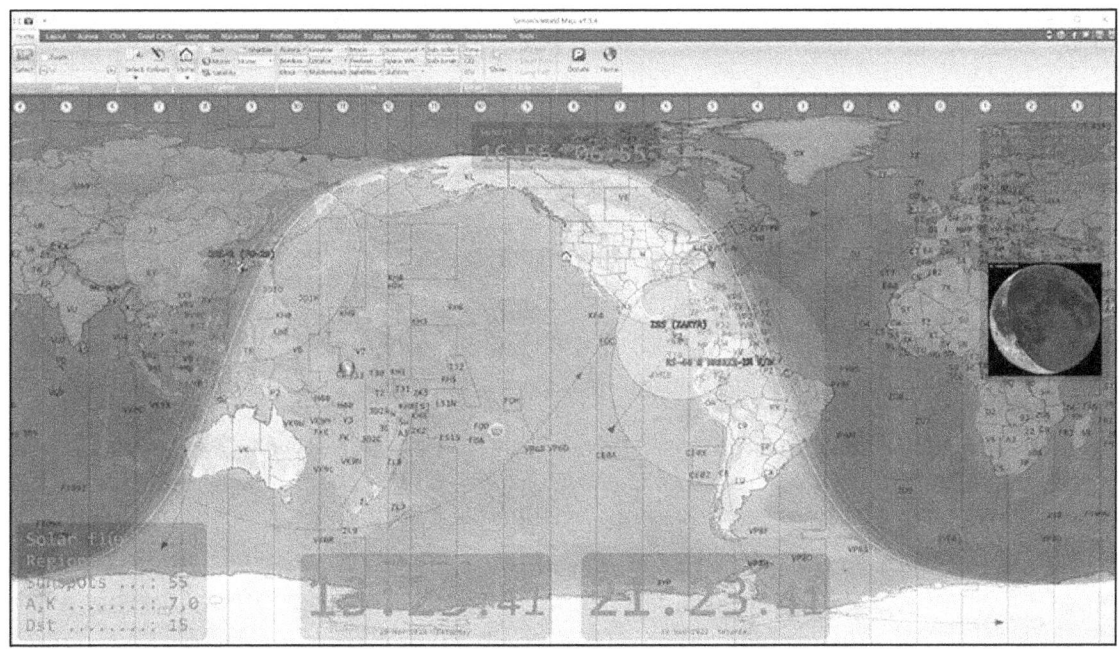

Simon Brown, G4ELI has written **Simon's World Map** which is an amazing piece of software at g4eli.com/world-map. This will not only give you super grayline map, UTC clock and propagation data but is getting better and better with each release with numerous features being added. Looks great on a 4K monitor and is a nice free alternative to the Geochron or the Raspberry Pi Ham Clock discussed earlier. Both of which require additional hardware; Simon's World Map does not.

Linux and Raspberry Pi

Let's get everyone on the same level first. Linux is an open-source operating system, meaning, it does not belong to or is maintained by a "*company*". Yes, power to the people. You have likely used it and not even known you did. Your Android devices, such as some smartphones are powered by Linux as are many eBook readers, smart and streaming devices and of course some ham shacks. It came to the scene during the early 90's and was developed initially by Linus Torvalds. Linux is a lot like Unix, yet another operating system. His name, Linus and Unix combined, gave us Linux.

If you want to run a radio specific Linux distribution check out **Skywave Linux** which can be found at skywavelinux.com as well as **Andy's Ham Radio Linux** which is at sourceforge.net/projects/kb1oiq-andysham.

Of course, do not forget about the ever so popular **Raspberry Pi.** These came to the scene initially as inexpensive teaching devices developed in the United Kingdom **(G)** in 2012. Basically, they are a single board, tiny computer which can fit in the palm of your hand. It took off with a vengeance and was embraced by tinkerers worldwide.

Raspberry Pi used to run Raspbian recently, now got renamed to Raspberry Pi OS. I think this was a good move, less confusing. And yes, this is also Linux based. You can find out more about these at raspberrypi.com and you can get started with some cool ham radio software at github.com/mawcg/ham-pi which has an auto installer to simplify things a bit.

A nice piece of software for the Raspberry Pi is the **Ham Clock** which can be found at clearskyinstitute.com/ham/HamClock. Those who have been to my shack know I run this on a dedicated screen when I am on the air. It provides a lot of goodies for both the DXer and the satellite operator alike.

Besides the above Ham Clock, there are many other things a DXer can use a RaspberryPi for. This includes antenna switching and other shack automation. These can be accomplished via external relays. Don't forget the **remote radio control. MFJ even sells a Raspberry Pi based solution**, just for this! I also use them as an SDR receiver head and have even used them to run WJST-X. I would recommend using the 3B series or above, though we are currently on version 4. You can do it with older models but with some compromises. Some hams even setup WSPR transmitters for propagation studies. All you need is the WSPR header, power, and an antenna. This can even be long wire. You can read up on this at tapr.org/product/wspr.

There are also headless models (no monitor out) such as the Zero series and the Pico series which are great for many applications as well. While not "*real*" DX, these are great for MMDVM (**M**ulti **M**ode **D**igital **V**oice **M**odem) hotspots for DMR (**D**igital **M**obile **R**adio). I also at one point ran an APRS iGate (**A**utomatic **P**acket **R**eporting **S**ystem **I**nternet **Gate**way) on 2m and an ADS-B (**A**utomatic **D**ependent **S**urveillance **B**roadcast) receiver for tracking airline traffic. I have many projects documented on my website at w6aer.com if curious. Though, these are beyond the scope of this book. If you are not sure what the above are, visit and find out. Lastly, there is an incredible list of all ham radio software for the Raspberry Pi at raspberryconnect.com/raspbian-packages/39-raspbian-hamradio.

Extra Software for FlexRadio Users

If you are a FlexRadio user, there are a lot of add-ons out there written for you by fellow hams. Below is a current list with links for the SmartSDR Software (Which I know of). As always, evaluate and use add-on software at your own risk.

- **Bob's meter** by WoodBoxRadio at flex-radio.nl/downloads.
- **FRStack** by MKCM Software at mkcmsoftware.com.
- **DDUtil** by K5FR at k5fr.com/DDUtilV3wiki.
- **Morse Keyer 6K, VoiceKeyer, SDR Memory 6K, SDR Monitor, SWR Plotter 6K, SDRSliceLabel** and **SDR Auto Center** by K9DUR at k9dur.info/smartsdr_utilities.html.
- **RigSync** and **SDR-Bridge** by W2RF / QRV Systems at qrv.com/rigsync.html.
- **Slice Master** by K1DBO at github.com/K1DBO/slice-master-6000.

There is also PowerSDR, which was written for an older version of FlexRadio hardware. This also has a good following still and many addons written for it. It has also been used on other SDR based radios as well. Apache Labs comes to mind.

Furthermore, there are "*Waveforms*" from FlexRadio Systems for **FreeDV** and **D-STAR** which are both digital voice modes. D-STAR does require a piece of hardware from **NW Digital Radio** but you can purchase it from their site directly at nwdigitalradio.com/product/thumbdv. These waveforms get installed alongside the SmartSDR Software suite instead of having to launch them independently once installed, can be initiated from the SmartSDR software.

Troubleshooting Ham Radio Software

I get asked a lot about the above, so figured I touch on the topic a bit. You will likely run into issues at some point. Sometimes this occurs when installing software, sometimes while running it, or both. The following will assume you are using current operating systems with the minimum hardware required to support the application in question. This seems to be a common issue. Folks trying to run software on ancient, underpowered computers or outdated operating systems. Alternatively, ancient software on new gear, but we already talked about that.

When installing on Windows, sometimes you will need to install a program as an administrator. You can do this by right clicking on the setup icon and clicking "*run as administrator*". This is not as much of an issue on the MacIntosh, but if you run into this use "*sudo*" (**S**uper **U**ser **DO**) command. This is the same on Linux where you are likely to encounter this more often. The latter two are a little more protective of the general user. Sometimes this is good, other times not so much. All depends.

On the MacIntosh, check gatekeeper. This may not allow the package to be installed if the developer is unidentified. To bypass in finder, control-click and open, then tell it to proceed. This will work most of the time.

You can also install applications in safe mode on both the Windows and the MacIntosh platforms. But if you must resort to this, it would be best to find an alternative application if possible.

At times in Windows, you may find that an application needs to be in the system path, or the registry is faulty. Sometimes a reinstallation of the program will cure this. If asked, it is good to always add the path for all users. Also, if asked, overwrite registry settings. This will repair the bad settings "*in theory*" if that was the issue to begin with. A reboot in-between is recommended. Turning the computer off and on is not the same. Without going into too much technical detail, do an actual reboot.

Also, there could be a dependency missing. On Windows this could be a .NET framework or a DLL (**D**ynamic **L**ink **L**ibrary) missing. These could contain functions needed to run an application. It is a very common Windows issue but can also occur on OSX. Somewhat less likely on Linux and/or the Raspberry Pi.

You can usually contact the author/vendor as a last resort, but prior to this see the FAQ (**F**requently **A**sked **Q**uestions) as likely you are not the only one with this issue. **Software user groups are also a great resource**, though some are managed better than others. Most popular ham applications have one and I do recommend joining them. Even if only to monitor.

Other things to check are firewall and anti-virus settings. You need both. If you do not have them, get them. You are living on the edge, trust me. You may however need to disable them for a few minutes if a specific installation is an issue though make sure your software download is from a trusted source. **Get the downloads directly from the author and be careful not to fall for fake websites, links, or even deceptive banners**. It is the wild west out there!

Some Windows (and even DOS) software for ham radio was written when dinosaurs roamed the earth as I had pointed this out already. Some titles still work fine though but may need to be executed in compatibility mode. Try and find an alternative or reach out to the author if possible. This may encourage them to maybe work on updated versions if they see there is still a demand. Ham radio software on the MacIntosh does not go back quite as far historically, therefore it is less on an issue on OSX. But if you see this, the same applies. Linux software in general falls somewhere in between the above two extremes. Same is true for the Raspberry Pi.

Lastly, **updates are important, and you should always try and do them.** However, updates sometimes can break things, move ports around, change sound card assignments and just wreak general havoc. If you have not yet experienced this, trust me, you will. This happened on both Windows and the Macintosh. I do not recommend skipping updates regardless, bad idea for many reasons. At times it is OK to wait a week or so. Sometimes this applies while patches are released by a given software author if issues arise for example. Always try and keep your computer current, if nothing else, for security reasons. Generally, updates also help with stability.

One software I have found to be very useful when troubleshooting is **Registry Changes View** by Nirsoft which can be downloaded at nirsoft.net. This software allows you to capture your registry settings and compare after updates and software installations. Allows a shadow copy as well to be saved and has some other very useful features. While this is a bit on the advanced side of troubleshooting, it can be a life saver and very handy for those who test a lot of software. I certainly fall into this category. As with everything else, use it with caution though if you are not sure what you are doing.

Save Time with Scripts

If you have a row of icons on your desktop which you double click, likely in order, every time you want to get on the air, this section is for you. I believe in working hard and working smart. Together, they are a powerful force. Hard work by itself pays off. Working smart will save you time and effort which can be spent elsewhere. In this case working DX. **May even help you reduce errors and stress.** Which brings us to automation via scripts.

You do not need to be an IT expert to do this. Using simple batch files on Windows, you can really speed things up when you are starting up your computer for the day to use with your radio. I have two scripts I run every morning. The first is used to start up my basic ham radio applications. The second is to start up the digital ones if I am going that route.

On your Windows PC create a blank text file which will have a "*txt*" extension. Rename it to a "*bat*" extension which is short for "*batch*". Now your computer will treat it like an executable instead of only showing you the contents. You may need to unhide your file extensions in Windows. You can do this, if needed, in the file explorer window under the view tab. Check the "*show file extension*" box.

Use your favorite text editor to add content using the example from below. Do not double click on the new file as it will execute it. Rather left click and select edit. Here is what I have in my personal script, I called this "*startup.bat*"

Start "DX Monitor" "C:\Program Files (x86)\DX Monitor 191\dxmon.exe"

Start "Antenna Genius" "C:\Program Files (x86)\4O3A Signature\Antenna Genius\AntennaGeniusDesktop.exe" -a1

Start "Smart SDR" "C:\Program Files\FlexRadio Systems\SmartSDR v3.3.32\SmartSDR.exe"

timeout 2

Start "Ham Radio Deluxe" "C:\Program Files (x86)\HRD Software LLC\Ham Radio Deluxe\HamRadioDeluxe.exe"

Start "Power Genius" "C:\Program Files (x86)\FlexRadio Systems\Power Genius Utility\PowerGeniusDesktop.exe"

Start "Tuner Genius" "C:\Program Files (x86)\FlexRadio Systems\Tuner Genius Utility\TunerGeniusDesktop.exe"

curl.exe --output C:\Users\lucas\AppData\Roaming\SatPC32\Kepler\nasabare.txt --url http://amsat.org/amsat/ftp/keps/current/nasabare.txt

curl.exe --output C:\Users\lucas\AppData\Roaming\SatPC32\Kepler\nasa.all --url http://amsat.org/amsat/ftp/keps/current/nasa.all

Start "KlaTrack" "C:\Program Files\KlaTrack_1.01d_windows\KlaTrack.exe"

Let me explain what you are looking at. It is not as bad as it looks! The "**start**" is just that. Followed by application name in quotes, mainly so I know what I am editing. Then the path to the executable. Pay attention to the quotes and spaces, these are important. If you do not know it, you can figure out the path to the executable by right clicking on the icon, then "*properties*". Here you can follow it to the actual folder if still not sure.

First my DX cluster application starts, followed by my antenna switch control software. The additional arguments are located after the last quote for the program if needed. This is just extra information being passed for each line to be executed. In this example the "*-a1*" just means it automatically connects to unit 1 on my AntennaGenius. I do not have to hit the connect button each time. Even more time saved.

The "***timeout***" command gives it a breather and lets the SmartSDR start properly before Ham Radio Deluxe loads, which in turn starts my logbook and rotator software. These are part of the suite. This is followed by the amplifier and the tuner software. If you have a slower computer, feel free to add more timeouts as needed between lines or even increase the timeout. The number after the timeout indicates seconds, in this case 2 seconds.

If you need to disable a line temporarily but do not want to delete it, just put a "*rem*" in front of it and a space. This computer will ignore this line when running the batch file. This is helpful for testing or if you do not always want to run something. I use this a lot. This is short for "*remark*" not "*remove*" like many people think. Leftover from the DOS days.

The last three are not necessarily HF DX related, so think of them as a **bonus**. The "*curl*" command allows you to download a file from the internet and copy it to a location on your computer. Since I operate satellites as well, every time I run this script at startup, I get the latest Keplerian elements data for nasabare.txt and nasa.all which are used by several of my applications. Including **KlaTrack** which I keep running to alert me visually of upcoming passes 6 hours in advance. You can get this software thanks to Christopher Thomspon, G0KLA/AC2CZ for free at g0kla.com/klatrack or via github at github.com/ac2cz/KlaTrack. You may find another use for this command if you often find yourself having to manually download a file. Perhaps a propagation report, callsign database and so on.

The above gets my basic ham radio applications started, minus the digital related goodies. For the digital startup script, I named it "*Digi_Start.bat*" and here are the contents. This script starts two instances of WSJT-X along with two JTAlert instances. These are tied to slices A and B on my FlexRadio. This method can be adapted to any other rig or rigs. The number of instances can also be decreased on increased as needed.

Start C:\WSJT\wsjtx\bin\wsjtx.exe /wsjtx --rig-name=Slice_A

timeout 2

Start C:\WSJT\wsjtx\bin\wsjtx.exe /wsjtx --rig-name=Slice_B

timeout 2

Start "" "C:\Program Files (x86)\HamApps\JTAlert\JTAlert.exe" /wsjtx

timeout 2

Start "" "C:\Program Files (x86)\HamApps\JTAlert\JTAlert.exe" /wsjtx

The above also includes some arguments and is a simplified version of the previous script. I did this to show alternative ways of doing it for this book. You can also leave the first set of quotes blank and use them later if needed. I left mine in for legacy reasons. No quotes needed in the first 2 lines as the "*–rig-name=Slice_X*" takes care of the identifier. The last argument on the JTalert starter just tells it to start the WSJT-X version. **I will cover these topics in more detail in the digital section.**

PowerShell can be an alternative to the above if you are comfortable with it. Windows Task Scheduler can also be used for some things you do every time you run your computer or at a regular interval. On the MacIntosh, you can use AppleScript to accomplish the same things if you like.

Bottom line is when I want to get on the air, I can accomplish this with one double click and everything I need is running. One more double click and I am on two bands ready to work digital DX. Startup time is way under a minute. If you install newer software which sits in a different directory, you may need to edit the paths. Besides that, this should be maintenance free.

Backup your Logs and Settings!

Yes, backup both and if possible, in the cloud. For sure not on the same drive or the same computer as this is a single point of failure. I know some very smart people who have lost years of logs, settings, etc. due to not having a proper backup. For those now saying, that is why I have a paper log, I got one word for you….Fire! Though, floods, tornadoes, hurricanes, earthquakes also come to mind. So, there are four more for you. I think you get the picture.

Since I know many like to have something physical, using a jump drive for backups is also fine but only in addition to having an **offsite backup**. Do not depend on a jump drive alone. They do get lost and can get damaged. On most computer-based logging programs you can export your backups, run them automatically and even have it saved to multiple places at once. Ham Radio Deluxe for example does all the above as well as many other logging programs out there. This is good to do and one of these locations should be something like Google Drive, iCloud, One Drive or Dropbox. The free space these services provide should be more than sufficient to store even multiple copies of your logs. Therefore, cost should not be an issue. Takes just minutes to set this up, a lot longer to recover lost data. That is if you even can.

Once you get your ham radio software setup to your liking, take notes and even screen captures on your computer for reference later. This will save you hours of work and frustration in the future.

As an added plus, you can share it with other hams who may need assistance as well. I keep a whole folder of settings and configurations I have screen captured. And guess what? Many of these made it into this book!

CHAPTER 6: Bagging the DX - Preparation

This section is going to be somewhat generalized and contains miscellaneous things which can be applied to all modes or did not fit anywhere else in the book. Consider this the potpourri section of the book. The chapters proceeding this one will go into more detail for various modes, broken down into the "*big three*".

There are operators out there who only work digital or CW for example. In fact, some hams do not even own a microphone, and I am not joking about this! Keep an open mind if you are used to or considering only operating a couple of modes only. You will be missing out! Even if you think that you are only going to work a certain mode, such as CW, I encourage you to read all the sections as there are overlaps and some information might be relevant to you.

Working Split

Despite of which mode you like to work, and you will likely have to operate at least 2 if not all 3 modes (phone, CW and digital) when working rare or semi-rare DX. One thing they all have in common is you must remember that most of the time they will be split. The rarer, the more likely this is the case. If you are uncertain, as always, start with listening first. Then listen some more. If you have panadapter, this will come in very handy here as well as an added visual guide. While I will cover in more detail, here are some basics. If they hear "UP" on CW, start 1-2 kHz up. If you are on SSB and you hear "listening up" without a specified number of kHz, try 5-10 kHz up. And lastly on FT8, try to pick a clear audio frequency which does not overlap the DX.

Split is sometimes called duplex incorrectly, as this means that you transmit and receive at the same time and not all radios can do this. Certainly not in the HF realm. True duplex is common in ham radio satellite communications when you hear your own signal coming back down on a different band with a slight delay. On HF you will not be working cross-band either. **Split is the correct terminology where one transmits on one frequency and receives on another.** This will always be on the same band when working HF DX. I will cover these more in detail for each specific operating mode.

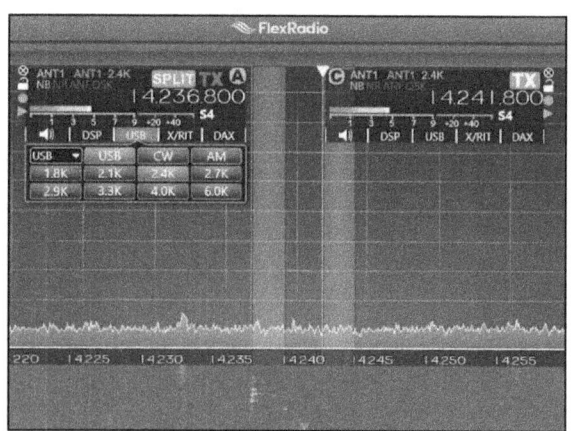

FlexRadio's SmartSDR software is pictured here with the **split mode** engaged on 20m USB. Slice A is setup for listening to the DX, in this case slice C is where I would be transmitting as slice B was on an entirely different band monitoring FT8 traffic. These do not have to be consecutive. This allows for incredible flexibility but may take a bit of time for some to get used to. I find it easier personally.

On Icom, Yaesu, Kenwood and Elecraft radios the concept is the same. These, however, would have a physical split button instead of using software on the computer. Though it can be engaged via software to if you have configured this. These buttons are usually very clearly labeled. Be sure to turn off split mode once you are done using it. Pictured here is my Icom radios split button.

However, working DX split is not generally true if you are working them during a contest. Although I have seen it. Especially on the 40m band where DX might be calling on the lower part of the band where the US for example cannot transmit phone. Generally announcing something like "…. *listening on 7.2*" or something of that nature. This is a good way for the DX to reduce the chances of stepped on, amongst other reasons.

At times when the station is not a rare DX, they may work simplex. They figure they can handle the number of callers. When in doubt, listen for something like "*up*" or other indictors as to where they are listening and who they are working. True for phone and CW. Also always listen to geographic regions which may be called such as "*West coast*" or "*6*" only. Others should be standing by…most do. All should!

Even once you figured out what is happening, as in the mode, this may suddenly change. DX may start out simplex and then go split. Alternatively, may start out on the preset FT8 calling frequency then announce that they are moving to somewhere else on the band and perhaps switch to fox and hound mode. More on this later. So always keep your eyes and ears open and watch what others do. Especially those with successful QSOs. If you see others moving up 5 to call, it is likely he is now split, and you may have missed the announcement. **As they say listen more, transmit less. And always double, even triple check your settings.**

Don't be a Dupe

One thing to note, again, despite what mode you work, try not to "*dupe*" or have a duplicate contact. Though the "u" might as well be an "o". This is very annoying not only to the DX at times but especially to those who are trying to get the DX in question for the first time. Be considerate!

It is OK to work DXpeditions once per band and mode, or combination of these. Certainly, a different callsign if they use multiples ones as is sometimes the case. This is also true with semi-rare DX to a degree especially if using a different station callsign.

Occasionally I hear hams, usually with a signal way stronger than needed for the QSO, get through and tell the semi rare DX that it is so nice to work them again on 20m sideband. Then proceed to and thank them for the 4th contact so far this year. Why!? And so far? Waste of both parties' time. Mind you these just contacts, not chats, and involve a pileup of others trying to get their contact. Maybe this is due to their boredom. Sounds like time to spend more time with the family instead. Or get a dog. Someone could have had that contact who really needed it. If nobody is responding to the DX, this is OK. But to be part of a pileup to just show off…Let's just say, don't be one of "*those hams*".

Same goes for FT8. I see some US stations working the same DX stations repeatedly. At times, the obviously annoyed DX responds with a "*QSO B4*". I will cover later how to avoid this using filter features of software such as JTAlert. If you have not worked a station for a year or two on the same band and mode, sure go for it. I have done this too. But several times a month is just…I think you get the point.

Sometimes you hear "*big gun*" stations, who know better, make more than one QSO within the span of an hour at times. Same band, same mode! The DX will at times respond with "*dupe*" but I have also heard "Oh, come on! Again?" I then every so often jump to ClubLog to check how the station is doing. I will see multiple QSOs listed for them on the same band and mode. Every time this happens, they take away a QSO from someone who really needed it. Not sure if they just cannot hear or are they unsure of themselves. I have given up trying to figure out most people.

This dupe rule, by the way, also applies in contests. Everyone can make a mistake here. I have! Generally, though since maybe I only catch part of a callsign. This is not unusual in crowded conditions. For contests, you can repeat the contact next year with the same station or during a different contest the same year. In some cases, on a different band or mode during the same contest. Rules vary from contest to contest so check them before doing so.

Your Callsign is not Written in Stone

To begin with, you will sometimes see the word "*callsign*" written as "*call sign*". Both are correct, I will write it as one word this this book for consistency. Having the "*proper*" callsign is something which can help you in all modes. A shorter callsign <u>can</u> be good but not always and there are other factors involved. You know, just to complicate things. However, a lower CW weight or an easily recognizable callsign can certainly help with DX. This also applies to phonetics, and less characters in your callsign may even speed up contesting if you are into that.

My call, W6AER has a low CW weight the. My call in CW is .-- -.... .- ..-. and if you note the characters for A and E are very short. Which is why I picked them. Other good characters are T, N, I and M for example. These are all very short. **See Table in CW section**. Something with a Q --.- or a J .--- for example, increase the CW length (or CW weight as it is referred to) as does usually having a longer callsign.

That is why many DXers and contesters opt for a short 1x2 (or 2x1) call versus my 1x3, but still more practical then the 2x3 if you are concerned with this issue. To get everyone on the same page, the first digit refers to the number of letter(s) before the actual number also known and the "**prefix**". The second digit refers to the number of letters after the number known as the "**suffix**". Therefore KK6TAP, Dema (my stepfather) is a 2x3 call, with six characters in his call. Two letters in the prefix, three in the suffix plus the number. Shorter calls are usually reserved for those with higher level licenses in the US, but many hams choose to keep their original callsigns. You can look for available vanity calls at radioqth.net/vanity/available. Dean Gibson, AE7Q also have a very nice online tool at ae7q.com/query to assist with this process.

If you do decide on changing your callsign, and it is available, you can apply at wireless2.fcc.gov/UlsApp/ApplicationSearch/searchVanity.jsp

The number 6 is a given to stations in California, although I could have opted for another number when getting a vanity callsign. There could be potential issues with this though. When DX is calling stations on the West coast and I was to come back with a 5 in my callsign for example, they may wonder if I am a not listening. Even worse, they might just choose to ignore me, and I never get in the log. All numbers are about the same CW weight, therefore unless you have some attachment to one versus another, I would not worry about that.

This weight issue also comes up when calling DX via phone mode. Whiskey Six is shorter length than Whiskey November Six plus 2-3 more characters past that. Less letters for the DX to get right with band noise, QRN and even possible QRM. I did however get logged as few times with rare DX as W6AE and W6TAR, so if the current callsign holders are reading this, you are welcome! All kidding aside, I had most of these corrected by the DX by proving the time and details of the contact but even with well thought out calls, things can and will, occasionally go wrong.

When using digital modes, having a shorter call also saves transmit time in some modes like PSK and RTTY. For modes on WSJT-X, this does not matter unless you exceed the character limit.

You may wonder if you can get an even shorter callsign like W6W or K1A. Sure, but only if you operate a special event station and only on a temporary basis. One of these calls cannot be assigned to you in the US permanently. You may see DX stations using super short 1x1 calls, like the DXpedition to Navassa Island in 2015 which used K1N. But unless you go on one of these DXpeditions or operate a special event station, you will only be calling a short callsign.

Some hams of course like to get something personal to them when it comes to callsigns. This can be their name, initials or perhaps even a callsign which was once held by their father or other relative. Go for it! **In summary, a callsign that is shorter and has a low CW weight MAY give you a bit of an edge in some situations.** Every little thing counts, just like with dB!

The Never-Ending Q-codes

To be frank, you only really need to know a few Q-codes for DX. I have seen these massive lists of Q codes and 70% or more I have never heard on the air and 90% I have never used in my life. I have been using digital modes for a while and rarely do I see Q-codes on FT8 other than a rare QRM, QRP or QSY. On CW and Phone, you may encounter these more, but only a handful really.

Q-codes are very useful with language barriers on the air. In case of limited English on the other end, using these can be a life saver. Here are the ones worth knowing for DX work in order of frequency heard, at least by me:

QSL – Received/acknowledged usually followed by a 73 meaning all the best. Kind of like the "*10-4 old buddy*" back in the CB days. At times this is way overused.

Alternatives are R (or RR) on CW or "*Roger*" on Phone when confirming. Some choose to skip it all together, and that is ok too.

QRZ – Who is calling? This might be a toss-up with QSL in frequency of use especially during contests.

QTH – Location. Can be a question as in what your QTH is or as in my QTH is X. Or you can just say "*I am in X*" like a normal person.

QSO – The contact in question, you may be thanked for the QSO.

QRP – Low power station calling, also can be a request to decrease power though rarely used as such. Sometimes used as a callsign/QRP when using low power.

QRM – Manmade noise. Usually someone stepping on someone else or splatter, interference basically. Remember **M** for **Manmade**.

QRN – Natural noise or interference, like lightning crashes and such. Remember **N** for **Natural**.

QSY – To change the frequency to or asking someone to change the TX location. Sometimes DX will announce this when there are too many callers on the main frequency so look for it. Can be "*QSY 14.080*" (frequency move) or "*QSY FT4*" (frequency and mode). This would the default FT4 frequency. Occasionally you may also see an "*FH*" somewhere in there, this means fox and hound. More on this later.

QSB – Signal fading, not due to QRM or QRN but propagation in most cases.

QRX – Stop transmitting or stand by. As an example, wait for the other station to finish if followed by another callsign.

QRT – When a DX expedition ends or stops on a given band or mode this may be heard or even seen on a DX cluster. Refers to stop(ed) transmitting. Not to be confused with QST.

QRL – Is this frequency in use? Yes, this is barely used, no pun intended. However, I do hear folks sometimes actually say "*Is this frequency in use?*" which I think is much better instead of the overused Q codes. You should be in the habit of always checking first anyway. Have I mentioned it is barely used?

QST – Announcement to all stations. Also, the name of the main ARRL periodical, which is very appropriate. This can be followed by an announcement that the station is moving to another band, operator change or up in frequency for example.

QRO – The opposite of QRP, also can be a question to increase power. I think I heard this once in my life perhaps.

QSK – Relates to CW and refers to listening between sent code. Frankly have not heard this come up in a QSO ever, but mainly seen in menus on radios.

QRV – This I only heard in the context of other hams (not DX) discussing when a station will be active, ready to make QSOs.

QRQ / QRS – Just remember the **Q** for **Q**uick and **S** for **S**low. This is used in CW very rarely and never heard DX use it but good to know. Refers to speeding up and slowing down CW speed either asking the other party or you are being asked to do so.

Some of these you have surely seen on ham radio exams, and I think that it is good to know at least the above. And no, **hams do not use the "*10 codes*",** such as 10-4 for acknowledged. Certainly not on HF and with DX. This is more CB and police lingo. Use QSL or roger if must. And speaking of roger, if you have a "*roger beep*" please do not use it on HF especially when DXing. If you are not sure what these are, this is the beep you hear sometimes when someone releases the PTT. Not for DXing!

Your Grid Square

If you decide to get active on 6m, satellites or work digital modes, like FT8, you will want to know your grid square. This is required when you setup your WSJT-X and other digital software as well as for exchanges on the above. Sometimes grid squares are also referred to as the maidenhead system.

A grid square measures 1° latitude by 2° longitude and measures approximately 70 × 100 miles. This is broken down into smaller, more accurate divisions. A specific grid square is indicated by two capital letters (the field) and two numbers (the square), followed by an optional two lower case letters for more precise location. For example, my location in Pacifica, CA has precise location is CM87sp.

Some software requires the use of the 4-digit version and others the 6-digit, more precise value. The 4-digit one is usually used when transmitting and in award applications, the 6-digit is often asked for by software at setup, even if it will only use the 4-digit version to send out during a QSO. Basically, I just need to use CM87, my general area when I say it verbally.

I recommend always using the 6-digit version of this in software, if possible, as is more accurate and will place you on the map correctly if being reported back. The software will only send 4 (as it should), the other 2 are for internal use only. For example, if I do not use 6 but rather 4 digits, PSKreporter puts me in the middle of the Pacific Ocean. Clearly, I am not a maritime mobile station. Therefore the 2 extra digits really make a difference, at least in spot reporting. You can look yours up, if you do not already know it, at levinecentral.com/ham/grid_square.php.

Know your Radio

Your radio manual is more than just something to put your coffee on. I do, however, understand its absorbent properties. Though getting harder and harder to do this as many now come in a PDF format and I think this discourages many hams from reading them even more. I prefer the PDF format, but I suspect I am in the minority. I recommend at least once reading it cover to cover. At minimum you can honestly say you did if tech support ever asks. However, I will say that while many manuals explain where to find a radio function and/or button, very few actually explain when and how to use them. Some of these things I will cover here apply to all modes. This is a good place to address them before diving into modes as I may later reference things seen here.

RF Gain / AGC-T

This is often overused or misused. In most situations this should not be fully open, meaning on max or all the way up to 10. Basically, you are adjusting your signal to noise ratio. Because you are trying to copy the signal, meaning hear it above the noise, this is important to understand. The setting you will use here will be different from band to band. I find that 40m and below is where I must play with this setting the most.

Since you are adjusting it down to noise level, on noisier bands (and days) you will need to touch this setting. Maybe even more than your AF (audio) gain. There are hams who use this as their RF gain as volume control on phone and CW modes. The dynamic range of the radio changes proportionately with the RF gain being adjusted. This is certainly something to keep in mind.

FlexRadio's SmartSDR software for example also has something called the AGC-T which stands for **A**utomatic **G**ain **C**ontrol **T**uning. The AGC-T is to be set to where the volume just starts to increase. You ideally should adjust this in a clear part of the waterfall. This is similar in concept to what you think of as an RF gain on a non-SDR radio. However, I find it works much better, at least for me. At times I can remove noise completely by just dialing it in perfectly. Radios will vary, so experiment. Try things and find what works for you. Some of this may take some time to get a feel for, do not let this discourage you. It will become second nature with practice.

Preamplifier

You will find two camps here. One says you need to leave them off due to noise being introduced, others feel their radio is deaf without them. At times they can both be technically correct. You need to learn when and how to use your preamps. Afterall, they are there for a reason. Just play it by "*ear*".

Preamplifiers are found on all major brand radios. Sometimes it is just called preamp or even "*P. Amp*" for short like on Icoms. On the FlexRadio it is found on the RF Gain slider. Down is attenuator (next topic) and up is preamp. This is the most logical design to me. On Yaesu radios, it is sometimes labeled as IPO (**I**ntercept **P**oint **O**ptimization). Bypass here is off and generally seen at 2 levels. Whatever it might be called or the method of implementation, it all essentially does the same thing.

Rule of thumb is on the higher bands, where there is less noise, it is more likely you will benefit from this. Below the 40m band you see me leave this off when using my vertical but at times on when using my magnetic loop. You just need to play around with this a bit. This also does vary from radio to radio somewhat. Certain transceivers really overdo the preamp, on others it barely makes any difference.

There are external preamps available to be mounted close to the antenna. Except for magnetic loops and Beverage antennas these are mainly found on VHF and above. These are not to be confused with internal preamplifiers found in HF rigs.

Attenuator

As they say, if the preamp is the gas pedal, the attenuator is the brakes. An attenuator is used to reduce the amount of incoming signal energy into the receiver section. Will not affect the transmit function. Now you are asking, why would I want to do this? You likely do not want to do this on the higher bands but for sure on the lower bands. I find myself using it sometimes on 80 and 160m bands where noise is higher. Sometimes this can overwhelm the receiver and by applying it, again play it by "*ear*", you might be able to dig out some signal you would not otherwise hear. This does take some practice and some patience. I will be honest, my first two or so years of DXing, I really did not touch this setting, and I wonder how much I missed out on. Likely I was flooding my receiver as things "*appeared*" louder, but the key is to dig things out of the noise so you can hear or decode them.

Signal Width

This is adjusted a couple of different ways. Sometimes via buttons with preset widths, sometimes with a knob, or both. Of course, nowadays by software as well. On SSB you likely want to set something like 2.4kHz or lower. SSB signal is 3kHz wide. On CW this setting is likely 600Hz or lower. These all depend on your radio and how it sounds to you as well as local QRM being one of the factors. Play around with this and see what works and sounds good to you. On digital you will find this to be all over the place but generally you want around 3kHz or more if you notice you are losing a bit of signal around the top and bottom edges due to your filter. This does not apply to RTTY though. On my SDR receivers I use a 6kHz filter for digital but only 1kHz for RTTY, 2.1kHz for phone and 400Hz for CW. There is no magic number, and you may find that you need to re-adjust them at times not just based on mode but also band conditions and even your ears.

DSP & Noise Blanker

The **D**igital **S**ignal **P**rocessing unit can really help you clean up a poor signal but again, it is not magic, though close. Still, do not expect miracles. **Noise Blanker falls into this category** and it can help remove some pops and certain types of interference very well. Usually this is just marked as "*NB*" on some transceivers. This works better on some units than others from my experience. Use caution and patience when adjusting if the level can be adjusted. Experiment with your radios DSP features and don't be afraid to use them but no need to overuse them either. If you have issues from local RFI, this sometimes can resolve it quite nicely. Some radios also have a **Wide Noise Blanker which may help you filter lightning pulse noise or other electronic "*spark type*" noise to at least some degree**. Regardless, always remember never to operate when there is a storm nearby. Not much point anyway as lightning strikes nearby will create a great deal of interference on the HF bands. Lightning will be more noticeable on the lower bands, though all bands can be affected. You may find more of a need for these filters on the lower bands but everyone's mileage varies.

DSP features and performance do vary a lot from model to model or make. You will need to get used to how it behaves on each radio. Transceivers I have owned in the past were all noticeably different in how they responded to various settings. DSP can sound very different from what might be used to hearing. Some say it sounds "*flat*" or even "*dull*". Yes, I would agree it can. Therefore, you must adjust the levels to your liking, maybe compensate a bit with the receive equalizer to adjust everything to your own personal taste. Also, as mentioned earlier, to your level of hearing loss.

To complicate things, there are two DSPs in some radios. The above DSP I just covered refers to audio DSP. There is also an IF DSP which is basically digitally adjusting your software filters and such instead of doing it mechanically or via crystal filters as has been traditionally done.

Notch Filter

Does exactly just that and can be used to remove a carrier or other unwanted signal. Sometimes these are generated by heterodynes inside your own radio, other times by wall warts (Power adapters) for example. This is sometimes great at removing other types of hash you need to remove to make the signal readable. Most useful for phone modes. An ANF (**A**utomatic **N**otch **F**ilter) is like magic to me. One push of a button and the offending carrier is (usually) gone. Can also be used for removing folks tuning up, therefore "LF" would also be an appropriate button as in "Lid Filter".

Roofing Filters

If this is adjustable on your radio, you want to adjust this according to what mode you are operating. Though, most transceivers switch this for you as needed automatically. Many radios are preset to where it "*should*" be. I have been known to modify radios to narrow the roofing filter and have gotten better performance. Generally, this is done by making them slightly narrower, at least in my case. Unwanted nearby signals are filtered out and the DSP has a lighter load to deal with as well. Of course, it you overdo it, your signal may become useless and unintelligible. Also, can backfire as is the case with digital modes where you may want a whole 3kHz. In the SDR world you will not have to worry about roofing filters.

AGC

Here is the most controversial of the bunch with some strong feelings and opinions. AGC (**A**utomatic **G**ain **C**ontrol) can create a heated conversation in ham circles. We have already covered RF gain and AGC-T. These are related, but not the same. AGC here has to do with "*recovery speed*". This phrase is used to describe how long the signal attenuation lasts once it is triggered. It is triggered by a strong received signal. After a given amount of time, the signal level will return to the initially set level via the RF gain.

It really is not as difficult or controversial as people make it out to be. Really comes down to what works for you. **Most radios have settings for slow, medium, fast and of course an off**. Off is just that, not inline. Some hams like to operate with it off and ride their RF gain. Meaning the gain will <u>not</u> be automatically adjusted even if some inconsiderate lid operator comes in tuning up their amp on the DX and blows out your ear drums.

Slow means that when applied by the circuit, it will recover slower to the level it was set to. Fast means just the opposite. It will recover much faster. Of course, this means that there could be large jumps in gain, and some find this distracting if you are not used to operating like that. Medium is as expected, a halfway point. Slow can be slow enough to where you may miss a DX station responding to you though, as in several seconds. Now you are likely starting to see where the debates start.

Here is what I do, and I encourage you to experiment. For phone mode I use the slow or medium setting. However, for CW and digital I use fast and occasionally off. This is just my preference. Now I know many will argue with this, but this works for me. I am more likely to miss something in CW or part of a RTTY or FT8 transmission being decoded if the AGC is not recovering fast enough. AGC performance does vary from radio to radio as well. Some digital operators leave AGC off completely when operating FT8 for example. Try both and see which yields the best results for you.

Flipside works better for me with SSB, but I have been known to operate using fast AGC with weak DX. I do want it to recover faster so I do not potentially miss my call being called even if a strong signal doubles with the DX. Many feel that using medium or slow is less exhausting for your ears, especially during contesting. I would agree with that.

RIT and XIT

Lastly, one that I sometimes wish some hams would not touch. RIT stands for **R**eceiver **I**ncremental **T**uning used for receiving frequency adjustment. Sometimes called a "*clarifier*". This is used when you do <u>not</u> want to change your transmit frequency but slightly adjust your receive frequency. Often doing so when a station calling you is off just a bit. So far, I get it and I agree with its intended usage.

Some would say it can be used in cases where you want to use it for split operation. I never do, I do not recommend this, in fact I never use it like that. But if you must use it for split operating due to not having a split button, please make sure you set these back to the default or you will be off frequency for your next simplex QSO. I have had folks tell me I was off frequency and just as soon as they realized they moved their RIT. For me it is much easier to just tap the split button again and go back to operating simplex. That is just the way I like to operate. I think some of this might be just habit in cases of operators who had to use it in the past. Kind of like the hams who still "*tune up*" their solid-state amps every time they change bands…. don't ask!

XIT is **Trans**ceiver **I**ncremental **T**uning, the pronunciation sometimes sounds like "*zit*" and is used to adjust the transmit signal frequency <u>without</u> adjusting the receive frequency. Essentially the opposite of RIT. Good to know what it is and some also use this for split operating. This makes more sense to do so with! If you are using a radio with a single VFO and no true split, you may have to use the above method. That is a different story, however. You may find this in modified military radios or older ham gear. **In modern ham gear I see absolutely no reason to use this as the method of operating split.**

Signal Overload, work your RF Gain

While the following may sound counterintuitive to some of you, trust me on this one. It is kind of like taking a shower. Too little water pressure will drive you nuts. Too much will act like a fire hose and blow you away. Though not very likely an issue in my drought ridden state of California…But I digress. Your radio goes through this too. The first step is to reduce your RF gain but do this slowly. Many newer hams have theirs fully open since this seems the "*loudest*". This varies from rig to rig depending on how slowly you will have to reduce it. Some radios are hyper-responsive, some you feel are barely changing anything. This will reduce the amount of noise coming in. **You must look for a sweet spot here.** The key is to reduce the noise as much as possible by reducing RF gain but not so much that the station can no longer be heard. I am not referring to background noise here, as that is what you are trying to cut. You likely will need to compensate with AF (audio) gain in some instances from reduced RF gain. To use the above analogy, you want a comfortable water pressure. This takes some practice so be patient. You can also try adding some attenuation slowly and systematically in some cases, though the above usually does the trick.

Make sure your pre-amplifiers are off if all you hear is noise. I know it sounds like the pre-amplifier is supposed to do just that, pre-amplify, but especially on the lower bands it just makes things worse in many cases. If you are using a magnetic loop with a weaker preamp to listen on, try the opposite. Depending on the loop, length of the cable run and its own preamp, this may be needed. Radios do vary, and some hams swear to me that on their rigs they hear nothing without the preamp on. I had different experiences but as always, test things for your given situation and working conditions.

While using my FlexRadio SmartSDR software I usually keep the RF gain at 0 or 8 for 20m and above. 0 or -8 for below 20m depending on conditions. Using your ears and the waterfall will help you find the sweet spot. If you are receiving on a magnetic loop antenna for example, these numbers will be different as I had mentioned. Same for beverage antennas is you have one.

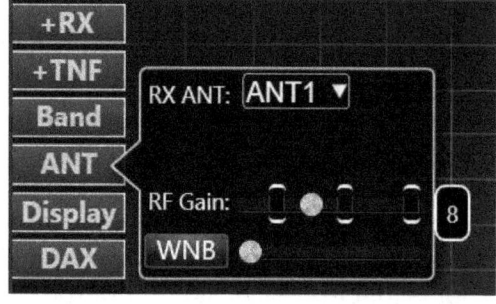

On my Yaesu radios, I used to keep the RF gain at around ¾ way up above 20m and ½ or thereabouts for below 20m. I did however usually keep my preamp on for the higher bands. Again, I adjusted for band conditions and additionally for the band I was on. Here I had to do it all by ear. If you have a panadapter, this really helps.

When using Icom radios in general, the RF gain at ½ seems to have worked for me and frankly barely ever touched it. But again, this varies based on many things per above. These are just my experiences. If you noticed the pattern above, **never have the RF gain fully open for weaker signal work.** As they say, if you can't hear them, you can't work them. At times though, this can be fixed with some minor adjustments, like the one above.

ESP and Insurance Contacts

There are times on phone and CW where signals are so faint, they are barely above the noise level. Often it is somewhat jokingly referred to as an ESP (**E**xtra **S**ensory **P**erception) contact. No, not really a ham radio term but we embraced it. Signals can be so weak that you cannot even be certain you were the one called. You may have heard a partial callsign or missed part of the report. May take several exchanges to finally get the 73. Or so you think! Sometimes you are not even sure you are in the log even after that. A few hams may log these ESP contacts anyway even if they never got (or at least heard) their 73 and hope or the best. Others do not consider this a valid contact and do not log them. I will leave this at your discretion. I <u>always</u> log if there is a chance the QSO was completed with a note. If the station is rare then say a small prayer, salt over your shoulder or whatever you are into. Been lucky most of the time, but when I am not, I move forward and there is always next time.

Insurance contacts refers to making an "*extra*" contact in case you are not sure you are really in the log. An example would be above. This is generally not recommended if you have a way of checking the logs first, such as online, and certainly not recommended via the same mode or band if avoidable. You do not want to take away someone else's QSO by being a DX hog. Or even worse, being called a "*duper*" as covered earlier. If needed, repeat the QSO for sure. If the DX is very rare, absolutely! But use common sense and common courtesy as with everything else when it comes to DXing…. or life.

QSB – Fading

This is the one that drives me the battiest and it often goes hand in hand with the above ESP we have just discussed. Basically, has to do with paths and signal subtractions and additions. It can really ruin a QSO and just requires patience and timing if you notice signals area fading in and out. Stick with it though. May take a lot of repeats no matter what mode you are in. If you are getting parts of the exchange, the signals may come up just long enough to complete the QSO. Or it may not! Always worth trying for a few minutes and checking back later. Propagation can shift dramatically in either direction at a moment's notice.

You will experience this sooner or later if you have not done so already. This affects all modes and bands, though some more than others. I have especially seen this on 6m and 160m bands, for different reasons, but can occur anywhere. It can also show up as "*flutter*" with sudden changes in signal strength.

While it can certainly ruin one, alternatively, it can also make a QSO. I have seen DX come up above the noise just long enough for a few stations to make contact. Within minutes there was no trace of them. Just a bunch of desperate callers hoping propagation would shift again.

Use What you Paid for

This comes with experience and with experimenting, but here are a few tips that can always be applied. I am sometimes taken back by how many do not use anything beyond the PPT and the volume control knob. OK, maybe the power knob, but still…you pay for the "*extras*" in your radio to get an advantage. Use them!

Let's start with filters. For Sideband some DXers find that reducing the filter down to 2.1kHz or lower helps them a lot in hearing. This will reduce the audio range of course but may help cut through the noise. I stay between 2.1kHz and 2.4kHz personally. If you go below 1.8kHz you may start to have issues with comprehension. Learn to walk that fine line between hearing better and not being able to understand anymore. There is a sweet spot there. This takes practice and will vary from band to band based on noise and even the tone of the voice of the DX. Here you may want to play with the RIT or IF shift, but just a little if you need to narrow the bandwidth as this will change the tone slightly. They may help with comprehension, though I find that usually not as much as getting your DSP adjusted right.

Some of the same settings can be applied for RTTY but not so much for FT8 since it requires you to theoretically monitor a 3kHz segment. The good news with FT8 is that the signal you are interested in is very narrow, therefore efficient. The bad news is you still need your filter open wider. On SDR radios, this is not so much an issue.

The same is true for CW as far as narrowing filters. You need to find a balance where you can decode but don't set it too narrow. See what works for you. We all hear a little differently and may prefer different tones or even hear different tones better.

DSP is your friend. There are at times two extreme camps here. Those who never use it, and those who adjust it to a degree where speech becomes intelligible. On some receivers this can be done easily. Use the DSP but also use caution. Play around with various settings of "*noise reduction*" if your radio has it. When used carefully, this can help you greatly. Refer to your user manual as all radios are slightly different. And a few are very different. **Noise blanker**, which we covered already, many feel is a must. I would mostly agree. And lastly a reminder about **adjusting your RF gain**. Use it wisely!

Low Band Noise Issues

Once you start getting more serious about working stations on 40m and below, you will notice that noise levels are much higher. If you live in a more populated area, it can be a true nightmare at times for many different reasons. This is especially true in the summer months and daytime. In fact, most of the lower bands are useless for serious DX during the day. Though in spite of popular belief, the lower bands are not all that dead in the summer. I have worked some great DX during summer nights.

There is a lot of man-made noise (QRM) as well as natural noise (QRN) present on these bands. Nearby lighting strikes can be heard on 40m and below quite often for example and can even be seen via a waterfall. Poorly built electronics tend to emit a lot of RFI. Especially noticeable for hams and I will cover this in more detail shortly.

Technically with FT8 you may not "*hear*" the other station as the signal can be so weak that may not be audible. It may still decode. Phone communication can be a little tricky with a lot of local noise. Even CW, while better, is not going to work as well as FT8.

Most of my new entities on the 160m band were obtained using FT8. I can practically count the number of DX phone and CW contacts on my hand.

Antennas specifically for receiving are helpful. I have covered beverage and magnetic loops already. Sometimes adding baluns help, though only to some degree. Directionality can also help, such as a dipole even. Verticals are noisier than dipoles. Dipoles and other directional antennas will have lower noise as there is directionality. Much as a magnetic loop, which has even sharper receive and rejection angles.

Low band DXing can take up a whole book, in fact it has. If you get serious about this topic, check out **ON4UN's Low Band DXing** book. I can't even begin here to compete here with the great information presented in this book. Much of that I know I learned from this book as well as by doing and trying things on my own, sometimes from ideas in this book. Check it out for sure.

RFI and EMI Should be a 4-Letter Word

Fittingly, this brings us to one of my pet peeves. RFI stands for **R**adio **F**requency **I**nterference and at one point I was considering creating a whole chapter on this but then this book would quickly turn into a multi-volume publication. Much as if I covered low band DXing in detail. I ran across quite a few books and articles that do a much better job of this than I ever could. I will, however, give you the basics you need to get going. If you want more theory and details, plenty out there.

EMI (**E**lectromagnetic **I**nterference) is real, and you cannot ever get away from it. Even if you operate away from society, your inverter or battery charger will likely still be an issue to at least some degree. So, what is the difference between RFI and EMI? Are they the same? Yes and no. EMI and RFI are used interchangeably in the world of ham radio. And I would say that is OK. But technically EMI is a much broader spectrum than RFI. Without getting too technical RFI is a subset. Meaning it is more specific. All RFI is EMI but not all EMI is RFI.

RFI is often misunderstood, and many do not think about the fact that it **works both ways, receive and transmit.** I am sure you have heard strange stories of Christmas tree lights glowing when they are actually turned off or garage doors opening

unexpectedly. These events are caused by nearby unintentional interference. Our radios receivers are also susceptive to such RFI, sometimes so faintly that we don't even realize. You may not have any RFI issues though or more accurately; don't realize you do. Sometimes it is all about connecting the dots.

Line isolators, such as the ones pictured here can also help with issues such as the one, I am about to describe. These are for USB and serial isolation, respectively. I recall talking to a fellow ham at a convention a few years back. He told a group of us that he does not ground anything, there is no need for ferrites ever, and this is just "*the man*" (who ever that is) trying to get more money from him. He stated that he never had RFI issues. Most of this is a myth, he stated. I said great, he is lucky he has no issues and was ready to move the conversation along.

But it was not over. Oh no! He then very angrily added that he only has computer issues because "*Windows is junk*". So, I (stupidly) asked for more details as he piqued my curiosity, working in IT and all. He described that every time he transmits his mouse moves around the screen or the computer freezes for a few minutes. Then he mentioned that sometimes his logging program also crashes. After a quick breath, he continued to tell me that his computer even reboots by itself sometimes when running an amplifier...Darn Windows! I have heard about these issues occurring before. I then resumed to verify with him that this never happens when he is off the air, and this happens only when he transmits. He confirmed by adding that yes and he is ready to toss his computer out and go back to paper logging. OK...So, I tried to explain to him that in fact this is an RFI issue, and this is not a coincidence nor is this normal Windows PC behavior. He would not listen to anyone and continued to tell us how "*Bill Gates is the devil*" and how he broke his barely 10-year-old PC to get more money from him. Seems everyone wants this guy's money. Yeah...I changed topics very quickly. Bottom line, many have RFI issues, but for one reason or another wish to not connect the dots. Perhaps it is easier to blame something else than apply a relatively easy fix. In this case there were several issues!

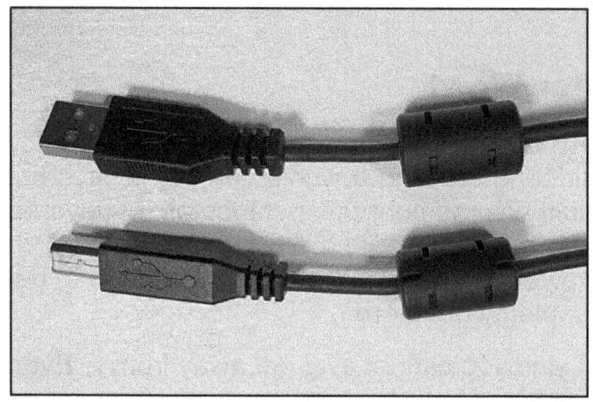

Another lesson here is to learn from others or at least hear them out. Likely, someone has seen the issues you are having and may (or may not) have a solution. Sometimes the dots are easier to connect when you just take a step back and look at the big picture. You know what they say about assumptions. Follow the recommended procedures and good engineering practice. You will be a happier ham. And likely your neighbors will thank you too as in this case. I am sure others around him had "*possessed*" electronics as well.

Sometimes there is a telltale sign to a really big issue at the desk. The dreaded microphone shock! **If you are getting zapped when transmitting, you likely have other issues going on. These need to be addressed.** Yes, it does refer to a surprise and not a pleasant one by any means. Those using boom microphones sometimes get a zap on the lip when transmitting if there are RFI issues present in the shack. Though other non-boomed microphones are not immune either. I have never experienced this but heard many stories.

This can be remedied usually easily. Start by finding the cause of the problem first. Might be obvious once you start to figure out when and where. Grounding, bonding and properly connected equipment is a start. Application of ferrites can usually help too. Remember those?

Circuit breakers can also trip when you transmit. Especially, certain brands are more susceptible to RF interference as I was told by my electrician. Older units, due to age might trip more easily already. Solution is to replace them with fuses. I am <u>totally</u> kidding! No. Just get a new, preferably a better-quality circuit breaker. Likely time anyway to replace them. I have never experienced this myself but heard of this happening more than once. There is a great website on RFI if you are interested in further reading about this topic at hamuniverse.com/rfi.html.

Finding Reception Noise Sources

There are a few ways of finding the culprit or more like culprits. Some are a little extreme for some tastes but work well. Others are simpler and, frankly, get the job done or at least well enough for an average DXer. If you are serious about DX though, like I am, you will want to do everything you can to cut the RF noise. Common sources of RFI include:

- Dimmer light switches, surely you even hear the buzzing by ear sometimes.

- Solar inverters, especially the budget ones. No, it is not the solar panels that generate the RF noise. Microinverters can be especially bad. Look around on forums to see what is quiet. I use Enphase IQ7+ microinverters, in fact 22 of them. They seem to be fine.

- Some switching power supplies such as DC adapters, including some phone chargers. This seems to correlate with cost.

- Plasma televisions are luckily going away. These were basically mini lightning generators. They looked good at the time, but hams hate them!

- Some monitors, especially older ones with VGA. Unshielded VGA cables are even worse. Even with a ferrite on them some are bad.

- Certain laptops and desktops and at times with their power supplies bricks. Easy fix for these. Ferrites! Battery power at times works too.

- Outdoor AC power transformers, such as those for garden lights. This is due to long cables running all over the yard. Ferrites can help. Keep them away from your radials if you are using any.

- Power utility lines and transformers. More on this one later. This one can be a tough one to address. Though I hear depends on where you live.

- Some air conditioners, mainly portable ones due to power cords.

- Furnace blower motors, this is likely due to the type of electrical motor used.

- Some LED lights or light strips, even "*higher quality*" ones.

- Washing machines, especially during spin cycle. Not too much you can do about this one. Laundry when you are not on the air maybe?

- Vacuum cleaners. Surely you recall from the analog TV days. Nothing much has changed here.
- Ethernet cables, especially longer runs or loops. Filters and ferrites work here as covered earlier.
- Now we have grow lights to worry about. Especially an issue in California. I don't think I need to spell out why.
- Anything with a very long cord or extension cord can be a source as well as this acts like a radiating or receiving antenna.

The most common way of doing this investigative work is using a portable AM radio to locate noise sources. I find that a portable shortwave radio is even better. You can get one for under $60. I have done this and works well. Some even suggest turning off all power and slowly turning things on, one circuit breaker at a time. This way theoretically the noise source is revealed when a circuit breaker is engaged by region. With this technique the source can be narrowed. If you have solar panels though, these can technically never be fully turned off during the day. The panels will produce power and the microinverters will still be engaged. Therefore, if there is RFI from these, you will have to address that first. Ferrites!

There are also gadgets one can buy online to assist with this RFI hunt. MFJ makes several of these and they perform this task rather well. The MFJ-805 has a snap on ferrite with a coil to enable you to see if there is RFI via a cable or cord. This unit does a great job detecting noise. I used this to test most of my house electronics. Now over 50% of my electronics have ferrites! Ideally the goal would be 100% though.

The MFJ-852 is great for narrowing down noise sources outside as well as inside. If you want to kick it up a notch the MFJ-856 has a 3-element Yagi for more directionality and gain. This unit is mainly for outdoor use unless you have lots of space inside. It is a little bit on the larger side.

Ultrasonic receivers have also been used to find power line noise as well as other sources of arcing interference. The drawback is that these receivers do need to be in line of sight. Meaning, will not work as well as RF location methods but they are great for pinpointing the exact location once the general noise source has been identified. Noise, like traffic or wind can make the above very difficult to use. Using RF based noise finders in conjunction with ultrasonic receivers, you can do some serious cleanup of your home (and street) as far as RFI. Your neighbors of course can be a whole other story!

Powerline Noise

Powerlines can be an issue for sure. Specifically, the arcing which may occur due to transformers which may be aging or are faulty. Additionally, the actual power lines with bare or loose connections can be an issue. More so when it rains or in my area because of heavy fog and the salty air. What a horrible combination!

You can contact the power company as well as look on the ARRL website to help assist you in resolving these issues. If you do a lot of the leg work, figure out where the issue is, this really could speed up the resolution. Just make sure you are certain as if you are wrong, next time when there is an issue, you might be the one *"who cried wolf"*. I have heard stories where it took years, yes years to resolve issues. So, patience is key here but so is persistence.

Antenna Phasing & Noise Canceling

One way to fight noise is to cancel it. If you studied any physics, you may recall that two equal waves at a different phase will cancel each other out. Radio waves are waves, as is interference. Now, if you can locate the local interference source, using a second antenna, this can be inverted and injected back into the signal. This in turn cancels it. Now, I really oversimplified the process and the fine adjustments required but this is the basic concept.

This method is very useful perhaps if you are dealing with a noise source that you cannot address with other methods we had covered. Such as a neighbor with grow lights or cheaply constructed LED lighting. Under counter lights and recessed lights with built in LEDs seem to be a reoccurring issue around me. These are the most noticeable on the 40m band from my experience. Can't exactly put ferrites on these!

DX Engineering and MFJ both make devices to assist you with this. The MFJ-1025 & MFJ-1026 do a great job with this and they work as advertised. Except for very severe instances, this will have no problem cleaning up local noise. I have tried these in the past and was very impressed with the results. If you are looking for even more bells and whistles or are dealing with unusually bad issues, the DX Engineering NCC2 is the one to get. This unit takes signal cleanup to the next level with massive expandability and customization. Slightly different approach but will do well.

Ferrites Are Your Friends

I have already touched on these, a lot, but let's learn more about them. If you are not familiar with ferrites, they are essentially ceramic mixes with different metals, called mixtures. The main metal is generally iron oxide (yes, rust) and mixed with other metals in smaller ratios, such as zinc, strontium, barium, manganese and even nickel. These ratios are what are referred to as mixtures. They perform differently and are used for various applications. There are a handful of mixtures useful for ham applications.

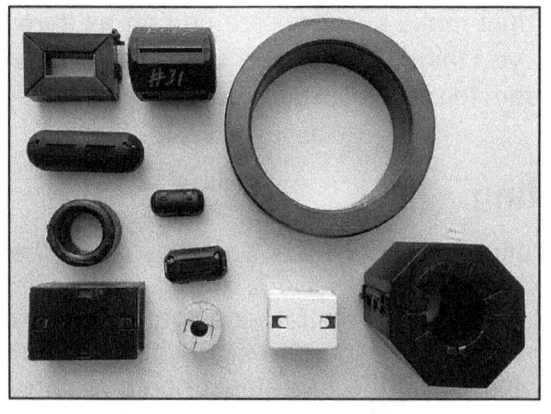

They are very fragile, so handle them with care. If they crack, they are usually garbage. Minor chips are perhaps OK for noise reduction purposes but <u>not</u> for use in baluns or at high power as there might be small fractures inside that can cause issues. Don't drop them or bang them against hard surfaces. I broke a few on the way home from ham fests, so trust me on this one. If you use them outside, they need to be protected from moisture and other harsh weather to at least some degree.

Here is why they rank so high on my list. They can not only help with RFI issues in the shack, but also with your HF reception. They are also used in some Baluns. **Ferrites are one of the most overlooked items in a ham shack.**

They come in two common forms: Snap-on and doughnut. Don't try and eat them though. Doughnut ring types do not have a split in them therefore, it is said they will work slightly better. I have not tested this myself but makes sense to me. The ferrites you likely will run into are generally round, but you can find some used on ribbon cables and other special applications where they are rectangular. Some of these can still be used in ham radio applications and are perfectly fine to loop cables through.

One problem with ferrites is that they are not well marked, if at all. I usually use a permanent marker (silver or gold are best) to write the type on them when I get them so I can reference that in the future. Paint based markers are best.

Since I only know enough about ferrites to be dangerous, please read articles by **Jim Brown, K9YC**. He is personally responsible for my ferrite education, addiction, and excessive spending on ferrites. Thanks, Jim!

All kidding aside, he has done more research on this topic than anyone I know or have ever run across, and I always check out his presentations in person. If you get a chance, you should too. He has many articles on his website at audiosystemsgroup.com/publish.htm. I would recommend spending some time there as some of this information presented is second to none in quality. Lucky for me Jim is local, so I get to hear him talk a lot in person. I likely owe him for my success on the lower bands as well as some of the very weak signals I have managed to dig out from the noise. To supplement the above, I also highly recommend checking out **QRM Guru's** article on ferrites as well for more information at qrm.guru/the-truth-about-ferrites.

Here is what you really need to know. The key thing to remember is **ferrite mix #31 for the HF bands and up to about 300MHz and mix #43 for the higher bands, 25Mhz and up to about 250-275Mhz**. These are useful for 10m and 6m mainly for HF operators. Most other common mixes will be for higher frequencies, but not all. For VHF and UHF DXers, mix #52 is above 200Mhz and up to 1GHz. Lastly, #61 covers 200Mhz to 2GHz.

I use both #31 and #43, often at the same time and everywhere I can think of. Then I add some more for good measure! **There is also a mix #77 which is great for noise suppression the 160m band as is mix #75 as this covers 150KHz to 10MHz.** This would cover the 160m to 10m bands!

If you can **loop the cable through a ferrite more than once**, do that! It will work even better than using 2 ferrites as **impedance gets squared**. See the next photo for a visual. For example, if you have a doughnut shaped ferrite and you loop it through 4 times, it will be equal to 16 snap-on ferrites. That is some magic! Well, not really, it's just science.

Mix #31 and #75 are the ones most used for wall warts. The problems with wall warts are numerous for a DXer. Specifically, with some newer style switching varieties.

Pulse with modulation creates a square wave. Square wave is a sine wave with fundamental frequencies. Even if this emission is not strong, it will likely, raise your noise floor…possibly a lot! Wall warts radiate via the DC lead as this acts as an antenna. This is also true for USB cables as they carry 5VDC. Even more so if they are plugged into a cheap charger. Cheap being the key word. upply technology is not bad when implemented and filtered properly.

Ethernet cables can also be guilty of not only causing QRM but can also easily pickup RF especially if it is a long run. Even more so if in a loop formation. The fix is relatively simple. Use shielded cables when possible and use ferrites. This may not work 100% but will certainly help if used consistently. You want the ferrites to give you about 25-30dB attenuation. This will do the trick for most hams when it comes to noise reduction. Though you can choke them more if you like, though your return on investments might not be as good. This is your call. I go overboard, but that is just my paranoia.

If you come across some mystery ferrites or do not trust the source of the ferrites you have purchased, there is a way to test them. Articles have been written about testing them on the fly using an MFJ-259 Antenna analyzer. This also involved 5 turns of 22-gauge wire. Set to 2MHz the inductive resistance should be 500 Ω if it is mix #43. Which is what you ideally want in many situations. There are a few articles on the internet about this as well as a demo video. You may see a reference to **soft and hard** ferrites. This has to do with their **ability to magnetic reversal**, not with their physical hardness.

A great place to purchase ferrites is **Palomar Engineers** at palomar-engineers.com. They carry everything most hams could want and if you are not sure what you need, they have you covered. Just ask! They also have many great articles to learn from on the website. Another reliable source is **Amidon** and they can be found at amidoncorp.com. **Digi-Key Electronics** at digikey.com is a favorite for many hams also. Additionally, try **JPM Supply** at jpmsupply.com or **Kits and Parts** at kitsandparts.com.

RF on the audio is common in some shacks. Make sure you rule out ground loop as this is a totally different animal. Speaker hum when transmitting is one that plagues many hams. But using ferrites does not always fix this issue, in fact rarely. If you operate VHF on 2 or 6 meters you likely know this. These are the most common bands with this issue from my experience.

One of the common solutions for speakers is to put a 0.01 uF capacitor between the speaker inputs leads to bypass the speaker coil. Other values may also work, check the internet for current research on this. The Speaker coil makes an excellent "*antenna*" if you will. I used to have issues with my computer speakers buzzing when I used any kind of power on the 6m band. This method in conjunction with shielded cables and ferrites, took care of it all.

Get Some Fresh Air

Yep, go outside! You read it correctly. I know some DXers are now saying, but I will miss all the good DX. Well, maybe the opposite! You ever wonder why sometimes remotely located DXpeditions hear you better than you hear them? Trust me, most DXpeditions do not travel with monobanders and massive towers, so it is not the antennas. They use the same or similar radios as you. So, what gives?

No local interference for one, though the saltwater helps as well if they are near it. So why not take the show on the road! You would be amazed at how much better you can hear when away from folks running poorly (and sometimes even not so poorly) designed electronics. With the RFI/EMI mostly gone, weak signals reception ability really come up. Your noise floor will be happy!

A nice bonus can be that if you find a rare grid to operate from, or perhaps a rare county (or even country) with a limited or non-existent ham population, you may just find yourself being the center of attention. Hope you are ready to handle a pileup!

While you are at it, you can participate in popular programs such as SOTA (**S**ummits **o**n **t**he **A**ir) at sota.org.uk or POTA (**P**arks **o**n **t**he **A**ir) at parksontheair.com. There is a chance, it might be both at once actually.

If you live near water or are willing to drive to it you may even be able to activate an IOTA (**I**slands **o**n **t**he **A**ir) at iota-world.org or even ARLHS (**A**mateur Lighthouses **o**n **t**he **A**ir) at arlhs.com. Or again, maybe both! The DX will be after you.

Be sure to do the necessary registrations for the above as required and get the word out beforehand for best results. There is a minimum number of contacts one needs for some of the above activations. Self-spotting for the above is generally acceptable.

If you reside in the US, you can also generate a lot of DX interest if you visit a wanted county as I had mentioned. You can read up on this at countyhunter.com. This is especially more so on the West coast where we have some counties with only a few hams and in some cases none!

Grid activations have been done for decades in the world of ham satellites since it is mainly about grid chasing for the VUCC award. HF can be just as rewarding and 6m activation during the summer sporadic E season can be sensational. In fact, once again, why not do both. If you get creative, you can kill many birds with one stone!

Photo provided by **Dr. Antonis Papatsaras, AA6PP.** This is a great SOTA/POTA setup (mobile shack) on a park bench using his Elecraft KX3, Elecraft KPA100 amplifier, portable antennas and 2 LiPo batteries. Logging is done on the iPad. Best part, he gets to look at scenic San Francisco without all the noise. Both RF and AF.

DQRM & Other Jammers

DQRM is short for **D**eliberate **QRM**, and QRM is man-made noise, as discussed earlier. This could be anything from digital transmissions, carriers and so on. Maybe this is why some hams think it stands for *"digital QRM"*, it does not. I have even heard jammers playing recordings of music over DX as well as obscenities and worse. If you even encounter this and can get a fix on the heading, it is a good idea to post to the cluster. With enough folks doing this, the location can be narrowed somewhat. These individuals think they are impossible to trace but this is not the case. They can be triangulated and identified by coordinated hams with actually very little effort. This has happened in the past, in fact in recent history.

As with Pirates, I will never be able to comprehend what possesses people to do this. And sadly, I have seen an increase over the years. If this is the best form of entertainment for some folks, I am truly sad for them. Do not respond to them or encourage them by acknowledging their behavior. Ignore them as attention feeds their fire and ego. They seem to move on when ignored.

Some believe these are people who are frustrated with not being able to make contact with DX in question and they want to ruin it for others. Others feel these are folks who already made a contact and want to limit others from also doing so. I think it is a little of each, but who knows…And who cares! Even more of a reason to just ignore them as if they think they are ineffective, hopefully they just move on and hopefully grow up.

Join a Wolfpack

Most DXers I run across outside of my clubs seem to be solitary DX hunters. For some that works and that is their preference or personality. You will find a much better success rate working in a *"pack"* much like in the animal kingdom. Mother nature figured this out long before DXers.

This can be done in many ways. For example, **using a 2m repeater to *"coordinate an attack"* and call out DX spots**. I know this is nothing new, but it is often forgotten. It works, and it is fun as well. This is something regularly being done by the NCDXC folks, which also has its own unofficial wolfpack. Members help each other, encourage each other, and take joy in the success of others. Kind of like how life really should be. Right? One member can be monitoring 6m and call out things while others are working on 20m. Take multi-tasking to the max.

Email lists also help, and many clubs do this. This can also be done with just a group of friends easily as you can create a *"group alias"* which is fast and easy. At times email list servers can be slow due to server congestion. This method does require one to be checking email often or at least be near it. I have set up an alert in outlook where if an email with a specific kind of heading shows up, for example *"DX Spot"*, it gives me a special sound alert. You can find this in the *"rules"* setting in outlook as well as many other email clients.

Using JTAlert to keep in touch with friends, or to call out spots to your closest ones is also a good option. I will cover this shortly. Sending a group **SMS text with a rare DX spot** also works and this is more commonly done than one can imagine. Yes, your cellphone can be used for other things than just social media. I often hear stories from back in the day before cell phones, with home phones ringing at night waking up the whole family with a rare DX spot alert. For my younger readers, phones had cords at one time.

Speaking of using modern technology, there are some Twitter feeds and Facebook groups as well that you might find helpful. Not sure where to start? As I had mentioned earlier, join a DX club! No wolfpack? Start one!

CHAPTER 7: Nice to Hear Your Voice - Mode

The phone modes include FM, AM and SSB for purposes of HF with minor exceptions which I will cover shortly. SSB is where you will be most. SSB (**S**ingle **S**ide **B**and) can be USB (**U**pper **S**ide **B**and) and LSB (**L**ower **S**ide **B**and). 40m, 80m and 160m will always be LSB. 6-20m and 60m will always be USB. There is no voice communication on the 30m band, so it is not applicable to this chapter.

FM is only found on certain segments of 6m and 10m and is not really used for DX in the US, nor is AM mode really though you will find folks using both modes where it is allowed to mix things up a little. Mainly AM on the 40m band at night. SSB is going to be a lot more efficient for DX than AM or FM on HF. However, using SSB still does take significantly more power than CW or most digital modes when it comes to beating the signal to noise ratio to get through to DX. It can be a challenge at times!

There are digital voice modes found on HF. One notable one is **FreeDV** at freedv.org. This involves software but you can also purchase a hardware solution (add-on) from Rowetel called the SM1000 FreeDV adapter at rowetel.com/?page_id=3902. One can also find **D-STAR** on HF and more on this at dstarinfo.com/DSTARHFNet.aspx.

DX exchanges will usually be very quick. It does take some practice to learn to properly tune in SSB signals. This is especially more so for weaker DX stations, so be patient. With practice come results so hang in there.

We have already covered split mode a touch. Generally, DX on phone will be up 5-10kHz for most but at times you will hear down 5, as in 5kHz below receiving frequency. This is common for Japanese **JA** stations depending on the band. You may hear *"5 down JA only"* in these cases. Always pay attention to the DX. If you cannot copy them clearly, stand by until you can.

Readability (R)	Strength (S)
1. Unreadable	1. Barely detectable
2. Barely readable	2. Very weak signals
3. Readable with difficulty	3. Weak signals
4. Readable with little difficulty	4. Fair signals
5. Perfectly readable	5. OK signals
	6. Good signals
59 is the best signal	7. Moderately strong signals
	8. Strong signals
	9. Very strong signals

On phone mode, signal reports will be in the RS format. When you are working rare DX they do not usually care for you to analyze their signal quality so the **exchange is always going to be 59**. Sometimes less rare DX will ask for an accurate report but when it is a rare one or a DXpedition, they just want to focus on getting as many in the log as possible.

You may occasionally get a 59 plus 20 or such report. This means you are booming in and are pegging the meter for someone. If I was to get this report from Oregon (**W7**) for example, given I am in California (**W6**), I would not be surprised. But from Bouvet Island (**3Y**), very much so.

However, if I get this report from Hungary (**HA**) for example, that means that either the propagation Gods are awake and good times are ahead or I am using way too much power. If the latter, tone it down a bit. I know some hams like to brag about how well they are received in the EU but with reports like this but technically not abiding by FCC rules. **Using the minimum amount of power required to complete the QSO.** Many DXers, seem to forget this. It's about making contact, not who can peg the S-meter across the globe more.

Speech Processors / Compressing

I was originally going to put this into the "*know your radio*" section but seems more logical here as this is only used in phone mode. Or at least I hope it is always off when you are in digital mode as it should be. Not all radios turn this off automatically even if you switch to a data mode, so be sure to check. Regardless, I do feel strongly that **at least some level of speech processing is very helpful for successful DXing.** Along with a good microphone and proper equalizer adjustments, it can really work for you.

Many tend to overdo audio compression and end up sounding like they are screaming into a paper bag, or worse. While it does help in many situations, by compressing the audio to make the speech pop through the noise, it can also backfire per above. I find that most of the time 30-50% compression works just fine. This is where I kept my Yaesu radios. Some radios, literally have a button for on and off. That takes the mystery out of adjusting it but will also drive the control freak DXer batty. Some control is nice. On Elecraft radios, this is called "*CMP*" and is highly adjustable. On the FlexRadio there are 3 settings, Normal, DX and DX+ and I find that these all work well, but DX (middle setting) seems to do the trick most of the time. On Icom and Yaesu rigs, I also set the compression to about half.

You may want to use two settings. One for rag chewing (chit chat mode) or local nets and one for DXing. For rag chewing, having to listen to a higher frequency, sharp voice for several minutes at the time will make all the dogs in the neighborhood howl. Much easier to listen to a natural sounding, less compressed or even non compressed audio when not in a contest or chasing DX. For DX, use it in moderation, but tone it down at times when not chasing the DX. Your local net and local ham friends will thank you.

Compression brings up your average output. 5-10dB increase is a good thing for the most part but use caution per above. Make sure you check your ALC after you apply compression. It will likely change. Then need to check and adjust your microphone gain accordingly and will be going over this in more detail.

Equalizers

Yes, plural for a reason. **There are two of these you need to know about. One for Receive and one for Transmit.** Use them both! Sometimes an Equalizer is just called an EQ for short. Same thing.

For the **receive equalizer** end, it is really a personal preference but there are some settings that help with understanding phonetics better. Though, this can vary if you have some degree of hearing loss or even if your native language is not English as your brain is trained differently. I can certainly relate to this.

When using the **transmit equalizer**, this will take some patience to do perfectly but will be worth it in the end. This is the harder of the two to adjust. There are some sounds which pop or certain letters that sound like others when the EQ is set incorrectly. The settings will be different for everyone, based on the tone of their voice as well as the microphone element used.

For me, I must be careful as my call "*Whisky Six Alpha Echo Romeo*" is sometimes heard as "*Papa Echo Romeo*". This is due to the pop on the "*P*". I must adjust my transmit equalizer to reduce this from happening. Easy enough once you are aware. Make notes on your settings for future reference. Do this for all your rigs, they might be different. Same applies if using more than one microphone.

You want the power of your voice and therefore the signal to be in the range where you get the best punch and are still understandable. This is where voice compression, per above, comes in. Keep in mind that you do not want to sound "*like you*" for DX but rather easy to copy without crossing the line to having annoying audio. This can be a little tricky and as with many things, may take a while to get just right.

The **EQplus** equalizer by W2IHY, pictured here, is widely popular and I have used one for many years.

Most human speech exists from 500Hz to 4000Hz. Can vary based on factors like age, sex and so on. Very little falls below 500Hz or above 4kHz if anything. You want to boost the highs and reduce the lows when compressing. These are what is perceived as bass. Here it is not needed, not to mention wastes a lot of energy. You want the emphasis to be around where most human speech is understood. 1500-2550Hz range is where the emphasis is, and this is your target goal.

While those with booming "*Radio Voices*" perhaps sound cooler, these consume more power and for DXing have less punch. It is ironic they call that the radio voice. I think the YLs (female operators) and kids have a clear advantage here even without using voice compression. They have less bass in their voices, and this really helps cut through the noise or even the pile up in many cases. They often get called before much stronger, deeper sounding stations.

As mentioned before, those who are aging and/or enjoyed loud music a little too much in our youth may have hearing loss on certain frequencies. A **receive equalizer** is great for boosting the areas where compensation is required. This may take some experimenting and readjusting but once you nail it, make note. Consider an online hearing test as a guideline for your adjustments. I often tweak my receive equalizer for CW to cut noise as well by reducing the lows and highs where not needed. For phone modes, I use an entirely different setting. We all hear differently, so see what works for you. Listening to a net is a good way to adjust your receive equalizer since most signals will be at least somewhat strong and you know it will be going on for a while.

To VOX or not to VOX

VOX is short for **V**oice **O**perated Transmit (**X**). I will get straight to the point, just don't use it please if you do not know how set it up correctly. Many do not, in fact I would argue that most do not. Sadly, these are the folks who sometimes think they do or do not realize that their environment is not ideal for its use. I know it's there, and I know it is calling your name to use it, but for the rest of us, I kindly ask that you refrain from using it if you are not going to do it properly.

There is nothing more annoying then hearing heavy *"serial killer like breathing"* as a friend describes it while trying to listen to a weak DX station calling on simplex. Some do not even realize when they go into transmit. Background noise often triggers VOX to transmit as well. Loud coughing, sneezing in one's ear is also quite annoying. So are kids playing in the back, as are dogs barking, doorbells, phones ringing, repeaters coming on in the background and your garden variety email and DX alerts. I have heard them all either during contests or during other simplex operating. I have even overheard phone calls being answered and a complete conversation during a contest. One time medical. So much for HIPAA laws. All of which I have experienced thanks to the wonderful world of VOX.

It should be set so it triggers with your normal speech but not at every breath. If your breath triggers it, the microphone could also be too close to your mouth, or the VOX is set to be too sensitive. If you do not set it to be sensitive enough, your transmission might be choppy. Find a good balance. You likely will need to readjust at times. **A quiet environment is a must for this to be used.**

Yes, I understand that some contesters find it is easier to use and they are usually not a problem. If the DX is simplex and you are calling using improperly configured VOX or in a noisy environment, you will make enemies. Trust me! Even if working split, I am sure the DX does not want to hear your dog barking in the back. Not to mention that when you transmit you do not hear. Sometimes I wonder about some hams, but I digress. Good alternatives worth series consideration is a handheld PTT switch or a PTT foot pedal. They are not expensive and your fellow DXers and contesters will thank you.

ALC & Microphone Gain

ALC stands for **A**utomatic **L**evel **C**ontrol, but is sometimes referred to as **L**imiting or even **L**inearity control in various literature. Basically, similar if not the same. Think of this as the *"police"* of the radio. To oversimplify it, it provides feedback to the amplifier to keep the signal clean. I will touch on this topic in the digital section as well as that is the other problem area when it comes to ALC. I find that overdriving is the biggest problem in signal quality with phone and digital modes.

ALC is used only when you transmit. It can change band to band on some rigs, so keep your eyes on it. You can have a *"clean"* signal on one band and splatter on another without even touching the ALC controls. ALC will kick in when needed if your microphone gain is too high, your SWR is too high, or your amplifier is unhappy with the output. This can be the internal or an external amplifier, hence the ALC connector on your external amplifier. But it will not be able to do much for you if you way over drive it.

Via phone mode, a very common issue is that some folks crank their microphone gain to the max (11 if they could) thinking they will be heard better. Many DXers as well as some contesters suffer from this illness coined *"splatteritis"* as some like to call it. The cure is easy, but usually one of the side effects is denial. This *"condition"* is accompanied with very distorted audio, sometimes impossible to understand. Not to

mention splatter, hence the name. These signals take up a tremendous amount of space compared to that of a typical sub 3kHz signal. Therefore, power is less focused (wider signal), and operator requires subtitles to understand. Those are not available to hams, but sure would make DXing easier! Throw heavy audio compression into the mix and you get a full-blown disaster.

How ALC is monitored by the operator varies from radio to radio but generally you want to make sure you do not *"trip it"* as it is often referred to. Though it does not function like a circuit breaker. On Yaesu line (top picture) the boom line generally there is an

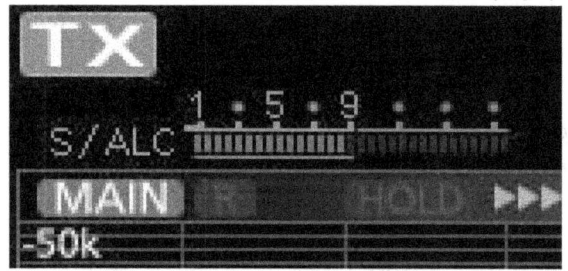

indicator where you do not want to go over this blue solid bar. This is usually the midway point. This is very similar to the Icom line (middle picture). If you look carefully, you will see a thin line below the indicator which runs from 1 to 9. I usually aim for around an 8 on my Icom 9700. On the FlexRadio for example monitor SmartSDR (bottom picture) to stay in the green on the ALC meter and possibly a bit in some yellow zone when talking. Here the radio shows -40dB to 0. Above 0 is very bad. I usually aim for around -10db in my rig. Consult your manual or your local Elmer.

When adjusting your ALC levels using your microphone, try and talk in your normal voice and/or as you do when operating. Refrain from saying *"Testing"* or *"Ho-o-o-o-la"*. Those who have been on the air for a while know what I am talking about. Instead **say your own callsign and an exchange like 59, 73, QSL** and so on as these are more realistic and therefore will give a more accurate final adjustment. This way it is more likely you will stay at the desired level. Repeat the above with compression applied if you desire to use it and make note of the settings.

Most modern transceivers, if not all, offer an ALC out which is to be used to connect to your external amplifier for essentially the same purpose as I had already touched on. Some radios achieve this via other methods, such as ethernet for example. As is the case with FlexRadio and the PowerGenius XL making this even simpler to connect and operate. Other manufacturers should take note.

Phonetic Alphabet, relearning your ABCs

There is a list of standards or rather recommendations if you want to avoid communication issues. You may have noticed these are not the same as what you may hear when listening to your local police on the scanner. My call is not **W**illiam **S**ix **A**dam **E**dward **R**obert. Though you will hear this if I get pulled over and my license plate gets checked. There are slight international variations on these, though most hams tend to stick to this list here.

A	Alpha	N	November
B	Bravo	O	Oscar
C	Charlie	P	Papa
D	Delta	Q	Quebec
E	Echo	R	Romeo
F	Foxtrot	S	Sierra
G	Golf	T	Tango
H	Hotel	U	Uniform
I	India	V	Victor
J	Juliet	W	Whiskey
K	Kilo	X	X-Ray
L	Lima	Y	Yankee
M	Mike	Z	Zulu

Numbers are all as normally spoken but do note that **the number 9 is often expressed as "*niner*" instead of "*nine*"**. But it is your call, both are fine. Also, if your callsign has a number zero, **say "*Zero*" not "O"** when using phonetics.

One trick that is nice if the DX has problems hearing you is count from 1 to the number they are not getting. This will help as likely they can hear peaks and maybe make out some of them helping reconstruct the rest. For example, using my call, if they got the AER part right, but not the 6, I may say: "*Whiskey 1, 2, 3, 4, 5, 6*" This usually does the job, for sure after 2 repeats they get it. Do this at a normal rhythm. Don't go too fast but you do not need to slow down too much either as not only is that annoying frankly but also harder to "*count*".

Your chances of getting through as best if you enunciate clearly. Use proper phonetics and speak at a normal speed and normal speaking volume. No need to shout, they will not hear you any better. Trust me! I tried this, and my wife asked who I was yelling at. In the excitement, we sometimes forget. May also trip your ALC!

I do not recommend getting too "*creative*" on the air with using variants or alternatives to the phonetics agreed upon. Some substitutions are fine, if they are commonly used, such as common city or country names. Good examples of OK substitutions are America for A, Victoria for V and Santiago for S. Bad ones would be Dingo for D, which can be heard as Bingo.

My favorite worse of all time Motel for M. Sounds a lot like Hotel for H and I have heard both on the air. The person's call ended in "*RM*" so he figured Roach Motel was cute. Funny but very annoying. Just don't do it!

I also sometimes do break habit and repeat my callsign with "*America*" instead of "*Alpha*" or "*Ecuador*" instead of "*Echo*" depending on where the difficulty is hearing it on the other end. Also, where the DX might be located. The above usage is a good one for South America for example. Don't deviate too much and **stick to the standard phonetics as much as possible.**

Especially if working DX, these substitutions might not be obvious to those whose native tongue is not English. Frankly, this really bugs me as well, since it throws me off, especially during a contest. But technically, it's your call if you understand the many drawbacks. Don't be surprised if some DX operators ignore you though or if you must repeat yourself several times. Can be quite a time waster for both parties. Here is a list of acceptable DX phonetics if you decide to substitute.

A	America, Amsterdam	N	Nicaragua, Norway
B	Brazil, Boston	O	Ontario, Ocean
C	Chile, Columbia, Canada	P	Pacific, Portugal
D	Denmark, Denver	Q	Queen, Quito
E	England, Egypt, Espana	R	Radio
F	France	S	Santiago, Sweden
G	Germany, Guetamala	T	Texas, Tokyo, Toronto
H	Hawaii, Honolu	U	United, Uruguay
I	Italy, Istambul	V	Venezuela, Victoria
J	Japan	W	Washington
K	Kentucky, King	X	Xilofono
L	London	Y	Yokohama, Yucatan
M	Mexico, Montreal	Z	Zanzibar, Zelandia

While in VHF circles where audio is clearer due to shorter distances and higher frequencies (less noise) it is fine I suppose it you really must deviate. I would refrain from it on HF as well a SSB operation during satellite passes where it is already difficult at times to make out calls due to doppler shift.

On HF the same is true due to atmospheric disturbances and even QRM. Stick to what people know, expect to hear and your chances of getting in the log just went way up. So will your QSO rate if you are a contester for example.

Working the Phone Pileup

So, what is the big secret to working pileups? I wish I could answer this simply in a few words. If I had to sum it up however, I would say knowing the **procedures, practice, experience, technique, and patience!** Doing is learning and is the only way to get experience. This will come with time. Just like you cannot learn to ride a bike from a book. And as with bikes, here you may fall too, but just get back up and try again.

Rule #1 to remember is that **the DX is in charge**. They dictate the rules not the callers, though some would love to. Examples are frequencies and who they are coming back to. If they are only calling Japan **(JA)**, respect that. Most DX operators when working split on phone, as they do most often, have a "*system*". They may not even realize it, but they do. This refers to how they operate and search for callers.

Meaning, they may go up and down the area of callers, 5-10kHz up or more usually. Once moved up and reached the end back to the 5kHz up mark. Alternatively, they may just start slowly tuning down back to 5kHz. At other times they try and tune in the signals with the least QRM. Both systems work at the end of the day. I have seen all the above with about equal distribution. There are even times when DX looks for the weakest signals on purpose. I have been told by actual DX that they have been known to do this. When listening, check to see if the DX is responding to stronger or weaker stations. Might be hard to judge, but you may get a general sense.

If the pileup just started, you have a much better chance of working it closer to 5kHz up or close to it. Once the word gets out, it will be rough. You will either need to increase power or find a new clear spot. Likely both! Many folks like to pile up on whole numbers, like 10kHz or 15kHz up. I see this on the waterfall all the time. Maybe this has to do something with human nature and liking things even. I have even heard DX announce, "*listening 5-10 up*" and guess what, most folks were at 5kHz or 10kHz up. 3-4kHz was clear. Go figure!

You may have a better chance when the DX is tuning around if you are at a more random number like 6.35kHz up as an example. I do this often. 98% of the time, I get in the log. Though that 2% kills me. But then I get over it and figure there is, hopefully, always next time.

I do recall during the K1N DXpedition to Navassa Island, you can hear stations up 40kHz! This required a lot of patience, but the payoff is worth it. Took me many hours to finally make one single contact. Some swear that they always get the DX 5kHz up. This has not been my experience unless it is some more common DX station. Does not hurt trying I suppose.

There is also a very good chance that the DX might listen way above the announced range. I often work DX above this and see DX cluster spots with folks posting information about the DX listening much higher than they initially announced. Don't be afraid to call 12kHz up, per the above example, assuming that the frequency is clear of course. If the DX has a visual aid, like a panadapter, they may see you. Even if not, they may tune until they hear you in the clear.

Using a waterfall to monitor the pileup and figure out where callers are being responded to almost feels like cheating. But it's not, so don't feel bad! If you have the capability, you would be almost insane not to use it to your advantage. But if you are one of those who likes to sleep on a bed of nails, I am sure many hams would love to take your panadapter off your hands.

Once you have made the contact, especially with a rarer DX and he is ready to move on to the next contact, so please let him. There is no need to tell him about your antenna setup and power used unless he is chatty and asks. He also does not need your name, your cat's name, or any medical aliments. Likely he has already tuned the dial and logged you.

Sometimes I listen to both sides of the QSO, and I have several times heard folks still spelling out their name to the DX followed by their distance to the nearest big city. This is all while the DX already moved on and made two other QSOs. For DXpeditions time is money, they want to get as many as possible in the log. So, it is best to help them by **keeping things short, friendly and to the point.** Do not repeat their callsign in a pileup. This not only slows them down but can even screw up their rhythm.

Also, lengthy filler words should be avoided. This includes contests and perhaps especially during contests. Examples such as "*My callsign is K8CHATTY*" or "*This is so and so 60 miles from Cleveland, Ohio and it is 79 degrees here*". This is OK for domestic contacts and perhaps for less rare DX looking for a real QSO. Normally the

DX will only need a callsign, the rest is surplus. You may throw off their timing and in case of contests their QSO rate. Many hams take contesting <u>very</u> seriously!

On the other hand, if you get through and give a report from an area where not many are getting through from, I feel it is OK to add a bit of information. But make it very quick. For example, when I respond with "*Five Nine*" as I usually would, I sometimes add "*Five Nine near San Francisco*" or "*Five Nine in California*". Why? Because they may not realize that there is good propagation to the West coast and may refocus. By doing so, may help my fellow hams on the forgotten coast. Notice I did not use "*niner*" in the "*five nine*" report since this is expected.

If the DX station did not copy your call correctly, say "*Negative*" and repeat your callsign. Slowly and only once, initially. Then if confirmed that they got your call, proceed as usual with a "*QSL 73*", "*Roger Roger 73*" or "*Thank you, 73*". Do not use "*No*" instead of "*Negative*" as I heard folks confuse this with "*Nine*". Now you have got a lot bigger mess to fix!

If the contact is just too hard to accomplish, you may hear "*Sorry no copy*" or even "*Try Later*" meaning in hopes that the propagation to your area may improve. Especially true for contests. Don't keep trying right away, come back in a bit. You may hear better than the DX, though I find that most DX hear better than me. I live in a city with lots of RFI noise, a remote island will likely only have to deal with AF (audio) noise from birds.

Some DX likely have limited English, so if you do strike up a conversation, keep it simple and limited if language seem to be an issue. Japanese **(JA)** DX seems to be different. Operators here seem to want to get to know you more and give their "*operating conditions*" which I find very nice and a good break from my usual response of "*73 from San Francisco*" and "*Thank you 73*".

Speaking of which, I often get a "*You are welcome, 73*" back as a confirmation to the above. Call me old fashioned, but it is nice to thank people for their efforts occasionally. Especially a DXpedition. Just make sure you keep it <u>short</u> and sweet to not hold them up too long. Always remember, there are others waiting. Be courteous!

Lastly, **slow down!** I have heard many hams calling so quickly with their callsigns that even with my 59 copy I had a hard time processing their callsign. I have heard this on all bands, even 2m repeaters and satellites. One time I even jokingly told someone to cut back on their coffee. Good phonetics helps but the speed may ruin it. This also applies to releasing the PTT too quickly. Sometimes chopping part of the last letter or even the whole last letter. Easy does it!

Timing, Tailgaters and Tail-enders

Timing is a large chunk of the skill you need to make the QSO. This also comes with experience. This is true for both phone mode and CW (next chapter). Sometimes they say that timing is everything. While not everything, in this case it is certainly also very important. Irregular timing and not understanding when to call seems to be a major issue for some. And not just newer DXers. Practice will make perfect and there is no substitute for it. I hear people often calling at the wrong time and they wonder why it takes them much longer to get in the log than others. Sometimes not at all! This is often one of the causes I see.

There are always a few Tail-enders and I am not going to lie, sometimes it does work, but try not to over use it as it can even annoy DX. Especially annoying if poorly timed when not split. Particularly so during a contest where split is not used. When split, and you really need the station, go for it. Always an option!

When I apply this technique, I generally wait until I hear 2-3 characters from other callers and then start mine. This will give the DX a better chance to pick out my call from the crowds….maybe. But I never do this exactly on top of someone else. This would be not only very counterproductive for both of us but also very poor operating practice. Also never call (in any mode) before the DX is done with his QSO with the other station. Do not ever do that!

Tailgating is used in reference to those who transmit right after the DX indicates they are ready for new callers such as "*QRZ?*" or if they gave a "*73*" or "*Thank you*". This may or may not work, depending on the operator. I had the best luck with somewhere in between tailgating and tail-ending personally. In other words, calling when I really should be.

Here is an example of what not to do. If you are calling the DX, give them a chance to respond. No need to repeat your call over and over with no pause. I have seen this mostly on phone mode where a station calls the DX. The DX responded to the caller, and they are still calling with their callsign and end up missing the DX stations response. I generally listen to both VFOs and hear this a lot. The DX largely moves on to someone else, someone who does listen.

I know a ham who does this a lot. He repeats his callsign twice like clockwork, every calling cycle. I have witnessed him being called while he is still saying his callsign phonetically and very slowly on at least five occasions. He does not get many in the log, needless to say. Usually, he gives up in frustration. I attempted to help but he said it works for him. I tried! It is just a hobby.

Timing is one of the most important keys to being successful in phone mode DX pileups. But as I said, it is not everything. You can also call it timing or tempo, kind of the same concept. Call it whatever if it gets you in the log.

Lastly, **take a breath occasionally and just listen**. Why? This is so that you can check if the calling pattern perhaps changed, check to see where most are being worked and a good time maybe even recheck your settings. You do not want to be the one who is not split, right?

Rule Breakers

Welcome to my soapbox section. Enjoy your stay! Examples of these are those calling with a partial callsigns, such as "*Alpha Echo*". There is always one who only throws out a partial call. And usually with horrible timing…I also notice how they often get ignored by the DX. Additionally, I also know from some DXpeditioners that they purposely ignore these callers sometimes. I also ignore these callers. Others could care less or may not even notice in the endless stream of callers.

Here is what irks me. Some well-known DXers recommend doing this to other DXers. I personally will not do so. I also ask that you think twice if you are one of those telling folks to use this "*method*". Two reasons for this. One, it wastes everyone's time. Likely the DX could have picked out your full call if timed properly and now they must and ask for your full callsign. That is a whole cycle lost when someone else could have had a QSO.

Secondly, it is not legal in many jurisdictions. **We are supposed to use our full calls to ID per FCC rules.** Not to mention sometimes these are the same folks who forget to split, and you hear something like "*Tango India……Tango India*" on top of the DX repeatedly. Please don't be one of those operators! People will get to know you on the air. Trust me, they will remember. Using a full callsign is your best bet in my opinion. The best legal option and the most time efficient option. I know someone is going to argue that you only have to ID once every 10 minutes. Technically true, but what if you do not get through? Humm…

Rule breaking also applies to some DX at times. **DX frequently do not ID often enough.** Most jurisdictions require an ID every 10 minutes, some 15 minutes but no longer than that. The worst part is, you eventually are going to have the "*band coppers*" (next topic) or other lids show up, on top of the station nevertheless asking the DX to ID. Of course, they don't ID themselves either, but don't get me started. This will cause a distraction (QRM) to all trying to call. I often see sought after DX stations working folks and not sending a callsign for long periods. In a recent contest I did not hear a station ID for over 30 minutes, and I just moved on. So, if <u>you</u> are the DX, make sure you ID and do so often. Don't assume the DX cluster will do it for you.

Sometimes there are stations who will start to call like sheep not knowing who the DX is. I know this as I often hear them ask the DX's call <u>after</u> contact, when I monitor both sides. Same goes for contests. If you are not sure who you are after, please refrain from calling.

I have already touched on this, but worth repeating:

Do not tune up on top of the DX. <u>Ever</u>!

I hear this a lot, sometimes every minute during a big pileup. You are being inconsiderate and/or lazy and could cost someone a QSO or an ear drum. How would you feel if that was you who could not hear your call being responded to by a rare DX? Bottom line is, be considerate, pay attention and follow the rules. Oh, and get a dummy load!

The HF Blue, Meet the Badge

This perhaps still falls under rule breakers, along with dupers, but I feel it does deserve its own heading. Ham radio has their own "*Policeman*", also sometimes called the "*Ham Cops*" or "*Band Cops*" by some. These are not terms of endearment, and these are not real cops obviously. Nor are endorsed or hired by anyone. In fact, they are some of the biggest and most annoying rule breakers.

They cause more of a mess at times than what they are trying to allegedly fix. These "*phony 5-0*" show up usually on top of the DX, lecturing a station who do not operate split, for example. The fact that someone is not split is likely accidental. If anything, a quick "*split*" or a "*5 up*" call to the station is fine if they have not realized for a while that they are QRMing. But usually, they take it too far. Way too far! If you want to be the fuzz, please do it outside of ham radio. And I am off my soap box now.

To sum up this chapter, practice, practice and then practice some more. Listen to those who get in the log. Learn from them. Listen to them on the air, watch them on the waterfall and be open to new ideas or suggestions. You do not have to take all the advice, but rather put it in your mental toolbox as you may one day end up reaching for something useful.

CHAPTER 8: Dits & Dahs - CW & Morse Code

This topic is the one many fear unnecessarily. I often get asked if you need to learn it or do you have to use it? The answer is technically no to both questions. It is no longer required for any of the US ham radio license tests and my understanding is, this is also true for most places outside the US as well. If you do not want to operate CW, that is your choice, but you might be missing out!

CW stands for **C**ontinuous **W**ave as it is basically an "*on and off*" signal with the "*on*" signal being 2 different lengths. The long is in theory 3 times the length of the short. Give or take. CW is also sometimes, and correctly so, referred to as the original digital mode. Basically, the signal is either on or off. Think of the "*dit*" (dot, short) as one and "*dah*" (dash, long) as three ones. Space is zero. You got yourself some digital!

There are many benefits to using CW mode, especially with a modest station. You will not need as much power here as you do on phone mode to be heard. Meaning, your 100W will go a lot further here than using SSB and a lot further than AM mode even. In comparison to digital modes, it does vary though depending on which mode we are talking about. Some may even be less efficient, though most newer modes will surpass CW performance. Sometimes considerably. More on this in the next chapter.

CW, for those new to the hobby, is basically sometimes called Morse code but you will never hear it called this in ham circles. Morse code is technically referring to the alphabet used or rather to the code created to send via CW. Morse is the last name of Samuel Morse who developed it.

A	.-	N	-.	0	-----
B	-...	O	---	1	.----
C	-.-.	P	.--.	2	..---
D	-..	Q	--.-	3	...--
E	.	R	.-.	4	....-
F	..-.	S	...	5	.....
G	--.	T	-	6	-....
H	....	U	..-	7	--...
I	..	V	...-	8	---..
J	.---	W	.--	9	----.
K	-.-	X	-..-	?	..--..
L	.-..	Y	-.--	/	-..-.
M	--	Z	--..		

I have caught myself calling it CW amongst non-ham friends and got the deer in the headlights look. More commonly used letters like "E" or "T" have a short code and less commonly used ones like "Z" and "Q" are longer. There was some thought put into it!

For CW, working split up 1 (or 1kHz) is the general starting point, and most of the time up 1-2 will do the trick. Of course, this can widen a lot when a rare DX station comes on. Rare DX will <u>always</u> be split and DXpeditions will be as well. Some rare DX like to work in the US extra class section of the bands, so be aware of this if you have not upgraded yet. This would be on the lower portion of most bands.

Most DX out there I find operates around 20-30WPM (**W**ords **P**er **M**inute) which is what my CW decoder at least seems to be measuring. Some contesters can be <u>significantly</u>

faster, though sometimes I feel like this is rather silly as many do not copy code that fast.

Words per minute? Now you are thinking, well, how long is a word? The answer is easy, 5 letters. But now you are thinking, some letters like E are short in CW and some like Q are longer. Yep, and that is why the word "*PARIS*" is used as a standard. Based on 25WPM, there can be several QSOs per minute in ideal conditions. Assuming no QRM, nobody forgot to split, and there is good propagation.

As with phone modes, **check your split and then check it again**. If DX is split, make sure you are also. Listen if not sure and do not always trust the DX cluster. This is especially so with CW. Not only do posters sometimes get the split mode wrong but often even the call. So always confirm it for yourself. And not to state the obvious, but if you do not hear, do not call.

By knowing certain basic things, you can work most any DX. You will need to know how to send 5NN for signal report (N is short for 9) and your own call. Stations sometimes use 73 or QSL at the end of an exchange. Occasionally the slash for various reasons such as maritime mobile or working outside their normal or licensed areas. No need to get too fancy! CW DX exchanges are short and to the point usually.

You may have noted that I have included the question mark in the code table. Well, this does not mean you have to keep always using it. I know a few hams personally who have an annoying habit of sending this constantly. Not sure if they just can't hear but if the DX called you and you did not respond they will try and call you at least one more time in general, if not more.

Luckily, DX is generally working split, so I only know this as I see it on the waterfall and the decoder. However, on simplex this can be very annoying. If you cannot clearly hear, might be best to wait until you can. CW mode uses the RST systems for signal reports.

Readability (R)	Strength (S)	Tone (T)
1. Unreadable	1. Barely detectable	1. Extremely rough note
2. Barely readable	2. Very weak signals	2. Very rough note
3. Readable with difficulty	3. Weak signals	3. Rough note
4. Readable with little difficulty	4. Fair signals	4. Fairly rough note
5. Perfectly readable	5. Fairly good signals	5. Note modulated, strong ripple
	6. Good signals	6. Modulated note
599 is the best signal	7. Moderately strong signals	7. Near DC note, smooth ripple
Expressed as 5NN	8. Strong signals	8. Near DC note, trace of ripple
	9. Very strong signals	9. Pure DC note

For reference, here is how the RST is broken down for CW signal strength. Sometimes you may see a reference to a "**C**" (chirp) or "**K**" (key clicks) on the end of the report, but these are rarely heard now with modern transceivers. Very unlikely you will ever encounter these in a DX pileup environment especially. Although in the unlikely event someone does point this out to you, check your settings and/or consult an Elmer.

Some of the other abbreviations used besides 5NN instead of 599 are as follows. Zero is sometimes a "T". 5 is sometimes an "E" so during contests especially you may hear ENN, this is 599. Though I do not hear this very often. The letter "K" can indicate 1000 and I get that one. But at times even "A" can mean 1 and putting the above together AT is 10 therefore ATT is 100. Confused yet? Takes a little getting used to and frankly I am not a big fan of these abbreviations outside of 5NN. But if you hear them in contests, that is what they mean. Others which may come up are U for 2, V for 3, B for 7 and lastly D for 8. Other numbers thankfully were left alone. One extra dit or dah in my opinion is not going to make a huge difference, especially if it throws off, and therefore slows the receiving station down.

Of course, everyone knows or at least heard of the SOS distress signal. Dit dit dit dah dah dah dit dit dit. This is what they call a "*prosing*", which are designations of combined characters sent without space. You likely covered these in your license exams, and you really do not need to have an in-depth knowledge of this to DX.

Shorthand is also very common in CW to reduce sending time. Many of these have been adapted and used in other modes such as phone and digital to various degrees. Of course, CW came first. Examples of these are 73, CQ and even DX.

You may run across the term "Zero Beat". **Zero beat is simply in reference to matching the signal of the other station when tuning it in.** This is just something you get good at with practice and/or a good panadapter! Those really help on all modes. Does not necessarily mean you will be transmitting there, in fact most likely you will be split. Some radios also have an auto zero beat feature where it will match the DX signal spot on. Two issues though. If the signal is weak, it will not work well at times. Also, if the signal is not split, special attention must be paid to QRM being present. Lastly, since much of DX is split, I find this somewhat useless for many, expect maybe during some contests.

As with phone, there are times when a CW op will call based on zones or call areas. An example would be like "*CQ USA P5DX*" (in my dreams) and this would mean North Korea is calling US stations only. Yes, North Korea **(P5)** is one I still need.

CQ JA or CQ EU are other common ones, rarely CQ 6 or similar, which would designate a call area. You do sometimes hear this when there is known propagation to the West coast or someone with local ties is on the other end. For North America for example this would mean California **(K6/W6/N6)** or Alberta, Canada **(VE6)** which is slightly more inland. Mexico gets a little screwed here as geographically they are not "6" on the West coast, (Baja California) but rather a "2". Also, much of British Columbia, Canada is "7". Hopefully all numbers get their turn. No, it is not a perfect system.

The DX calling by numbers is more prevalent on phone mode, and this is more so due to sideband taking more bandwidth. This will cut down on the overlapping signals. A lot more CW callers can fit into the same space due to a much narrower signal, but it can still be a tremendous challenge.

There are a few abbreviations I would avoid in CW as well as digital as they can be confusing. One is DE which normally means "*from*", so CQ DX DE W6AER Meaning I am calling any DX station from my callsign. I feel this is wordy as I would just CQ DX W6AER for short. It is clear I am the one calling. But if I am calling CQ DE W6AER, well, am I calling the state of Delaware looking to finish my WAS or just being formal?

Alternatively, if I am calling CQ QRP W6AER versus CQ W6AER/QRP. The latter is obvious, I am QRP, meaning sending at 5W or less. But the first instance is not. Am I QRP or am I LOOKING for a QRP station? Humm…Again, if you ask 2 hams you will get 3 answers. I tried asking both questions for real. 20 minutes into it, I had to change topics on them. True story! Open to interpretation. My point is things can get confusing!

Related to the above, **CQ DX is just that. Do not respond to this if you are in the same country.** This also applies to sideband and digital modes too. Our friends up North in Canada when calling CQ DX are also generally looking for stations outside the continent or past the Southern US border. On the flip side, some US stations also do not consider Canada to be DX for us. Technically it is, but you decide how you interpret it.

You can also send CW using light, as sometimes ships close to each other do. Even your eyes. Say what? Some may recall hearing about the case of the captured naval aviator, Jeremiah Denton Jr. who after being captured in by the North Vietnamese appeared in a 1966 TV interview. He was being asked about his treatment. While being forced to lie, he blinked "*torture*" with his eyes. He was later freed after 8 years in captivity and went on to become a US Senator.

You can watch the clip at youtube.com/watch?v=rufnWLVQcKg.

A quick word about **CW pitch**. CW pitch frequency can be adjusted for your preference. This can be as low as 300Hz to as high as 1kHz. I personally leave this usually at default. Generally, this is 6-800Hz and I find it works just fine for my tone-deaf ears. Some ops tell me they prefer a bit lower pitch. If it helps you, adjust it. On most rigs you can adjust this to your liking per above.

CW Keys & Paddles

Keys & Paddles both come in a variety of flavors. It can be a little intimidating at first. A CW key is basically something that allows you to send CW. It refers to the actual hardware used.

A **straight key** is one you may think of from the early days which requires the up and down motion of a lever to make a contact to send code. This is the simplest variety and the oldest design though still very much in use today. Though I will be honest, I cannot get used to them for the life of me.

One very popular straight key, if you want to try one, is the **J-38** from World War II. This was used by the US Army. There are also many Soviet era and Eastern European made straight keys which one can pick up on eBay cheaply and most are very well made. The most sought after are the ones from the former East Germany (Stamped DDR) but pretty much anything cold war associated is in demand, or so I am told. **Nye-Viking** keys are also popular with some. **Vibroplex** has some beautiful straight keys as well. There is a speed limit on how fast one can send with a straight key. Getting past 20 WPM for most is tough using these. Though I am sure there are those who can, I am not one of them.

Pictured are two excellent CW keys. The Begali Sculpture (left) which I own myself, is a paddle and a Begali Spark (right), is a straight key model. Both are exceptionally beautiful units.

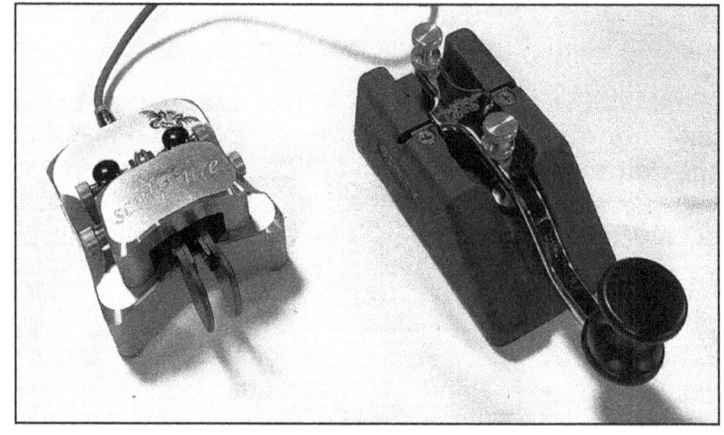

Think of a **Paddle** as a variant which allows one to move the keys (most have two) left or right to create tones. One side being short tone, the other long tone. Paddles allow for faster speed and in my opinion much more comfortable for operating due to ergonomics. Though as I always say, everyone is different. You may find otherwise.

The **Bencher BY-1** line Is perhaps one of the most popular paddles out there due to reasonable cost and a good build quality balance. Usually, you do get what you pay for as with almost everything else. Here you get both.

There is **Vibroplex**, a brand which I have already mentioned, is also known for making bugs or mechanical keys. They are also known to be the oldest name in amateur radio. Bugs are not as popular now, but some still enjoy using them. They have made some beautiful units! Bugs work using adjustable weights and springs. They are a marvel to watch in action. I used to own one myself. They are not exactly what I could call compact!

Begali is often considered by many DXers and contesters as top of the line when it comes to keys. Pietro, I2RTF is a true craftsman. I would tend to agree that it is one of the best made out there. I also own one and I know many serious DXers who only use these. The price tag is a bit higher than some, but the quality is top notch and well worth it. All units are machined, have excellent weight and are incredibly adjustable. Frankly, I cannot say anything even remotely bad about them and those who know me know I have very high standards with certain things. CW keys are one. You can visit them at i2rtf.com.

Other paddles to check out are from **Kent, N3NZ** and the **Hungarian HA8KF** paddles look rather nice! With so many options something will work for you, so try a few. Visit vendor booths at your next conference and ask around. You may just end up with more than one. Most DXers do!

For the most part, you will need to find something that works for you. We are all different, so it is hard to suggest just one. Although a suggestion I do have, as something that is important to me is to get something heavy. **You do not want your CW key to move around, so a solid and immobile base is a must**.

You may have heard the term iambic paddle or Iambic keying before. They both refer to a paddle with two finger pieces (elements) that produce continuous alternating dits and dots when squeezed for a longer time. The one squeezed first starts off the pattern of alternating lengths. Using this method, the sending speed can be increased dramatically for some users. The start order can be adjusted by selecting the "*iambic mode*". There are two modes, A and B. A is the default mode for most and the most used of the two modes. Mode B will send an additional element, mode A stops when you stop. That is also my preference, but some swear by mode B. Try them and see what you like. I would be very curious to hear!

One last thing. A "*key*" is not to be confused with a "*keyer*" and I have even been known to mix this up in casual conversations. A keyer is a device or a circuit that is built into modern radios (though can also be standalone) and allows one to send CW to put it simply. The key is what is used to trigger this and is the external component you see. These are the ones we have just covered.

Using Memory

If you are a contester, you likely are very familiar with this topic. Using pre-recorded memory for CW and for other modes is commonplace. It is also very handy for calling a DX station and certainly a lot less exhausting. I can pretty much work a DX station with my **F1** and **F2** keys in a pileup.

Above is my keyboard used for ham radio applications. Here you see two rows on the top with labels. One is labeled N1MM and the other FLEX. These are the two programs, Flex referring to SmartSDR, where I use preset memories. These memory buttons correspond to the F-keys below.

I created the labels for the above with ¼ inch tape using a Dymo labeler, same as with the I-Mate keyer. The keys will not be mapped the same program to program. As you can see above as I stayed with the program default mapping for N1MM. For example, on the FlexRadio SmartSDR software, F1 is my callsign. Some N1MM presets get adjusted for whichever contest I partake in, within reasons.

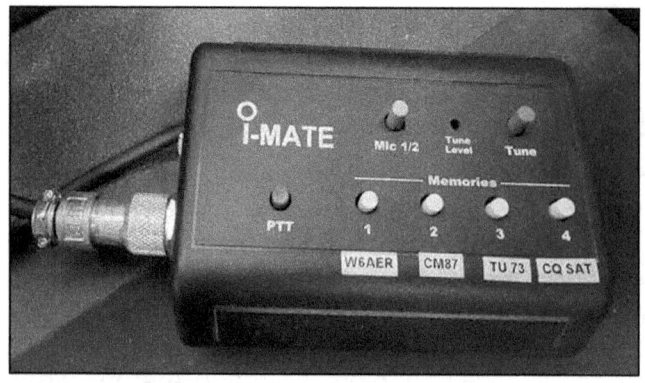

Pictured here is an **I-Mate** transceiver memory keyer which is sadly no longer made. If you find one used, get it! But if you cannot, there is another option which is just as good. The **SOTAbeams ContestConsole Switching Unit** is a very similar product and will do the job just fine. These generally cost around $100. Using this device, the actual radio hardware memory is played back, not from software. Therefore, the buttons will always be the same no matter what application you are using, unlike the keyboard solution above. I use this setup with my Icom IC-9700 hence the "*CQ Satellite*" reference. Though it works with most Icom radios, including HF, as an easy external way of memory keying. I have this setup for both CW and phone modes. I can even connect a 2nd microphone to it, but I just use it with my Heil Pro 7 headset.

CW Filters and Bandwidth

Your radio will have a few preset filter settings for CW use. These can be physical filters or software ones. Both can usually be adjusted nicely, though the roofing filter for CW is generally just one per radio if applicable. I usually like to start with a 4-500Hz width for CW when trying to tune a signal in. I also find this range works well with CW decoders, though do some experimenting as radios can vary.

If I go too **wide**, I hear too much side noise and possibly other stations. I also have issues not copying it as well in my head, perhaps due to excess static distracting me. For sure this will be the case in a contest as stations tend to be tightly packed. Being too wide will be next to impossible. Some however, outside of contests, like going a bit wide as they claim it gives them "*higher-fidelity*". That is not my experience and I personally reserve that for music.

If I go too **narrow**, I find that I may miss something if I tune around too fast, though watching the waterfall solves much of that issue. Also, some software decoders decrease in accuracy, some even start to spew out garbage. This is likely because they need some reference noise in given situations. Some hams also find the sound too "*sharp*" as it is sometimes described.

Finding a midpoint that works for you is key. Your ears, your call. Once I find a station I am interested in, I narrow the filter more if needed to find a workable balance. This will take some time to learn and will differ from person to person, radio to radio and certainly band to band as some can get noisy.

If you are tuning using the aid of a waterfall or any sort of panadapter, you will see quite a bit at once. Narrow the view a bit if possible so it is not sensory overload. The viewing area should not be nearly as wide as when you are hunting for phone signals. CW signals are a fraction of an SSB signal. You can stay on a 500Hz bandwidth filter while viewing much wider and just move to where you see or hear the signal.

If you are tuning by ear, you will need to move very slowly or temporarily jump to something like a 2kHz filter, so you do not miss anything. Then narrow if needed to zero beat. I had to constantly do this before I went to an SDR radio, and this is how I always operated all my older Yaesu radios when in CW hunting mode. Frankly seeing the signals is amazing and would have a hard time going back to not having a waterfall or panadapter now.

There are add-on hardware CW filters and signal processors available which are treasured by some CW enthusiasts. I used one about a decade ago, just to give it a try. I find that I can get pretty much the same results with more modern radio or an SDR today, therefore no longer use one. But if curious, or depending on your setup, give them a try. You can usually find these used cheaply and they are not too hard to homebrew.

I recommend using APF (**A**uto **P**eak **F**iltering) for CW to help with band noise, and frankly even without. This will really help you with decoding (head or software) as well as fatigue. APF can really bring signals out of the noise and has helped me to bag several band fillers. If your radio has this, get familiar with it and use it!

Some radios have just a straightforward CW mode. Simple enough. But some may instead have CW-L and CW-U like on Yaesu radios, CW and CW-Reverse on Kenwood. This is not related to phone sideband modes. Selecting the offset (swapping) modes can help with QRM by making competing signals that were higher in pitch than the desired signal, lower pitch in the CW-R mode. Try these and see if it makes a difference for you. Some CW ops swear by this method.

Do you Really need to Master CW?

My honest answer is…No, you do not. I can see the angry emails coming now from CW aficionados. I know some will disagree with me on this but hear me out. As long as you can recognize and send the following you are in OK to work DX:

1. **Your Callsign** - Know this one well!
2. **5NN** - Short for 599, signal report
3. **73** – Best wishes, end of QSO
4. **TU** - Short for thank you
5. Understanding of the **DX Call** you are trying to work
6. **AGN** and/or **?** (question mark) in case of repeat needed

If you are going to try to work DX during a contest you will need to be able to send the required exchange as well, so this kicks it up a notch in complexity. This could be your CQ zone, a **serial** (contact) number, or possibly other criteria depending on the contest in question. You can always look this up online if uncertain. If you do not know your **CQ zone**, you can look it up at cqww.com/resources.htm. This might be the most asked additional information for contests. Some contests ask for **grid squares** which we will cover in more detail later. For contests, you will need to know the numbers as well if it is a serial contest or anything which requires a numerical response. During contests, you may hear **NR**. This is short for number, and it may or may not be followed by a question mark. This means the other end needs a number from you per above.

You also will want also to be able to recognize **TEST** and this is often used. Example being CQ TEST W6AER, meaning that I am in a conTEST (Abbreviation) and looking for stations. These stations will likely want more than a 5NN from you per above.

There are other things which might come up, but not often. **RR** for Roger Roger (yes, twice) is one. This is a confirmation, sort of like a **QSL** which you may also hear time to time. Also, **FB** for Fine Business. This is kind of like an RR. Additionally, GM, GA, GE and GN which are good morning, afternoon, evening, and night. Though not as common, might be heard from very casual contesters. Most serious contesters get to the point and move on quickly.

I have already discussed CW related software earlier, including CW decoders. **Using one of these CW decoders in conjunction with a memory keyer and being able to recognize your own call when heard, you can pretty much get the DX in the log via CW without having to master CW.** How do I know? I got CW DXCC and WAS doing just that, after which I got more serious about re-learning CW. I fell in love with it and now especially enjoy CW contests. I am not going to claim I am great at it, but good enough to hold my own and go after bigger fish. At the end of the day, that is what matters the most for DXers.

Learning to Pound the Brass

Sometimes you hear hams say that they are "*pounding the brass*". No, nobody is angry at you or is threatening to beat up the police. This basically means to work CW and is a reference to the telegraph days in the 19th century when keys were all made of brass.

So, yes you can do it the "*easy way*" as above with memory keyers and decoders. However, with just a little more effort the rewards will be greater. Just take the time to learn it. You ask again, how do I know? Same as above, I have been there.

Go about studying it slowly. No more than 20 minutes at the time and take at least a 30-minute break between sessions. Studies show you lose focus studying anything for more than 20-minute blocks at the time. I am more in the 15-minute range personally. Do not study for too long in one day. Taking it all in using shorter blocks of time is much better for long term memory. Your brain will need to process it all and it is not like you are cramming for an exam. Time is on your side. You will get there eventually if you just stick to it. If you are in a situation where you want to re-learn CW, it should be much easier the 2nd time around, though I would follow the same system. This was my experience.

There are also studies showing certain times of the day are best for long term memory, some claim right before bedtime is best. Give it a shot, maybe it works for you. First thing in the morning, after coffee of course, is my preference. CW can be forgotten and again, I am living proof. I learned CW in my teenage years, way before my license from a toy 49MHz handy talkie. This unit had a Morse code cheat sheet printed on it and in addition to the PTT had a second button to send code. I used to think it was the coolest thing ever as a kid. Fast forward many decades, I was limited to SOS until I took it upon myself to improve. Funny side note, the handy talkie only had letters printed on the unit. I guess numbers were not important back then!

How long will it take? There are two ways of looking at it. To get to a point of being comfortable is something which may take longer for some than others. This is hard to answer. I know folks who claimed to have learned it in a month, others have been slowly learning it for over a year, trying to build up speed. I would argue that you will be learning it or rather improving over your lifetime.

What is the best method to learn to pound brass? There are many schools of thought on this, and I have tried them all. Seriously, I have tried them all! Many feel that CW should be learned by sound, almost musically if you will. Others swear that if you are writing it down versus decoding in your head, you will never learn it. Works for some, I find that this slows me down a bit, but it does work. We all learn differently. Something that works for me, does not work for others at all. I find that using all the above in conjunction worked best for me. Some learn only by one method, some by multiples, like me.

One would think that just reading and memorizing the characters works for everyone. It does for some, but you are learning visually, and some folks have a hard time applying what they learned visually to something they need to process auditorily. I am not going to say don't do this, but make sure it works for you. **The way to attack learning CW is by determining what works for you.** So, when you hear someone say, you can't learn it by such and such method, take it with a grain of salt. Maybe you can since that is how you learn. This is something for you to figure out on your own.

Currently one of the most popular methods of learning is called the **Farnsworth method**. Farnsworth method is where you are sending average speed, but you increase the space between letters and words. This gives your brain a little time to

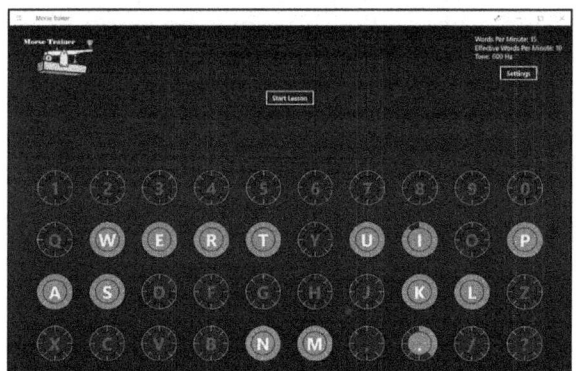

process but you get used to the character speed rate. Once you improve, reduce the gaps. The Farnsworth method was developed by Donald Farnsworth (Former W6TTB) in the late 1950s. I rather like this system myself.

In the Windows Store you can find **Morse Trainer** which is pictured here. This is perfect for initial learning, for testing yourself as well as reinforcing what you have learned. This is one of my favorite programs! It is visual, has sound and is interactive. You are using many senses to learn. This was written by Shane M. DeSeranno and retails for only $1.99 at the writing of this book.

Koch method CW Trainer by Ray Burlingame-Goff, G4FON (SK 2021) is also worth checking out. This and other CW related items on this website can be obtained at g4fon.net. The **Koch method** teaches at normal CW speed, which is generally around 25WPM. It was developed in the 1930s by Ludwig Koch, a German psychologist. It works by adding characters as speed builds up.

There are many **CW audio courses** out there available on CD. The ones from **Gordon West, WB6NOA** are very popular, and I find them rather good. He has these available in different speed ranges, blocks of WPM, to help build up your speed after you have learned the basics. As a side note, he also has excellent books to assist you with upgrading to a higher license class if you wish to do so. Visit gordonwestradioschool.com for more information.

The **ARRL** has also published a set of CDs though not as in depth as the ones from Gordon. I have listened and worked through all of them, and they all have a slightly different approach. Consider checking them out.

For those who are already up to speed with their CW there is also a piece of software called **Morse Runner** by Alex Shovkoplyas, VE3NEA which simulates pileups for practice among other things. You can find this for free at dxatlas.com/morserunner.

Another piece of training software worth looking into is **RufzXP-Tancredi** by Mathias Kolpe, DL4MM and Alessandro Vitiello, IV3XYM. This can be found at rufzxp.net.

There are also apps out there for both Android and the iPhone. Search for "*Morse Code*" as results that come up will be more relevant. **CWops** has an academy worth exploring, and it is completely free! Many feel this has helped them tremendously in improving their CW skills. You can read more about it at cwops.org.

Lastly, there is the International Morse Preservation Society I had mentioned earlier called **FISTS**. You can find them at fists.org. I encourage you to check them out if you get serious about CW operating.

Working the CW Pileup

Now the part you have been likely waiting for. In theory, CW pileups are very similar to other types of pileups. There are a few key differences though worth covering here. You will need to try and **match the operator's speed** or maybe even slow down a bit more, especially if there is a lot of QRM, QRN or QSB. No need to show off your CW skills. Honestly, nobody really cares, and the DX might not be able to copy at your speed. I know this hurts the feelings of those who pride themselves on speed, but the point is to get the DX in the log. If the DX is slow at sending, you should be slow as well. He may copy faster than he sends, like I do, but you do not know this for sure. If you are faster than he can copy, he will move on to someone that he can copy. You likely would do the same as well. So, this is "key"!

Watch your speed difference if you are using a memory keyer and/or key. I often see in contests where the DX calls CQ at a normal speed. Then when they respond with the 5NN at a lighting fast speed or the exchange (usually numbers) at a super slow speed. Maybe it is just me, but this really throws me off! I even wonder if it is the same station who just got back to me. It would be like as if I called CQ on sideband in a deep voice, then responded to your call with a high-pitched voice. That would mess with your head, I am sure. I do not want to do this to others. I know the 5NN is expected and always the same, but still. You get the point.

The CW operator will likely indicate if he is working split. He should, but not always indicated so or at least not often enough. You may hear the **DX call followed by an "up" only and it is safe to assume it will be up 1kHz or more.** If the pileup just started, it is likely not yet spread out too much. Meaning closer to 1kHz up to maybe 2kHz up. Try and work the front or the back here in a clear spot of course. If you have a waterfall, this helps to find a clear spot, though be aware that just because you are not hearing (or seeing) a station, there still one might be there. Think of dead spots and directionality! As the word gets out, depending on the rarity of the station, this can grow wide!

The DX can work the pileup in a few ways. Likely sweeping across the area where they are listening. Moving from one side to the other and possibly back slowly. Alternatively, moving to one side and jumping back to where they started. We already covered this in the phone section. Every operator is different, so this is where listening and observing come in. Of course, others might be doing the same thing and will try to jump where they think the DX might be listening next. The trick is to "*outsmart*" them. You can do this by moving a bit further up or you can always just find a nice clear spot and wait for your turn. Frankly, this is what I often do. However, do **check regularly to make sure you are not doubling with someone** on the transmit frequency. Neither of you will likely get through with ease at least.

Remember, many DX stations like us will choose the easiest and most convenient way to work the next station. Put your subreceiver on the same spot as the DX's current split contact can work beautifully unless everyone else is using the same trick as I had mentioned. Wait for the contact to finish and call them. Make it easy for the DX station to work you. If it is a huge pileup, move just a bit from the current station's frequency. The direction is best determined by listening.

Note that **there are instances when the DX is listening down**. Sometimes both up and down. I have seen this and even worked DX this way. There are times they announce "*down*" but not always. I could never put my finger around this one. Check your waterfall if you have one as well as the DX cluster as someone may have figured this out and posted it. Always confirm what you read on the DX cluster though as I had already mentioned earlier. It is posted by people and people can make mistakes.

Furthermore, keep your **CW DX exchange short and simple when working DX**. If they call CQ, you answer with your call, he responds with his call and 5NN. You then send a 5NN, he responds with a TU 73 or QSL and you log it. Then do your happy dance and post it to the DX cluster if nobody else has yet, or at least not from your area. Do not add anything extra, do not deviate. Also, just as with sideband DX, no need to repeat their callsign. Slows things down and they know who they are. I hope!

I have seen folks send extras like their name and location. In both examples the DX came back doubting he had the right callsign as this was not expected and they thought it might be a correction. This can be especially difficult if you have noise on the bands. Don't break their rhythm. Trust me, just don't. Keep it simple! Simple is good and it is more likely to get you in the log.

What is QSK

This refers to a radio's and therefore your ability to hear while you are sending CW. Essentially it allows you to listen between tones. Normally when you send CW, the radio goes into transmit mode and stays there until done. And this is OK. This can be due to using memory or due to delay switching back to receive while you are sending. Radio brands and even some models do vary a little in how they behave when sending CW. Some give you more ability to adjust these parameters, some less. Some transceivers have QSK but not all. Also, some amplifiers do not support this feature so always check!

This can be useful in contests or high noise situations as well as hearing the DX if you transmit and they transmit as well. It does help with timing for sure for many. I personally find it a bit stressful to use when sending by hand, but perhaps that is just me. I am also not a big fan of seeing the radio looking as if it is having a seizure. Not so bad though if I am sending using the memory keyer. Again, we are all different. Some would not even consider operating CW without it. Try it if <u>all</u> your hardware supports it and see if it is for you. Might just be! Clearly it exists for a reason.

CHAPTER 9: Digital Modes

This section will be a lot longer than the previous two dealing with CW and phone modes as digital is a lot more complex. There is also more than just one digital mode, in fact way more than I can cover. Luckily, you only need to know a few to DX. A computer is involved in most cases. Though some digital modes can now be directly decoded and even sent on a select number of modern radios without a computer. Digital really is a whole new set of skills to learn. For DXing even more so. Having some pre-existing IT skills helps for sure. Part of the reason I had spent so much time on computers early on in this book. It is the world we live in now; things have changed dramatically even in the past few years. Certainly, a lot from the early days of digital.

JT9	-26dB
JT65	-24dB
FT8	**-20dB**
FT4	**-18dB**
CW	-15dB
RTTY	-9dB
SSB	+10dB

The phrase "*digital modes*" is an umbrella term for many modes used in ham radio. Many of these modes have very different characteristics, user base, popularity, and efficiencies. Speaking of efficiency, I have mentioned earlier that 100W of CW will get you a lot further then 100W using SSB. However, 100W of FT8 for example will outperform CW. And JT9 will do even better than FT8! However, digital modes like RTTY, not so much. In fact, it is somewhat closer to CW in performance. Perhaps this sheds some light on why some of these modes have become so popular. If you want to read more about how some digital modes outperform voice and CW, check out qsl.net/k4fk/presentations/Mode-sensitivity-2013-Dec-QST-Siwiak-Pontius-1.pdf. This article is from a few years back but does an excellent job in covering this topic. It does not cover some of the newer modes due to when it was written. FT4/8 but as you can see from my chart, fall just below JT9/65 in the "*punch*" you get with them. More on these modes later.

When working DX on digital modes you will be operating most likely split. Though exceptions exist and do not always involve having to change the frequency of your radio when you transmit. You might even be using some other special methods which I will discuss in more detail below as it relates to specific modes.

Signal reporting methods will also be very different in some of the modes. When using WSJT-X or JTDX for example, the software does this for you. You do not need to figure out the signal strength. As with the other modes, we lie to DX. Everyone has "*great signal*" and a 599 or 5NN is in order. Though I do sometimes see honest reports given.

In RTTY for example, the RST system is used. Therefore a 599 is given to a good signal when reporting, much as in CW but you do not write 5NN. Sending 5NN takes longer to send via RTTY. See RTTY section for more details.

PSK31 and its variants also use the RST system. Though this mode has started to fall out of favor with DXers when JT65 appeared and has much less traffic now due to FT4/8 popularity. Really too bad!

On your radio, you will likely be engaging the digital mode, though this could be labeled "*Data*", "*DIG*" or "*Digital*" on many units. On FlexRadio's SmartSDR it will be "*DIGU*" or "*DIGL*" but there is also a "*RTTY*" specific mode too as on many other transceivers. Some hams use USB mode to transmit digital via the radio. I have also done this at one time when I used a rig with no built-in soundcard. This is OK too, but there are some special things to pay attention to. Use the designated mode on your radio for best results if available. All the digital modes will require the use of a sound card which can be in your radio or external as we discussed earlier in the hardware section. Both are fine, though internal is much easier to get started with.

I will be focusing mostly on FT4/8 as frankly, this is where you will be likely working DX as of the writing of this book. This has changed over the years. FT8 seems to be the watering hole as of right now for DX in the digital realm. But when the "*numbers*" are good (I will cover this later in propagation) FT4 traffic seems to really pickup as well.

The Ingenious FTs

In the year 2017 a revolution occurred. This all took place without an uprising or any demonstrations. It was rather peaceful. Was not widely covered; the media did not jump on it and only a few select folks even understood what it was about. These folks were ham radio operators, and the revolution was the "*FT8 revolution*". And yes, it was a rather big deal for hams, but even we did not understand at the time just how big. WSJT-X and Joe Taylor, K1JT changed things for ever in ham radio.

Like so many topics in ham radio and this book, this section may also create some controversy. Some think of these FT modes as something that is trying to kill CW and phone. Others think of it as just computers talking to computers. Some think both.

And that is their right, we live in a free country. However, I will disagree with everything above. None of this is actual reality if one looks at broader data. FT8 has brough many hams back to the shack. Now they can operate when conditions are not ideal or if they have compromised antenna systems due to HOAs, funds or time.

I feel operating FT8 requires a different set of technical talents than CW or phone operating. This is also true for other digital modes. While yes, computers are involved on both ends, as are with RTTY or PSK31, there is operator involvement, and they are all valid QSOs. In fact, I would argue that there is more technical skill required to operate digital than CW and phone. I know, for some these are some fighting words.

So why work FT8? Because simply put, that is where the DX is at times. Few will disagree with the fact that most DX is to be found on FT8, especially in the past few years. When band conditions are not the greatest, this mode generally still works. When conditions improve, many CW and SSB stations show up out of the blue. And during CW and phone contests, there is certainly no shortage of non-digital mode operators. They tend to come out of hibernation.

These FT modes can be broken down into FT8 and FT4 mainly. FT8 being the most used and FT4 having been developed as a contest mode technically but should not be ruled out by any means for DXing. I have worked great DX on FT4 on many occasions. There is operating found on FT4 outside of contests and this is especially true for the 20m band, with 30, 40 and 17m getting more active on FT4 as well. I hope to see this expand to other bands soon. I have even made contact via FT4 on the RS-44 ham radio satellite as of late. FT4/8 are not conversational modes, but rather they focus on quick, preset exchanges. FT8 is not channelized, except for the 60m band where everything is. There you will find FT8 on CH3 at 5.357MHz in the US. Though, there are some presets or rather recommended operating frequencies.

Basically, within a 3kHz window you can pick any spot (preferably a clear one) to transmit. The recommended frequencies are preset in the software, but others can be added as needed. And these will be needed if you want the more in demand DX!

FT8 has a T/R cycle of 15 seconds, and each signal takes up 50Hz

FT4 has a T/R cycle of 7.5 seconds, and each signal takes up 83.8Hz

Both transmit a bit shorter time than indicated above but this is the allocated time or timing window if you will, to allow for clocks being off slightly and to make things nice and even I suppose. Buffers are always good.

Setting up your WSJT-X Software

WSJT-X is a piece of software which allows digital communications via several modes. This is by far the most installed and used digital software out there now. There are alternatives and I will touch on these in a bit as well. The supported modes are FT8, FT4, MSK144, JT4, JT9, JT65, QRA64, ISCAT and WSPR. **For DX purposes, 90% of the usage is FT8.** The rest of the usage is primarily FT4 and WSPR. WSPR, pronounced "*Whisper*" is only used for propagation studies and does not count for DXCC credit as it is not a 2-way contact. I will talk more about this later as it can be used to help you figure out what areas of the world are open.

A few years back JT65 and JT9 modes were king, though much slower and much less automated. I recall working stations all over the world using these modes for many years, especially on the lower bands. Though nowhere near to the degree of current digital traffic. This has changed dramatically after the introduction of FT8 where seems everyone has moved to…for now. Until the next big thing. Just watch!

Now you might be wondering what WSJT stands for. The "*WS*" is for **W**eak **S**ignal and the "*JT*" is for **J**oe **T**aylor, K1JT, the original creator. The FT in FT8 and FT4 stand for **F**ranke-**T**aylor. Which is a combination of the names Taylor, per above and Steve Franke, K9AN. Therefore FT-8 is simply Franke-Taylor 8-FSK modulation. Note that others have since also contributed to the WSJT-X software. FT4/FT8 has enabled weak-signal DX during poor sunspot years, leveled the playing field for "little pistols", and expanded the reach of QRP. The guy who invented it should get the Nobel Prize. Oh wait, he already did!

Originally WSJT was introduced for **EME** (**E**arth **M**oon **E**arth) contacts, better known as moon bounce. Essentially using the moon as the reflector instead of the ionosphere or ham radio satellites. This is usually done via VHF and UHF frequencies. If you think digital is technical, this adds a whole new level of difficulty but clearly can be done and many hams have done so very successfully.

To begin with, make sure you have already installed some way of syncing your computer clock accurately. We covered this in chapter 5. This will be very important to stay on top of once you start operating. For some modes, such as those found in the WSJT-X suite (FT8, JT65, etc.) having an accurate clock is essential. I am sometimes amazed at how some folks are off over a second or more, but we still manage to decode them. Of course, there are those you may never decode and may never even know they were there. The more accurate your time, the better the results/decodes, equals more QSOs. If your transmit cycle does not match most others on the waterfall, you are the one off. Please check your clock.

Visit time.is and make sure you are running something additional to the built-in computer clock sync as you do not want to miss DX!

FT8 also uses the clock to know when to decode and transmit, as in establishing cycles. For a successful QSO, which takes about 90 seconds, decodes occur 4 times a minute, 2 second break between cycles. More than 2 seconds difference, no decode! The **DT** in the window of the software shows time difference. DT is short for "*Delta Time*". So, if you are getting a lot of -1.6 as an example across the board, it is you who is off. Generally, should average to about +/- 0.4 or so as not everyone is going to be super accurate. If you do not see anything above or below, you are likely good. You can check this though once you get everything up and running.

Once you figure out your computer clock management, you will need to download a fresh copy of WSJT-X from physics.princeton.edu/pulsar/K1JT. WSJT-X is available for Windows, OS X as well as Linux. All versions perform equally well with no difference in features to my knowledge.

I always recommend that you **use the latest production version of the software, especially if you are a new user**. If there is a beta or release candidate (marked with RC) and you are comfortable with using it, go for it. Versions older than 2.0 are not going to be backwards compatible with anything before them. There was a major change in how the software operates at the version 2.0 release. But you will likely no longer run into this issue on the air as most everyone has switched.

After you installed it, you will need to do some basic configuring. To begin with you will need to know a few items which you will be asked about during the setup process. These can be visited via the main menu under "*file*" select "*settings*" or by pressing the F2 key.

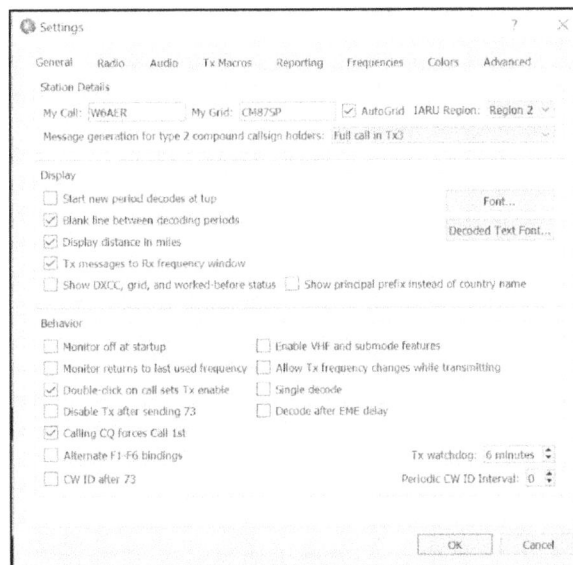

Your **callsign**, **grid** and **ITU region** are the first items you will need. These are the on the general tab, very top. You will need to **know your IARU Region**. IARU stands for International Amateur Radio Union.

Region 1: Europe, Africa, and North Asia. **Region 2:** North or South America. **Region 3:** Southern Asia, Australia & Oceania. So, for example, if you are in **the US, you are Region 2**. Select this option. If you fill in your grid, this can be auto filled for you though.

You will need to know your grid square. Use the 6-digit version of this to be more precise, not the 4-digit version as I had previously mentioned. It will provide more accuracy. This can be looked up online at levinecentral.com/ham/grid_square.php if you do not already know it. Likely you will need this a lot. So, write it down if you have not yet committed it to memory.

Next you will need to setup your **radio settings**. This is the next tab over under radio. This is the hardest part for most users and may require a bit of experimenting and research if you are new to this. You will need to look up your radio settings from either the drop down or online to see what the closest usable setting is for your setup. At times when your radio is brand new to the market or it is a little more obscure, you may need to find a closest match. Many times, the Kenwood radio setting seem to work for off brand units. This varies a lot from user to user and radio specifics are beyond the scope of this book. There is a ton of radio specific information online on this topic.

Radios themselves also vary in which mode they need to be in to function properly with WSJT-X as I had already mentioned this. Some older radios need to be in USB, though I do not recommend this if avoidable due to filters on the sides. This might also be the case if using an external sound card, though see if you can make it work in a "*digital*" or "*data*" setting. Consult your manual if unsure, ask an Elmer familiar with your radio or there is always the internet. I would literally take 50 more pages in this book to go over all possible scenarios. Once you figure this out, make a note for future reference. These settings can likely be used fine with other software as well.

The **COM port** via your computer can be real or virtual as on some SDRs where there is no actual serial or USB cable used but rather done via the network like on the FlexRadio for example. This makes things a lot simpler. The example here shows just that.

For COM ports, **pay attention to the Baud rate**, this can be found in your control panel, device manager in Windows. Once you get these, **Test CAT** and then **Test PTT**. If both are good, you are halfway there! They will be lit green and red respectively if you are all good to go. You just got through the hardest part!

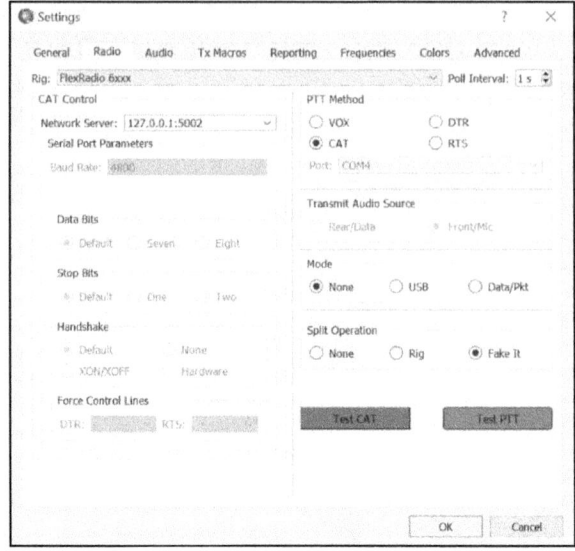

You will need to **leave the split at "None"** unless you are in fox and hound mode, in which case you need to change this. But more on that later. The mode can work in None, USB or Data, this is all about how you set your rig up. "None" has worked for me with several rigs, as have the others. I can manually switch this on the unit. But again, test it and see for your specific situation. The rest you can leave for now in this section.

Next, you will need a sound card selected correctly. Chances are by default it will be wrong. As one can see, here I have multiple radio related audio devices installed. This will likely be the case for most hams with more than one radio. Here you see an Icom 9700, FlexRadio and an SDRplay, all with their own sound drivers. The rest are ones

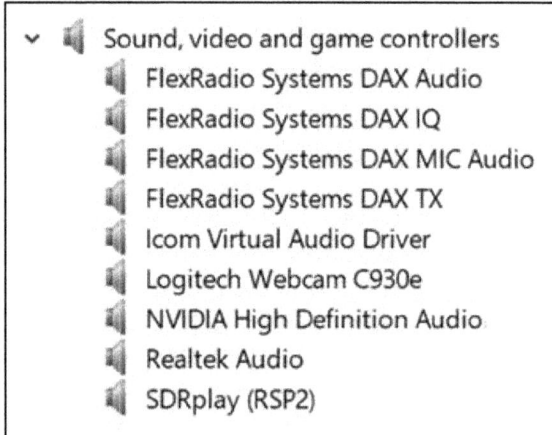

you may find on most average computers such as webcams and internal audio. My system generally defaults to Realtek Audio, which is what is used by the operating system to drive my speakers. This is a very common driver on many Windows PCs. You do not want this sound card, but rather the one to be used for receiving and transmitting via the radio.

These can still get rearranged and reassigned even after proper configuration. I have heard the AOL "*you got mail*" (recently mind you) as well as parts of conversations due to folks engaging their microphones accidentally, likely those on webcams, while transmitting. **Select the correct audio in and audio out device here. Don't forget to do both.** Check these first if you run into issues, followed by your COM ports. These two are a very likely the cause. Windows is especially notorious for adjusting your COM ports and audio devices especially after updates, so it is good to keep your eyes on this. Macs have been known to do this as well though not as much. If you see something like "*Realtek Audio*" or "*Camera*" for the WSJT-X microphone you for sure will need to re-adjust it. Most if not all webcams have a microphone as well. I have heard stories of hams accidentally listening via a webcam as it picked up the transceiver speaker audio. Again...**These may shift after driver or operating system updates!**

You do not want to use the same output as you use for your computer normally, unless you take special precautions and know what you are doing. The same is especially true for the microphone! If you must use only one sound card, be sure to turn off your system sounds and dedicate the sound card to digital mode operating only. I know some tell say that you can use your existing sound card, and this is true with proper isolation. But unless you are very comfortable with the operating system and really understand what is involved, please do not. You will just create possible QRM.

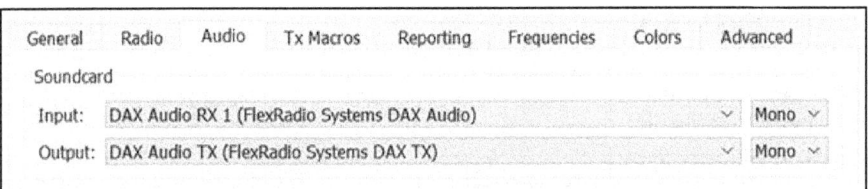

Above is an example of what you see if using a FlexRadio. It will have RX 1 correspond usually to Slice A and RX 2 Slice B and so on. These can be changed, but to keep it simple stick with an order you can remember if you have multiple radio options. TX will also be the same on the FlexRadio 6000 series, at least if you are using current software. You will see some IQ channels for some radios, which are not what you want here. So, pay special attention.

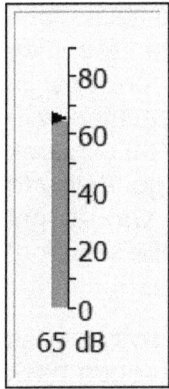

The Audio level can be seen on the bottom left of the application. **This should be <u>green</u> in color and generally around 60-65dB.** When not in the proper range, you may encounter decoding issues. Some keep it lower, but the key is for it to be green.

If it is indicating yellow or red, you will need to adjust your audio input level (microphone or line in). Red is way too much, meaning it is overdriven and clipping. Yellow is too low, meaning there is not enough audio data coming in for the software to do its magic. Both bad and will need your attention. This level is not to be confused with the transmit level, though you always need to pay attention to both. Maybe even more so to transmit levels.

Often Forgotten but a Must for DXers

There are two things which I feel everyone should really do! One will benefit you and as well as DXers. To begin with, in the settings of WSJT-X on the **Reporting Tab**, please select the checkbox options described in the paragraph below. By doing so, the program will send your reception reports to PSK Reporter, and in turn this will help with tracking propagation and your radio system's performance.

These are located under the Network Services area, mid-section, on the reporting tab. The default values are fine for UDP, just **check Enable PSK Reporter Spotting, use TCP/IP as well as the 3 boxes on the right**. Done!

If you are using more than one WSJT-X profile as in if **you are using more than one radio (or slice) all using WSJT-X at once, increase the UDP port number by one for each instance.**

In the above case, these would be 2238, 2239, etc. You will need to set these in each WSJT-X profile, but the good news is, only once. You will only need more than one profile if you use more the one callsign or have more than one receiver. I personally ask that if nothing else, you set this.

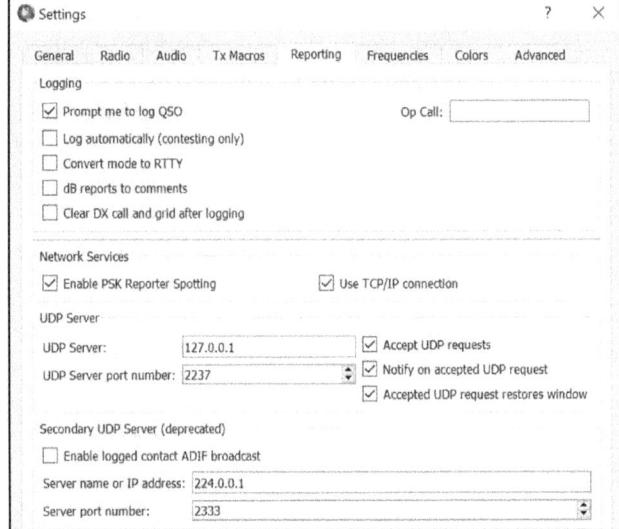

By all DXers working together and reporting back, we get a much better picture of propagation and how far we can work DX from a given point by analyzing data from a certain grid or region. PSK Reporter is a good way to judge where your signal is heard as well. It also keeps a visual log of what you heard. You can find it at pskreporter.info/pskmap.html. Many hams also use an alternate site called Hamspots. I personally go back and forth between the two, they are equally good, and I feel they complement each other very well. You can find more information on this at Hamspots.net. These settings are especially helpful if you leave your machine with WSJT-X running overnight and want to see what stations you decoded. You can look

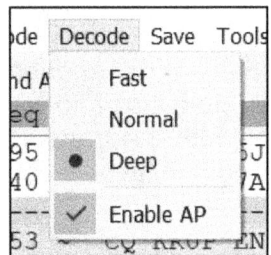

back for up to 24 hours with PSK reporter. This is a good way to learn propagation patterns as well. Also, a good way to get frustrated with what you may have missed out on while sleeping! But as I always say, there is always the next time.

I highly recommend configuring your other digital software as well for spotting feedback, when available, so it can report back to the PSK reporter servers. When I see a station blindly calling someone, they clearly can't hear, I sometimes check to see whether they decoded anything if they have reporting turned on. Usually, the answer is no, and this just makes me chuckle. Also makes me very puzzled. Not sure how they expect to complete a QSO if they cannot hear the DX. Magic perhaps?

The second setting is also on the menu but not where you would expect it. This is under the "*Decode*" menu dropdown. Select "*Deep*" if not already set there and try to initially set the "*Enable AP*". Some report that when you enable AP it gives a lot of false decodes but the tradeoff is a possible 4dB gain. I am not so sure about this 4dB number, but it does help. False decodes have not been a major issue in my experience, though the above is an example of one I saw recently. Still better than missing a valid decode! This is up to you, though for DX, I would use the above settings. May take some experimenting on your part.

Multiple WSJT-X Profiles

I have already briefly touched on this, but you can set up multiple instances of WSJT-X very easily. This may be needed if you have multiple receivers you want to use as well as if you have a radio with multiple receivers or slices such as a FlexRadio. This will work on all radios. Here is an example of what you may end up with on your desktop:

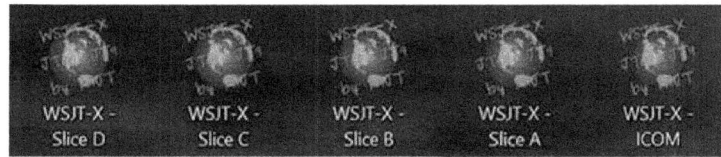

Much of this is accomplished by just following these simple steps. To use multiple instances for slices or radios you need to basically **create a desktop shortcut** to WSJT-X if one does not already exist. Then replicate a shortcut for WSJT-X so you have a copy and end up with a *"shortcut to WSJT-X"* as many times as needed. Next, **rename the shortcuts** so they make sense to you, such as the picture above for example. They can be called anything really but keep it simple and short.

Then one by one **right-click** and **select properties**, click the shortcut tab, and rename this following the example below. Repeat as many times as needed.

The **BOLD** below is what you are adding, please note the **two dashes, not one**. Just copy exactly, name can change if you like but **no spaces except for the space after the exe**. Should look something like these:

C:\WSJT\wsjtx\bin\wsjtx.exe ***--rig-name=Slice_A***

C:\WSJT\wsjtx\bin\wsjtx.exe ***--rig-name=Slice_B***

C:\WSJT\wsjtx\bin\wsjtx.exe ***--rig-name=Elecraft_K4***

The rig name can be anything after the equal sign if it makes sense to you and has no spaces in the name itself. I have used an underscore to compensate for this. You will need to reconfigure each profile the first time you start it so **pay attention to setting your COM ports and audio sources on each instance**. Once you set them up, it will give you years of joy. You will not need to reconfigure any of this when new versions come out as you edited the shortcut only. Therefore, if you install WSJT-X in the same folder location you should be good, software will upgrade for all of your instances since they are pointing to the same location. The process is essentially the same for the MacIntosh, just edit your shortcuts.

In the above example I have set up all 4 slices for my Flex-6600M as well as my Icom IC-9700 which I use for satellite work and VHF/UHF DXing. When running 2 large 4K displays I had at times 5 instances of WSJT-X running at once. No problems! You can also use this same method to create a different WSJT-X instance to use a different callsign if you ever operate a special event station, are maritime or outside your area.

Using the WSJT-X Software

While I will go over the basics, I am looking to add value to DXers with things which may not be widely covered or understood. Use this section in conjunction with other resources I will mention if so desired.

The FT and JT modes are not conversational modes, meaning no chit-chat. It will be a minimal exchange for a valid contact. A typical FT8 QSO flow is basically very similar to what you might see in other digital modes, except for the signal reports. Here is an example using me working Igor, R0JF:

1. **CQ** W6AER **CM87**
2. W6AER R0JF **PO30**
3. R0JF W6AER **-23**
4. W6AER R0JF **R-19**
5. R0JF W6AER **RR73**
6. W6AER R0JF **73**

This will be the same or similar for FT4, JT65 and JT9 as well. **There is a character limit with FT8 as well as FT4 and other similar modes.**

Basically, start with a CQ, your call with your grid square. These are already preset if you followed the instructions correctly. Someone responds with their grid square and call, then exchange reports. The first person sends just the signal report. ***NOT the R or it screws up the cycle, many screw this up when they go manual.*** Then the return signal report has the R meaning Roger. Next response confirms received report with RR73, then 73 to confirm all received in the exchange. Log it!

Some also like to send an RRR before the 73 exchanges. This is very redundant. You can even skip the grid square if you are in your usual QTH. These speeds up the QSO by a significant amount. The DX likely does not care about your grid square or need "*Roger*" twice. For some possible shorter openings like on 6m and 160m as well as in congested conditions, I do recommend not even sending the grid and starting with the signal report. This shaves time off the QSO and really expedites things on both ends. Less chance of an incomplete QSO.

The left side of the WSJT-X window is the activity overview, and the right side is for a selected RX frequency, generally where **you** are working. Wherever the WSJT-X waterfall **RX** green bracket (more on this in a bit) sits is what will show in the right decoder window. This marks 50Hz wide space which is the width of the FT8 signal. You want to find an open space on the FT8 waterfall and transmit there. Do this after listening for 2 cycles to decrease your chances of creating QRM. Bands can be changed from the dropdown menu on the left and if you set everything up correctly, the radio should follow.

The columns you see here left to right are as follows. The **UTC time** is the universal coordinated time, this is far left. Followed by **signal report** in dB. Next is the DT which is **delta time**, or actual time deviation from your clock as I had mentioned earlier. Then shown is the actual **Frequency.** This is the tuned frequency plus this number in Hz, so where you are sending. Example 10.136MHz is where you are tuned, working someone at the 1000hz (1kHz) mark, you are 1kHz up so at 10.137MHz technically. Keep in mind the signal is very narrow, but you are listening in a 3kHz window (or close to it), assuming you set things up correctly. Therefore, you can decode many stations at once.

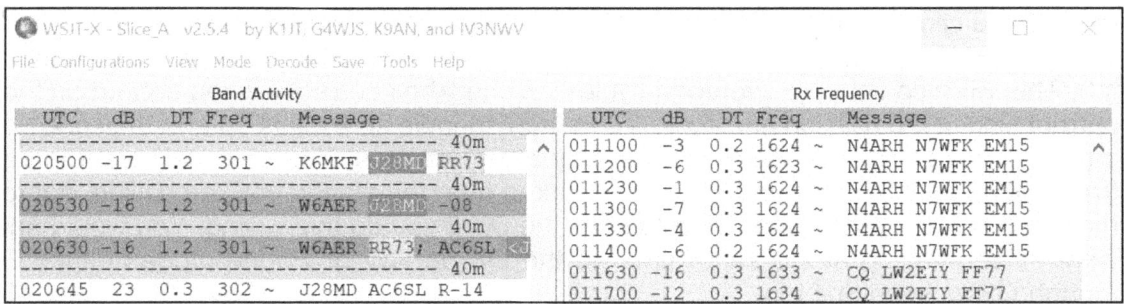

You may have noticed the Little "~" symbol. This indicates the FT8 mode. In case you are wondering JT65 is marked by "**#**" JT9 is "**@**" and finally FT4 is "**+**". You will see these symbols if you switch to those modes. I would encourage you to try them all at some point. There are preset frequencies for the other modes in the software for the common or rather recommended watering holes. You can use some WSJT-X modes outside of HF and all the way into UHF and even microwave! It is not uncommon to see FT8 activity during VHF contests. And as I have mentioned, FT4 is also used on ham radio satellites cross band. At times you can find FT4 on the RS-44 satellite, might even be me sending a CQ. The last column on the left side is the actual **message** or exchange. I recommend operating with the "*Hold Tx Freq*" checked.

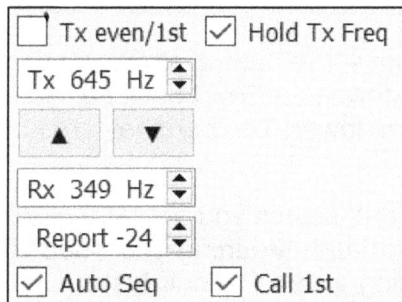

As far as the WSJT-X **bottom controls**, if you are new to digital, it may this look like a bloody mess. Though, I think it is very well designed once you understand it. I will not go over this in detail, just enough to get you going, the website manual does a much better job than I could ever do. The most important controls are **Enable TX** and **Halt TX** on the top portion of the controls. They do just what it says. **Erase** clears the right screen click it twice, clears both sides. Nice if you want to keep it clear or just changed bands, etc. **Log QSO** does just that, but you may need to configure it to work with your software used to log. Tons of information about this online, including some great videos. **Decode** basically processes the last receive cycle again. This corresponds to where the green pointer is on the waterfall. Likely you will not need this much. It will flash when automatically decoding, twice per cycle.

Tune is one I wish people did not use unless done properly. Not because some do not need it, but because it often creates QRM. Frankly, some are just lazy or do not know any better. They tune up in the wrong spot (move over slightly) or with way too much power wiping out QSOs in progress. This is generally followed by them coming in blasting on top of someone else. Yes, digital modes have lids too sadly. Please don't be one!

The band drop down is for **band selection**. These are preset recommended watering hole frequencies. The **even/odd switch** is a little tricky. More on this later. The right-side buttons are where you initiate a CQ followed by enabling the TX. Alternatively, you can double click on a callsign calling CQ and the software will do the rest if all goes well and if you checked the "*Double-click on call sets TX enable*" in the general settings tab. This will also set your even/odd cycle and checkbox correctly if you clicked on the correct side of the callsign and cycle. I see many people screw this up and they end up calling at the same time as the DX, so **pay attention this**! You obviously want to be transmitting on the opposite cycle if you want to hear the station you are calling, not to mention you will QRM those trying to work the DX if you do this wrong. You would not want to be sending at the same time as the DX on CW or Phone modes either…. though I have seen that too.

The **power slider** on the right is where you can pull the output power back a little. I frankly do this on the radio, but I am in the minority. See what works best for you. A little experimenting and experience will guide you best here. FT8 and FT4 do not use reports like 59 on phone or 599 (PSK, RTTY) or 5NN like in CW. **FT modes can be NEGATIVE** numbers such as -18 **or POSITIVE** like +02

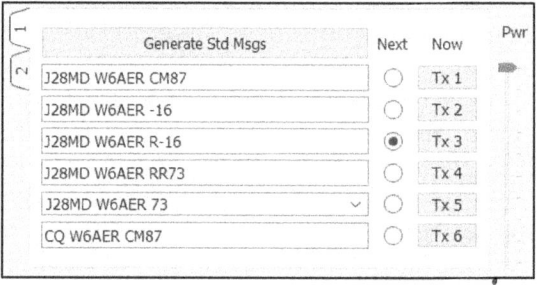

FT8 will decode down to about -24 and I have seen as high as +26 but I suspect sometimes this is due to 2 stations on the same frequency adding up to appear to be stronger unless you have a strong station very close to you. I have seen both. FT4 will not decode down to as low but close. Note that modes like JT65 and especially JT9 can decode even lower! Too bad they are not as widely used anymore.

If you are getting a +10 or higher from a different continent, unless you are very close to one or propagation is outstanding, you are likely too loud. Meaning you need to reduce power out. Getting a strong report back is nothing to brag about, just means you have your amp cranked to 11 unnecessary. Lower numbers are always weaker.

Do note that contest modes, such as the VHF contest usually skip the signal reports just exchange grid squares. The same is true during some contests using FT4 mode. Nothing wrong with this, still a valid QSO and can be submitted for awards.

I highly recommend bookmarking wsjtx.groups.io/g/main and maybe even subscribing to the daily summary as keeping current here will not only provide you with tremendous education but also possibly save you a lot of headaches in the future if you are active on digital modes.

Tweak the WSJT-X Waterfall/Spectrum Display

Once you get the basics down, there are a few more things I would recommend adjusting. This is all optional and your call. I find that the default settings of the WSJT-X waterfall speed and width are not ideal for me, at least not for DXing. For one, I like to see all 3kHz at once. Why you ask? In crowded conditions some, especially rarer stations, will transmit on the lower portion or upper end of 3kHz. I have even seen DX above 3kHz and had them all to myself. You want to be able to see it all if they are on the edges of your display. Also, some radios have digital presets where you will need to widen the default filter as they might be preset for RTTY and this can be much narrower. This varies from receiver to receiver but generally can be adjusted easily. Adjust your setting so you can see traffic 3kHz wide or slightly more is possible. When using SmartSDR I set this to 4-6kHz but keep the waterfall at 3.5kHz or so. When stations start showing up on the edge, I pay special attention.

I also like to adjust my own waterfall speed. I often find that the default speed is too slow, and I may miss some things. The settings I find work for me are as pictured here. I even go as far as setting the Bins/Pixel to 4 to get 3.5KHz under my WSJT-X. These settings are found on the bottom left of your waterfall window. I have the "*flatten*" checkbox checked and most other settings are default but feel free to experiment to what works for you.

Generally, I have this running per above so I can make sure I can find a clear spot to transmit on **after observing for 2 cycles**. I recommend you do the same. Not just as a courtesy so you do not QRM but also for your own benefit. If you are doubling with someone, while it may still decode, you decrease your chances of a QSO. By speeding up the waterfall you can also see at times a hair before you transmit if you are doubling. Yes, another reason to speed it up!

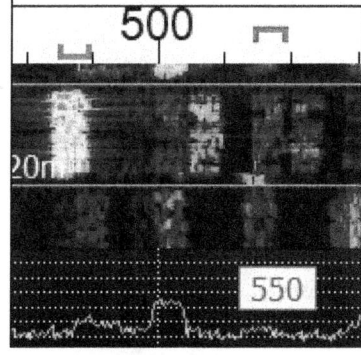

The **receive** bracket is moved with a **left click** and the **transmit** bracket can be moved with a **left click while holding the SHIFT key**. These can also be adjusted in the main software window by typing into the Tx and Rx windows, though I find this clunky. Here I show my receive indicated by the bracket pointing up around 350Hz. This would be green in real life. The bracket pointing down is red in the software and here located around 650Hz. You count from the left side of both brackets. The software will also show this in the center of your main window. I prefer having the waterfall below the main application but some like it above. Up to you, not written in stone. I do, however, recommend not turning it off.

One last thing. You may not see the waterfall traffic and the same stations as someone 200 miles away does. You may not even see the same thing as your neighbor! I have proof as I have a fellow ham a few blocks from me and he sometimes decodes things which others, including me, do not. The reverse has also occurred. This seems to be especially true for the 6m band. The reasons for this can be endless. From antennas to receiver setup to even local propagation and elevation, obstructions (trees, rooflines, metal objects like gutters, etc.) and while not very scientific, I would also include luck. Never assume anything.

How Much Power to Use

The simple and legal answer to this question is the minimum required to make the contact. Remember this in your licensing exam? But how much power is this? Seems a lot less than most are pushing nowadays. Many really should tone it down on FT8 as some DXers go max by default. No need to run 1.5kW. I have worked all around the world with usually no more than 400W with very minor exceptions. I sometimes run 100W or less for a day and don't even notice much difference. While writing this book, I had my amplifier in for repairs. Frankly, I barely noticed the difference outside of the 80m and 160m bands where it is much noisier. There is very little to be gained from going above 6-800W in my opinion except more QRM and a higher power bill. If you find you cannot make QSOs with this much power using FT8, time to re-evaluate the antenna system, your settings, and your operating style.

As I had mentioned, if you are getting consistent +10 reports from the East coast from the center of the US you are running entirely way too much power! Even from a nearby state that is too high of a report at times. Again, PSKreporter will give you a good feel for where you are heard as well as what you hear. Set a balance! If you see that you are only decoding from the US as far Portugal **(CT)** and Spain **(EA)** but you are being heard at +5 in Ukraine **(UR)** then you are doing things wrong. They may hear you; you will not hear them. This is a good way to figure out the required power. Try and hover no more than around zero for your target area, I usually shoot for -5 or so. If they hear you but you do not hear them, then you are just QRM and are wasting their time if they try and call you. **Just because you can run more power does not mean you will hear better.** However, you will be called an alligator even if you are not from Florida. Bigmouth small ears and is not a term of endearment either. Don't compensate for a poor antenna or poor operating with too much power as a QSO is a 2-way contact.

Please observe 60m and 30m power limits. I have heard "*seasoned operators*" brag about how they got a nearby DX with only 500W on 30m. There are **several** things wrong with that statement! Besides being illegal to run that much power on the 30m band.

Many are running QRP. Meaning using 5W or less. I have done this, the first time by accident frankly (forgot to turn on the amp) and made a dozen QSOs with perfectly good reports! I often run the amp with only 5-10W from my radio. This way both my radio and amp are happy and have to work less. This gives me the right amount of power out in most situations, except for a few hard to work spots from the West coast.

On the lower bands, you may need a little more power due to higher noise which you will need to overcome. Also, you likely will be using a compromised (reduced size) antenna. But, you need to hear at least somewhat well to start with. Often this can be done by using two antennas as we had already covered. One for reception only, such as a magnetic loop (directional and low noise) and a vertical to transmit. I have recently seen a nearby station on the 160m band from here on the West coast calling CQ. He was super strong, many EU and US stations responded to him, but he never heard any of them. He was calling CQ for 15 minutes with no QSOs. I could copy both sides fine. Clearly, he was an alligator. Don't be one! People do notice…I did!

Working Split with WSJT-X

I have touched on this briefly before, but let's go over these in more detail as they are rather important to DXing success. To be clear there are two different splits used in WSJT-X. For normal WSJT-X mode, left mouse click to move RX and **hold the shift key to move TX**. Just remember that **green is RX**, **red is TX**. You are on the same 3kHz frequency range but shifting your "*audio*" to a different spot in this range just as I have already covered. Split does not necessarily change the recorded frequency, just moves inside your audio bandpass. Since your signal only occupies 50Hz signal on FT8 for example, inside a 3000Hz domain (3KHz) this can be easily done, and many signals can be accommodated. Find a clear spot, **check the "*Hold TX Freq*"** and stay put unless someone parks on the same spot and cycle as you. Meaning transmits at the same time as you. You should check occasionally by taking breaks after every few transmissions.

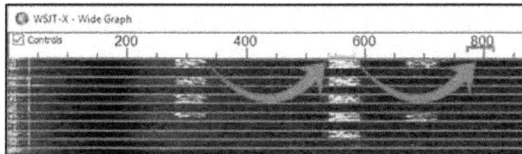

One reason folks might choose to not work split and leave the "Hold TX Freq" off is due to their search and pounce style FT8 operating. When the station you are calling has the "Call 1st" checked with multiple callers present, this will put the ones on the same frequency as the calling station first in line. In theory! Does not actually always work out as if everyone does this, nobody wins. This is one of the reasons I always work split. You really should too! You can do search and pounce operating from a held transmit frequency.

I have heard of hams using the RIT/XIT (called a clarifier on some radios) to work FT8 split. Don't do this please! Leave those darn buttons alone. You do not know exactly where you will transmit due to the shift created and likely will be on top of someone else. I feel strongly anyway that the above knobs are way overused by some. I think this is due to some legacy operating practices perhaps, but time to let it go at least in digital operating.

The above split method is to be used in normal operating mode, not Fox and Hound mode which is very different. This uses frequency shifting split mode (real or faked) and is automatic. **Fox and Hound mode is ALWAYS going to be split.** Speaking of which…

Fox & Hound, No Animals were Harmed

Understanding the Fox and Hound mode is rather important and using it properly is even more so for DXers. You will need to learn this mode and learn it well if you get serious about DX. Period!

Fox & Hound is what I think of as the "*automatic split mode*" for digital with some added software controls. While it is good operating practice to work FT8 split anyway as I had mentioned above, this really forces you to do so by forcing callers up. There will always be a few who forget to switch modes or somehow manage to call in the wrong spot, but it is a start.

This is a mode which is used often by more exotic DX and for sure DXpeditions. For the receiving end, that would be you if you are reading this, you will always be the hound. **The "*hound*" is the one calling the DX and the "*fox*" is the DX.** So, make double sure you set this correctly. You will create chaos if you do not and likely will get hate mail if you bust QSOs for others. Or even worse have your callsign posted on spotting networks with "*interesting*" comments which I will not mention in this book to keep it PG rated. Never do that by the way!

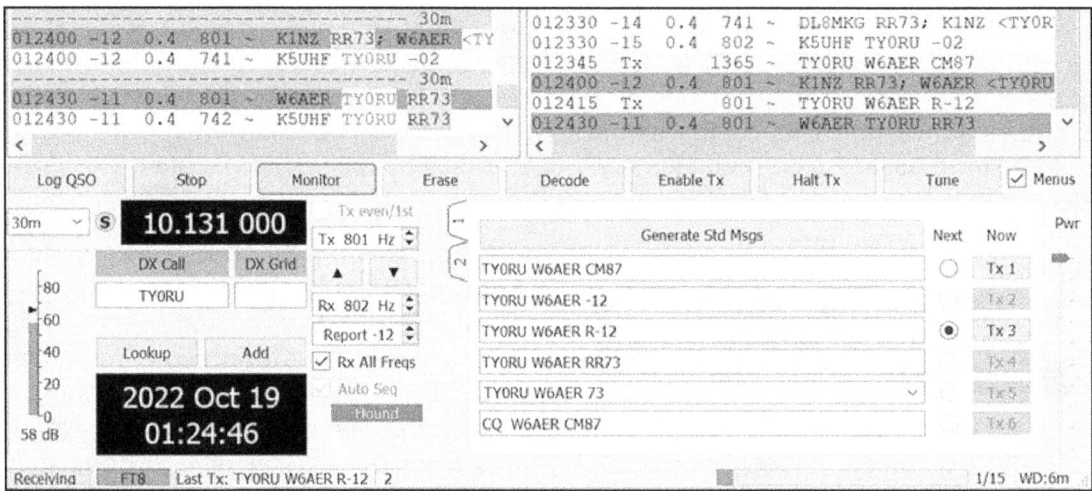

In the image is an example of multiple callers being responded to and what it looks like in the readout window. Here there are 2 streams with one stream containing 2 QSOs. In this example me, W6AER is working TY0RU, Benin. The DX (fox) is going to only attempt to complete the QSO with you (the hound) three times. After which, if he does receive your response, you will have to start the process over.

Note again, as this is very important: Once you get responded to, the software automatically moves you to the below 1kHz marker where you transmit. In the example pictured, I got moved to 801Hz, even though I was calling at around 2kHz. The DX will always be below 1kHz. There is nothing you need to do on your part, software will shift for you.

You will **always call above 1kHz** (or even 1.1kHz minimum in case you or the DX are off a bit) when in this mode and try and spread out. RIT and XIT should be off or at zero. You will not only not get replied to, but you will QRM others possibly who are on the 2nd cycle of their exchange, so make sure these are not engaged. The 2nd cycle, once the DX responds to you, will occur under 1kHz and that is why you should not be under the 1kHz marker otherwise. Software will do this for you, just stand by and watch the magic happen.

The DX can and likely will be operating multiple streams below the 1kHz marker. Meaning responding to multiple parties at once. One drawback is that then the stream count goes up, signal strengths go down. So, if signal is weak already with 1-2 streams,

3 streams will likely not be consistently decoded. I have seen 4-5 streams used reliably for hours when conditions were favorable. There were also times I could not even decode one stream consistently, so you just never know. This will vary a lot based on time of day as well as location and sunspots. Be patient and monitor until you can copy consistently. Likely at some point the conditions will be in your favor. A side note on the number of streams to be aware of is that when there are 2 steams running, they will be -6dB weaker, meaning ¼ power. At 5 streams this will be -14dB and therefore conditions must be amazing if only running 100W on the transmitter end as the effective output is only going to be 25W at 2 streams and 4W at 5 streams! The use of an amplifier of course will dramatically assist with the above situation.

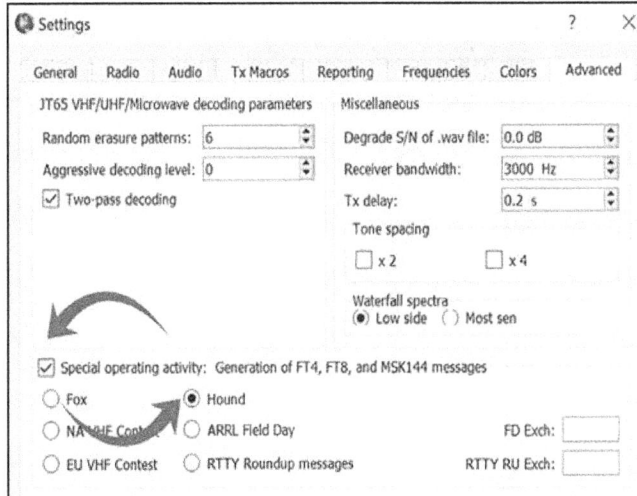

So now that you have got the general idea, let's look at settings on WSJT-X. **There are 3 steps that will be required to get going in F/H mode.**

You will also need to **set your radio to split but in the software** (not the hardware) via the WSJT-X general tab under settings. **This is likely going to be the "*Fake it*" button for you**. The rest of the settings we have already covered in the initial setup section, you likely will not need to change any of these. You will however need to uncheck this when going back to "*normal*" mode as well as the other changes you make for operating in F/H mode. In the same settings box, switch to the advanced tab now and **check the "Special operating activity" section check box as well as select "Hound"**. You can now close this. You will now see on the left of your frequency readout in the green dot a letter "S". If so, you are split and if you followed the above so far, you are almost ready to go.

You will now need to **generate the messages based on the callsign of the DX**. Do not make a mistake here as the software will not double check for you. Whatever you generate is going to be sent. Enter it in the DX call box and click generate. This will pre-populate all fields as needed. Once you start to decode the DX consistently, go for it! **If you do not hear them, please do not call them. It will not work!** Listen to two transmit cycles to make sure you have a clear spot in the caller area above the 1000 marker (1000Hz).

It is a big no-no to call when your region is not being called. This is also true for other modes of course as I have already covered. Look for CQ JA, CQ EU and call accordingly. If you only see CQ, anyone can call. If you see QRM or LID, pause and make sure it is not directed at you. Hopefully not, but everyone makes mistakes, just check your settings. Correct it if needed and you are good to go again.

When you are done operating F/H mode, and after you set the above two settings back to normal, specifically uncheck the special activity checkbox and put split back to "*none*" if that is where you normally have it set to.

Also please be sure to **Check Hold Frequency and Call 1st as well as Auto Sequence**. These may get turned off when you revert to "*normal*" mode then you may end up calling on the receive frequency of the DX and we already talked about why that is rarely a good idea. I have been known to forget this as well, so keep your eyes on it.

Adding Working Frequencies

You recommend you do this if you are planning on running Fox and Hound mode, though some operators just manually move to the DX frequency and that is also fine. Though if you do this manually do it after switching to F/H mode or you will need to move the frequency again.

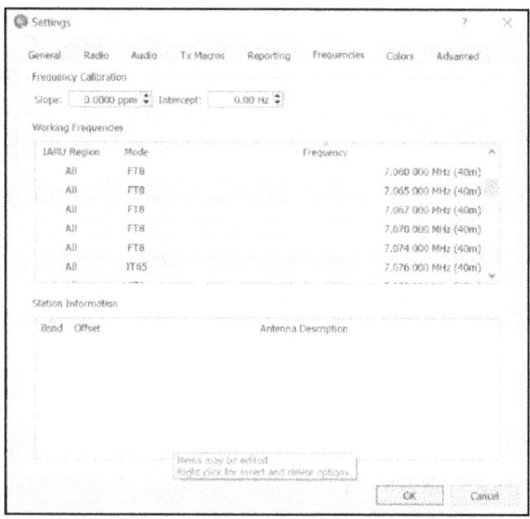

I would say 99% of the time, F/H operations are not done on the main, preset frequencies in WSJT-X. I have seen exceptions, but they really should move off the general frequencies.

If you are running a multi-instance install, you will need to do this for each install, but only once per instance. Although you can export and import these as well. This will save you a lot of time in the future.

In **Settings**, go to the **Frequencies Tab** and **right-click in the window**. A pop-up window will appear as seen here in the second image. **Set region (optional)**, **set mode**, likely FT8 and type in the **frequency** you want to add. This is in MHz so pay attention to the number of digits. Click **OK**, done! Check to make sure you have entered it correctly. Easy to make a mistake here. Now if you did the above correctly, you could now select it via the dropdown in the main window when needed.

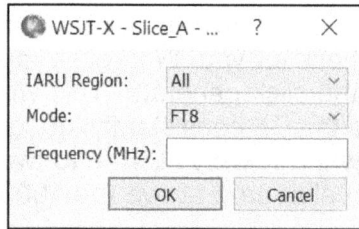

The frequency you need will most likely be the one published on the DXpeditions website. At times you will also see these spotted on the DX clusters and possibly published in DX newsletters. On the 6m band, there are two frequencies to monitor for DX. Make sure 50.313MHz and 50.323MHz are both on your frequency list. Some DX operators will only show up on the latter.

Check your ALC…Again and Often

Please watch your ALC! Yes, this also applies to all digital modes as well, in fact perhaps even more so. Over-driving your outgoing signal will distort, splatter and QRM others. Chances are it may not decode on the other end if the signal is dirty. Keep your eye on the ALC, don't overdrive! **Please make sure you set this right and check it regularly.** It may even vary by band on some radios, as weird as this sounds. I have personally experienced this with a modern transceiver before.

Basically more "*Audio out*" does not mean more power out, as in louder is not better. In fact, it can backfire badly. This is the same as when you are working SSB mode. Always get your audio out adjusted for a clean signal. Then you can amplify by increasing the power out or kicking in the amplifier, if needed.

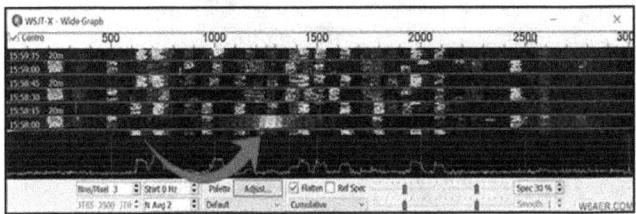

Very few things are worse than seeing an over driven signal on the waterfall. Not only is it annoying to watch but often QRM (interferes) with neighboring signals even if it can be decoded. This can be especially frustrating when some exotic DX is on. Often these signals are so distorted they cover other signals. You would think someone calling CQ for 10 minutes with no response would figure out something is wrong! Example of what not to do above.

If ignore the ALC settings or if you step on the DX, you may experience something like what is shown here. This screen capture has been redacted. The callsign and some of the exchanges were removed to protect the guilty party. Don't be one of these guys please.

Below is how it is seen on the FlexRadio Waterfall in SmartSDR.

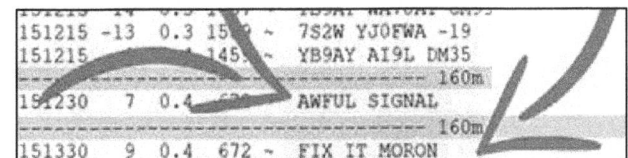

Needless to say, that it is impossible to decode the signals near it as it is interfering with just about everyone on the cycle. The next cycle (above it) you can see how the waterfall should look with clean FT8 signals and on the right, you can see perfect FT4 signals starting at 14.080MHz.

And here is another example of a poorly adjusted signal. Yes, QRM central! Look how many stations are being interfered with. This QRM was nearly 3kHz wide.

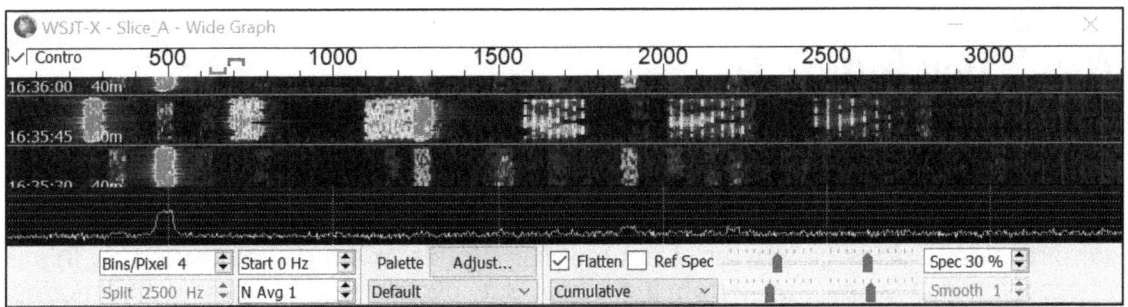

A similarly bad signal as shown on the SmartSDR waterfall seen below. If you are not sure how to do this correctly, ask for some help at your local club, help online, check out videos and of course always try your radio manufacturers manual as well. Generally, due to its importance they cover this well in every manual I have run across.

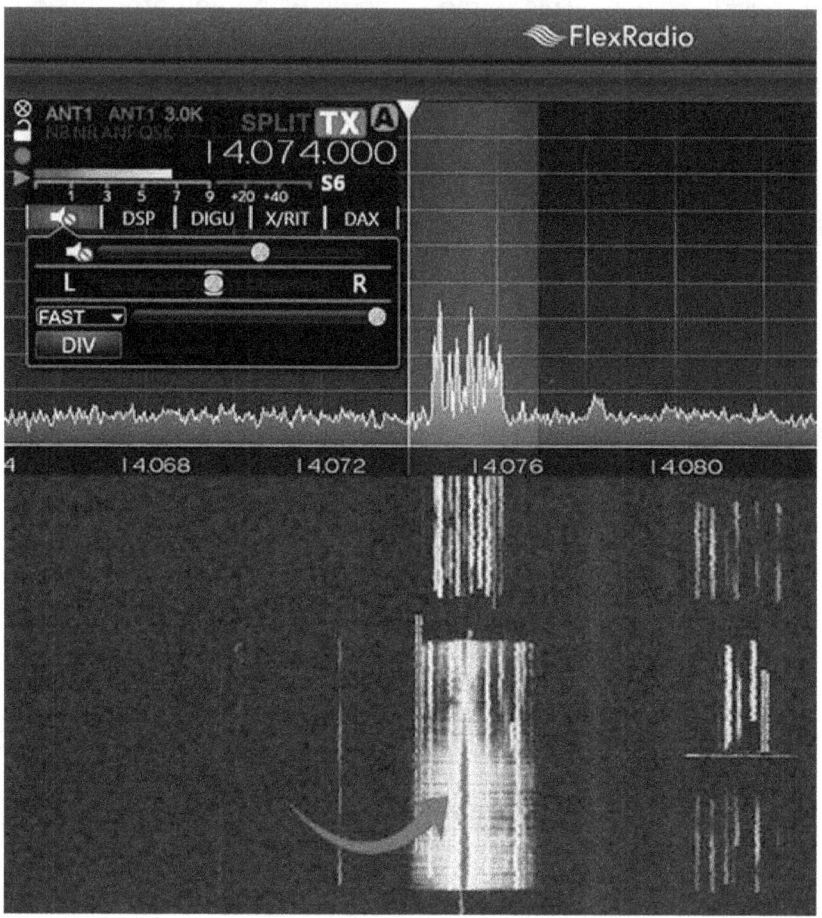

Watch your Duty Cycle

This is a silent or sometimes quite noisy killer of radio gear. This basically is in reference to on and off transmit cycles and the *"cool off"* time in between. But it does get more complex with modes which do not always use full power output, like SSB where it can vary even based on your compression level. When using voice compression, it does go way up. I am mentioning this in the digital section as this is where most issues could arise from the higher duty cycles used by some digital modes. RTTY being likely one of the worst offenders.

Why is this important? FT8 and FT4 are nearly full duty cycle modes, meaning when they transmit, they are putting out 100% power based on whatever you set the radio/amplifier to. CW is the same, just much shorter cycles. Yes, there is a break but if you noticed, this usually does not fully allow for gear to cool back to the initial temperatures.

In general, be careful with overheating. This applies to both your amplifier and your radio. When running barefoot (no amplifier) do not run 100W in a full duty cycle mode if your radio does not cool well. Most do not and this will slowly kill them. Around 60-70W is good to run long term without issues. 80W is OK for short periods. 50W is even enough to drive most amplifiers to 1.5kW. But this does not mean your amplifier can handle this much power output it not designed for it. Just watch your temperature and temperature meter if your radio and/or amplifier has one. If not, listen for a fan. If it has been on for a while, you are cooking. I have visited a ham once in the early days of FT8 where I could literally smell components overheating when he was operating FT8. His fans were so loud I was waiting for his radio to take off like a drone. I would never buy a used radio from this person. Likely that radio has surrendered by now.

There is one other factor relating to the above which likely will not affect modern gear, or at least not on HF. Heating and cooling cycles may shift your frequency slightly. This becomes an issue on UHF and above especially if you DX up there. Something to keep in mind if you experience this.

Digital Etiquette

If you read nothing else, please read this section. Two key takeaways here are **not calling on top of the caller frequency and listening before calling.**

There is a legacy tradition in North America about starting times for transmissions. Though I must say it is a bit dated and usually now reserved for the 6m band when it comes to DX for the purposes of this book at least. This has to do with Western stations starting even and Eastern station starting odd on the clock. So, being a W6 in California, would need to start at :00 or :30 on the clock. The East coast at :15 and :45. Needless to say, folks in the center of the US are confused by this, as would I. Sometimes the Mississippi river was used as the divider, but this was not really ever followed. When hunting DX it can show up on either side, so just answer their call, log them and don't worry about it. If I decide to call CQ, I check to see where most of the DX has shown up and call opposite to this. Has been working well for me so far. You will find some folks possibly very frazzled by this. You have been warned!

Lastly, once you start operating, take a little break during transmissions. **Stop after a few cycles and check to make sure there is nobody else sending on the same cycle and location as you.** Possibly you might be canceling each other out and nobody wins. Don't get mad, just move over a bit if this is the case and try again. Likely was not intentional. You may also be missing some great DX on the opposite cycle too so yet another reason to listen, listen and listen some more. See what is going on, who is working who and how the propagation might be shifting in your favor, hopefully. You will likely find this to be very rewarding in the long run.

Final Adjustments & Checks

Be sure to **turn off all audio compression** you normally would use for phone. Will not only not help here but will not work. Most radios handle this for you now automatically when you move to a digital mode, but if yours does not or if you are uncertain, check it. Make a note!

Turn off other radio DSP settings used for CW/Voice if not needed, some radios do not do this for you if they are not mode aware. Meaning, some radios retain settings by mode, or even band. Others are set the same across the board regardless of mode or band change. Therefore, it is up to you to check this as well. Though according to some hams, on some radios, using the digital "*noise reduction*" on a low setting may be helpful for FT8.

Check your bandwidth. A dead giveaway is dark spots on your waterfall display. Make sure you are at least 3kHz wide or as close as you can get.

I often get asked about AGC (**A**utomatic **G**ain **C**ontrol) on digital modes. Should it be AGC OFF or ON, SLOW or FAST? I recommend you play around with different settings and see what works best for you. Radios are different and so are operators. I am usually on FAST when in digital modes, but that is just me. Some feel AGC OFF works best for them. This can also vary a bit based on the radio you are using as they do behave differently even within the same make.

Digital Mode DX Clock Trick

Sometimes you will encounter, or at least see a station spotted whose clock is off. Way off! At times so much that you cannot even decode them, but you can see the telltale sign on the waterfall where they are out of sync based on their transmit cycles from everyone else. Sometimes you may decode them but may see a big difference in their DT versus yours causing sporadic decodes.

You can still work them! How? There are actually a few ways. If this happens, try manually adjusting your time by a second or two to compensate for them being off. Alternatively, with **TimeFudge** by Mike Black, W9MDB. This software will do that trick for you as well. Be sure to turn this off when done or you will be off.

It is a touch hard to find online, but try here:
dropbox.com/s/b39x5zismr5pv2v/TimeFudgeSetup1_3.exe?dl=1

The other option is the software I already use myself to keep my computer time accurate. Mania Radio **BkTimeSync** by Mauro Capelli, IZ2BKT has a button for "*Manual Set*" which brings up a "*Manual DT*" window. Just enter the time in fractions of a second and you are good to go.

You can obtain this at maniaradio.it/en/bkttimesync.html. I have managed to score a band filler once using this method and after I made the QSO sent him a "*CHK CLCK*". He did and soon the rest of the planet was calling him.

JTAlert – A Must for the Successful DXer

WSJT-X is an amazing piece of software, but it can be made even better with a piece of complementary software called JTAlert. This software was written and is maintained by Laurie Cowcher, VK3AMA in Australia. I will go over the basics of the setup as well as the settings I use, and I recommend in this section. Your preferences and needs may vary, but these are good guidelines to get you going at least.

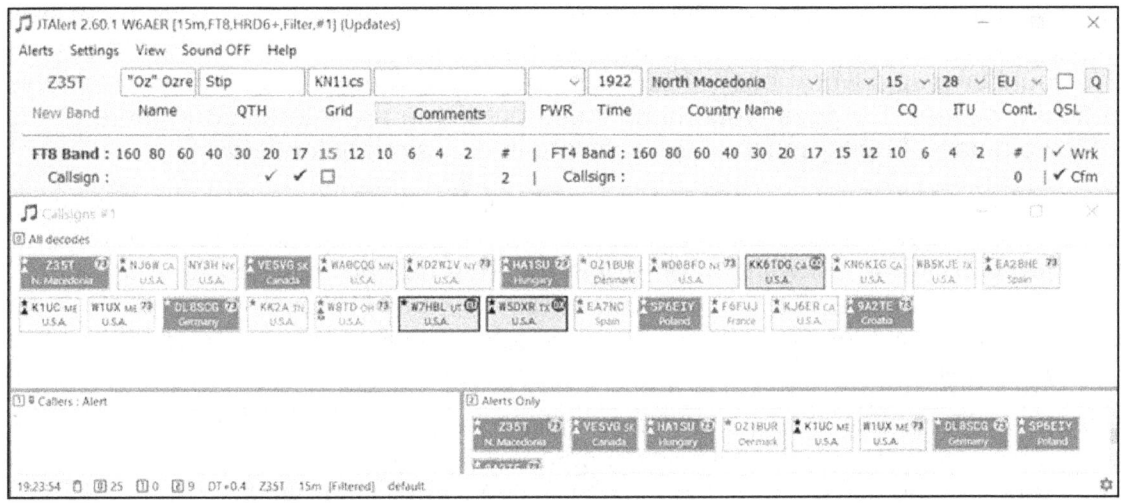

JTAlert is an add-on if you will (pictured above) to WSJT-X and JTDX, an alternative package. It will give you many great additional features. For one, alerts for needed WAS, DXCC, WPX, and many other awards. Alerts for wanted station callsigns, CQ and "*Calling You*". Check for LOTW and eQSL users via flags. Activity report by bands, see what is happening before changing bands. Also includes simplified integrated logging for many 3rd party software. The "*B4*" alert tied to the above saves a lot of duplicate QSOs, you can even hide already worked stations. Wish more stations used this feature. See my dupe section as to why. It also gives you chat capability with other JTAlert users online. Also, you can see what bands you worked a specific station on already. Many additional features with more added all the time! **Basically, this will give you an edge in DX hunting and save you time.**

Referring to the image above, there are multiple windows. The top display, which is optionally turned on also in **view**, shows the last caller I clicked on or worked. The information is looked up from QRZ though this does require at least an XML lookup account. There are other options too. The very bottom shows what other bands I may have worked the station on indicated above. The bottom window is generally the default one, though highly customizable. Adjust this based on your preferences.

To begin with, you will need to obtain this wonderful piece of software. You can always find the latest version at hamapps.com. This software is a little different, as it comes in three parts. There is the **main application**, the **Callsign DataBase** file, and an **Audio add-on** for announcements which is very handy, but optional. Use caution if you opt to install the audio package. Do not enable all voice alerts, especially if you are very new to DX, as it can get very annoying as it may not shut up...like, ever! So don't overdo it like I once did. Only select the things you are interested in being alerted about. If you are a visually impaired operator, this is also a very useful feature.

Updates happen often and there are nice features added at times. The above image shows updates available on the top of the window near the version number. Hard to believe this is still <u>free</u> software so for that a **big thank you to the author Laurie Cowcher, VK3AMA.**

Once installed, you need to configure it. Go to **settings and click manage settings or hit F2.** I will only go over a few selected things I feel are useful to DXers. Otherwise, this could take up a whole book. There are many resources online which dive into this software deeper. I suggest you check these out if you are interested in more of the intricate details. Frankly much of the software is very straight forward.

You will need to enter your Callsign and First Name. This is under the "*Own call*" section. This is to alert you when someone is "*calling you*". Fields with these items will be lit up **RED** in the software when triggered. This can be adjusted to your liking. I would also enable the CQ alert one down from it. You will repeat this, sort of, under station callsign towards the bottom where you also need to give your 6-digit grid square (told you it will be needed a lot) as well as your CQ zone and ITU zone.

I disabled the display of "*B4*" stations to indicate duplicates. This is what I do, and it really reduced clutter, more room on my screen for things I do need to see. On a busy 20m band it could use up more the maximum of 4 rows if left enabled. I do not go past 7 in a row as the font size drops too much to make it fit. This can be adjusted from the **view** menu as well to your preference. Larger monitors and lower resolutions come in <u>very</u> helpful here.

I am a long time Ham Radio Deluxe user, so I configured JT Alert for the HRD logbook as seen below. Also please note I use a SQL DataBase to speed it up due to log size. I touched on this earlier in the software section, here you see it in action.

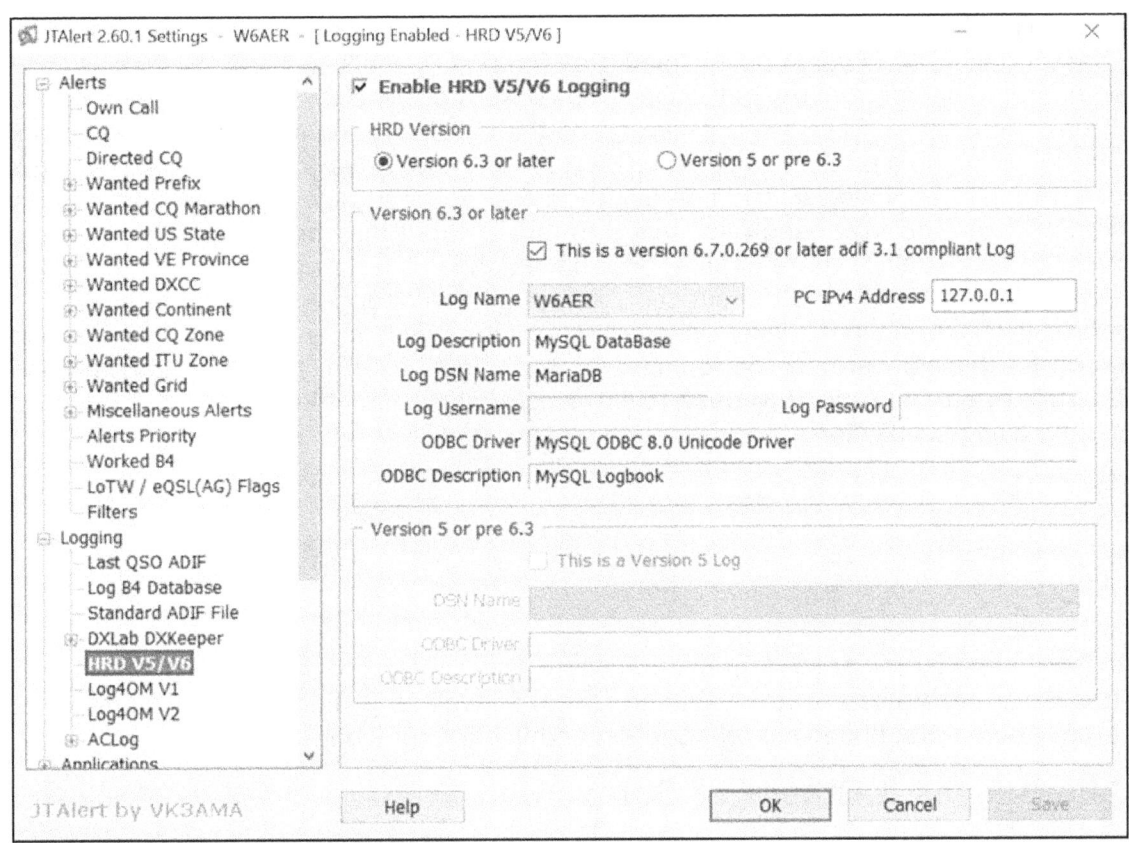

The audio alert is set on the bottom, under the "sound card" and requires the audio file package installation from the website. The sound card to selected here is not the one used in WSJT-X but the one used for your regular computer audio. This needs to be specifically selected in case you want audio alerts.

Next are the "*Wanted*" sections. Here you select what you, as a DXer, want to be alerted to. Wanted DX is what I always want to be alerted to, independent of band as I am interested in band fillers as well. I do not alert for WAS award, so I keep it off, though if you still need to obtain it and have an interest in it, this is handy to use. I like to see who uploads to LoTW and eQSL for example as if I do not really need them and they do not do LOTW, I may move on to someone else. These are all up to you. The software is very configurable.

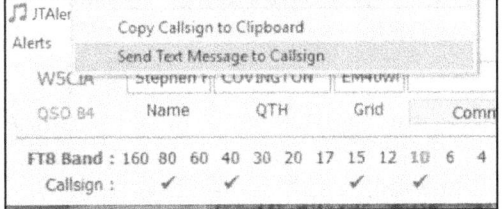

You can also configure how the LoTW and the eQSL indicators look on in your window or turn it off all together if not interested in knowing.

The way I configured my colors is **Blue** for new Prefix, **Green** is CQ, **Red** around the **GREEN box** is a conditional CQ. So, this would be a CQ DX or another region specific CQ such as CQ EU (Europe) or CQ VT (Vermont). **Yellow** is for a needed entity.

One of the features I use a lot is the text message ability, which is basically a private chat window. This is very helpful if you are sharing DX spots with friends, need to ask someone to get off from top of the DX, fix their ALC or send QSL information once the QSO is completed. Start by right clicking on the callsign in the main window, left side. This will bring up a menu, select "*Send Text Message to Callsign*". Now, assuming that they are also a JTAlert user, hopefully they are, you can now communicate. I know some DXers who may send a message to rare DX that they are being copied on the West coast, usually with a signal report. I have seen this work and shortly the DX must have turned their antenna our direction and with even stronger signal reports, worked a bunch of W6 stations. Don't underestimate the power of persuasion. The Support group for this software can be found at hamapps.groups.io/g/Support. Lastly, **please be sure to enable reporting back to** Hamspots. This is found under the web services in the settings of JTAlert.

```
┌─ HamSpots.net - Send Decode Spots ────────────────────────────┐
│ ☑ JTAlert can send new decodes as spots to HamSpots.net for viewing on the WSJT │
│    and Band pages. This is 2 to 3 minutes faster and more reliable than using │
│    PSKReporter.                                                │
└────────────────────────────────────────────────────────────────┘
```

This is something that is very useful to DXers as well as you. Much like PSK reporter it really helps with observing propagation and filtering for what you need.

Alternatives & Other FT8 Software

Alternatives to WSJT-X are available. Though WSJT-X is what is used by most hams and is the original software that started it all. That is the reason I choose to focus on it in this book. This does not mean there is anything wrong with having other options. In fact, I think it is wonderful. The basic concepts presented earlier are the same and can be easily applied to alternatives. Feel free to experiment with others.

One well-known and popular alternative is **JTDX** and it has some nice additional features. Some find the interfaces easier to use on some alternatives, particularly JTDX, and several even claim better decodes. As far as adjustments and personalization are concerned, much of it can be adjusted in both. This includes some visuals too, therefore the main difference is the button layout. This might also come down to what you are used to, screen real estate and esthetics. JTDX is also available for the MacIntosh platform. Do note that at the writing of this book the 16-bit audio version is recommended for most installations under OSX.

When I used JTDX in the past, I even tried to run it in parallel on my radio to see if I get better decodes on the same cycle using the same antenna, receiver and at the same exact time. I got mixed results but the results slightly favored WSJT-X but only after configured as noted earlier. Others have tried this as well; some think it favored JTDX and others agreed with me. You'll be the judge! In my opinion, half the fun in the hobby is experimenting and trying new things. One thing to note is that as a DXer you **want to enable the "*SWL mode*" on JTDX** to enable more possible decodes. This may provide you with a few additional weaker stations in your decodes but also a few possible false decodes. I find this though to be not a serious issue for me. I feel, the more the better, I can decide for myself which are bogus.

JTDX is an excellent piece of software and has many users with good support groups. If nothing else give it a try and use it as an alternative to spice things up. You can download it at sourceforge.net/projects/jtdx or jtdx.sourceforge.io.

There is another a piece of software based on WSJT-X called **MSHV** by Christo Hristov, LZ2HV based in Bulgaria **(LZ)**. His website can be found at lz2hv.org. This software has some other great features worth checking out. Some who have tested it claim this software provides better decodes under some conditions. Are you starting to see a common theme here? I have not tested this myself side by side with others, but I encourage you to test this for yourself. I would be curious to hear your feedback!

MSHV also does something very neat. It allows multi-stream capability as seen in fox and hound mode <u>without</u> using fox and hound mode. Meaning, one does not need to transmit above 1kHz and then have the software move down to below 1kHz to complete the QSO. It also does not require a forced split on transmit. Here it is not needed, and you do not enable F/H when working a station on the other end using MSHV. In fact, when the other end is using this software, no software reconfiguring is required on your end. You go about your business as usual; the other end does the hard work and may respond to multiple callers at once. I have seen as many as 4 streams working a pileup very effectively, theoretically there can be up to 5 streams.

A dead giveaway that the station is using MSHV instead of F/H is that stations might be calling below the DX and getting responses. A second clue is that stations being worked are not moved to a different frequency but rather complete the QSO from one spot. Lastly, F/H can transmit two different messages to 2 different stations on the same audio frequency. MSHV, currently at least, can only transmit to one station per frequency. MSHV tends to make more appearances on the preset FT8 frequencies. Though, I do have mixed feelings about this. Sometimes it can clutter things up.

WSJT-Z is an *"enhanced"* version of WSJT-X which can be found at sourceforge.net/projects/wsjt-z. This software has extra features and layouts and is a bit controversial due to the *"Auto CQ"* feature. I will leave it at that.

WSJT-X S52D version by Iztok Saje, S52D is one which allows for diversity reception. What this means is that you can use two receivers, and this will allow you to use 2 different antennas. The software will combine the results and reportedly can decode about 15% more decodes. If you have the capability to use this, for sure check it out. You can download it and find out more at lea.hamradio.si/~s52d/ft8div.

For those interested in more the just basic signal and grid exchanges, there is now an option. It is called **JS8Call** by Jordan Sherer, KN4CRD and can be found at js8call.com. There is DX here and if you are looking to chat a bit more, this is for you. Chances are, unlike FT8 and FT4, if the DX is using this, they are also looking for more. As in a conversation! So, for those complaining about FT8 being *"soulless"* this one is for you.

Another software I would highly recommend is called **WSJT-X Monitor** for the Android by Feo Tec. This software will only work on the same Wi-Fi network as your computer receiving FT8, but this will allow you to roam around your house and still be in the loop on what your WSJT-X software is decoding. You can find this software in your Google Play Store. I have tested this and loved it!

This software even allows you to set up alerts so you can be alerted when something you need comes, such as a new entity. This is useful during a rare DXpedition without having to be glued to your computer. Kind of like having JTAlert in your pocket! There is also a paid PRO version if you find it useful, support the author and get the extra features. The software also works well on Android tablets and is much easier to view.

RTTY

RTTY (**R**adio **T**ele**Ty**pe, sometimes written as one word) is one of the oldest digital modes, if you do not count CW which is technically digital. In fact, RTTY is so old it used a mechanical system when it was first introduced. Now your radio likely has a built-in sound card for piping the audio to your decoder software and many radios even have built in decoders for RTTY. Before the "*FT8 revolution*", we used to see a lot more

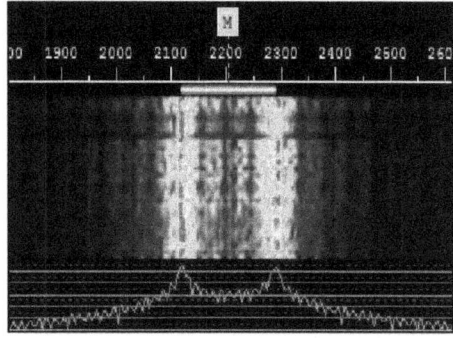

RTTY signals when a DXpedition hit the airwaves. The sad fact is, now many DXpeditions do not even operate RTTY. Kind of a shame, but I do understand it does come down to numbers. One can work many more stations using fox and hound mode via FT8 than RTTY. Not to mention it does not perform nearly as well as FT8 when conditions are not ideal. RTTY is usually operated at 45 baud which is about 60 wpm (**W**ords **P**er **M**inute). A "*baud*" simply refers to the number of times a signal changes its state per second. Meaning, flips from zero to one or the reverse in case of digital. If you are old enough to recall using a modem for the internet, that is exactly what it meant. And yes, even your slowest 300 baud modem back in the day was faster than this mode. This is fine for a simple text exchange.

There is no lower or upper case in RTTY just as in the WSJT-X modes. Using the shift enables the use of more characters but not enough for case changes. There is also no error correction present. RTTY requires more bandwidth than FT8 and many other digital modes as well as more power out for a more reliable decode on the other end. This can be very taxing on your radio and amplifier. **Important to pay attention to duty cycle with RTTY.**

In this mode, frequency shift keying is used. The amateur radio standard mark and space tones are 2125 Hz and 2295 Hz. This creates a 170 Hz shift. Theoretically, others can be used but this is what you will find in use by all hams. Most programs I mention here will have the above preset, therefore you will not have to worry about it much. You may run across signals outside the ham bands and even inside the ham bands as well sometimes, which are also RTTY but with different widths or baud rates. Most software out there can be adjusted to decode some of these if you are interested. Though there are a few which are encrypted. Due to the nature of RTTY shift, drift, and propagation both can play havoc with it though. Additionally, overlapping signals or other types of interference are not nearly as forgiving as with the modes found in WSJT-X. Now you are likely starting to see reasons for its decline in popularity.

You may also encounter references to a RTTY signal being inverted. This just simply means that the mark and space tone are swapped. Likely due to you being in the wrong sideband mode. The signal will look OK but cannot be decoded. Easy to fix!

Be sure to check out RTTY contests as DX does show up and while not as common as, let's say on FT8 nowadays, it is still used. Contests are where one is most likely to see RTTY activity as of late. Besides the usual garden variety DX, I have worked many Caribbean stations during contests as well as along the African coast regularly. This is a good time to grab these entities on RTTY. Seems many visiting operators like to go someplace warm for RTTY contests and operate. And if not, I am sure their amplifiers will keep them warm given how hard RTTY will drive them.

AFSK vs. FSK

If you have already set up RTTY, there is a very good chance you are using AFSK and that is fine, at least for working DX. Maybe not for contesting, according to some, but this is a DX book. I know there are those who will disagree and that is OK. People for some reason have strong feelings for one or the other. **Both get the job done, both do it well.** They are nearly indistinguishable from each other when done right. The most important thing is to get the DX in the log after all.

So, what I am talking about here? There are two ways of sending RTTY. FSK (**F**requency **S**hift **K**eying) and AFSK (**A**udio **F**requency **S**hift **K**eying). AFSK is by default setup on many modern receivers, though most support both. If you are using something like a SignaLink USB Interface, radios built in sound card or your computer's internal sound card, you are using AFSK. The main differences we care about between AFSK vs FSK are follows.

FSK is basically going to be **keying your radio directly.** You will not have to worry about audio levels here and once you set it up, it is pretty much a set it a forget it method. ALC levels are also not something you need to watch as much.

AFSK, hence the name, **uses audio**. The generated signal is sent as audio to the transceiver. AFSK will require you to pay extra attention to the selected sound card as well as audio out levels. Remember the section on ALC? Applies here as well. You also need to make sure all signal processing is off. If you can manage the above, you are likely OK. AFSK stays keyed down for the length of the transmission, unlike FSK. Most digital users will find AFSK easier and faster to setup initially.

FSK requires your radio to work a bit harder, and some think this is unnecessary wear and tear. It does require your radio to switch transmission on and off during a transmit cycle at a very fast rate. Some claim FSK can wear out some radio gear and some do not even fully support it, in part, for this reason. I would take this with a grain of salt. On the flip side, some radios have filters available that might not be available otherwise if using AFSK. This is not the case lately but something to consider. Bottom line, depending on what you are doing, you may decide to set up one versus the other. As I say many times in this book, try both and see what works best for you.

RTTY Software

There are many options here to choose from, likely since RTTY has been around since the dinosaurs, well at least for a long time. It was king of digital modes for quite a while before the "*revolution*" so that certainly helps. While not new by any stretch, perhaps one of the most popular software for RTTY out there is **MMTTY** by Makoto Mori, JE3HHT. Not the easiest to set up and learn initially, I find it to be one of the best out there still. You can download it at hamsoft.ca/pages/mmtty.php.

Hamsoft's MMTTY has not been updated for a while but works fine under Windows 10 and 11 as far as I can tell from my brief testing. Has excellent help files on the website to get you going. MMTTY Is perhaps the best at decoding when conditions are less than optimal and therefore great for DX! It is also a contester favorite, though there is a steep learning curve as I had mentioned. Some contesters even run multiple instances of it with different settings to gain an advantage.

Other recommended software options for RTTY are the **Ham Radio Deluxe** digital suite, specifically the **DM780** component used for digital modes. This is very easy to set up and use, therefore very beginner friendly. I sometimes use this to work DX instead of the MMTTY when I find the signal to be strong enough.

If you are interested in trying out some other great software, **FLDIGI** can be found at w1hkj.com and it stands for "**F**ast **L**ight **DIGI**tal". Is completely free and quite good. And yes, it is both fast and light. You may need to download and install FLRIG as well for radio CAT in addition to configuring your soundcard input and output. Give it a try and see if this fits your style of operating.

There is also a piece of software called **GRITTY** by Afreet Software which you may also want to check out. I often use this as a secondary decoder to check my decodes in sketchy conditions. Easily accomplished by simply directing your receive audio to this secondary application. This can be found at dxatlas.com/gritty.

WinWarbler is an application from **DXLab** and is comparable to DM780. It is also very user friendly and easy to setup. Supports other digital modes as well. Again, you can find it at dxlabsuite.com. **DigiPan** is one of the easiest to use and has a nice user base though it is now considered somewhat outdated. DigiPan stands for **Digi**tal **Pan**oramic Tuning. You can find it at digipan.net.

MixW from Ukraine **(UR)** is another program worth looking at. This one even has some additional features besides a digital module. This one is a paid application but does have a free trial. It also supports a few digital modes which most other software in its class does not. Though likely you will not need these for DX purposes. You can find MixW at mixw.net. **MultipPSK** from France **(F)** is free and is available from f6cte.free.fr. This has an interesting interface which may take some getting used to, but many superb features. I have used this a lot when experimenting with more exotic modes. The last two mentioned are likely your best bet if you are into hunting exotic signals, including those found on VHF and above.

CocoaModem is a great option for the MacIntosh and is also free. This can be obtained from w7ay.net/site/Applications/cocoaModem.

All the above software will need to have the sound card and radio control configured for your specific setup. I will not go into details here as there are endless variations though the good news is, if you have configured and recalled your settings for WSJT-X, most are going to be the same here. The most important things to remember are to watch your audio levels and use what you are comfortable with. I encourage you to try them all. We are all different, so what works for me might not be for you.

Working RTTY and RTTY Pileups

Most RTTY DX will be on the 20m and 40m bands in my experience, though as conditions improve, 15m tends to be the watering hole at times as well as 10m during contests. You may see RTTY on lower bands but due to noise levels there it will not perform as well on those bands except for maybe winter nights.

RTTY is somewhat of an oddity when it some to digital modes. There are times, such as during a contest, when you will respond to a RTTY signal at the same spot where they are calling. You will likely not be the only one calling, much as while using SSB or CW when the DX operator is not split. It will be a battle of amplifiers and timing. The same skills may come in handy here as discussed for those modes, mainly learning by practicing and perfecting your technique.

However, there are times, such as during DXpeditions, when DX will be listening up and sometimes even down. Here you will need to engage your split on the radio just as you would with phone or CW. The DX will be monitoring a segment outside of where they are transmitting and responding in the same spot, always on a fixed frequency. If you work a DX RTTY station working split, you do not ever want to transmit on their frequency just as you would never do that via SSB and CW split operations. The QSO format will be different for DX and for Contests in addition can vary from operator to operator. Here is a DX QSO example with a split operation using Temotu Province, *H40GC DXpedition from 2016.*

CQ CQ H40GC H40GC UP

W6AER W6AER

W6AER 599 H40GC

H40GC 599 W6AER

TU QRZ H40GC UP

In contest QSOs for example, the state and signal report can be used as an exchange. These can vary a bit from contest to contest. The state, for example, is repeated twice to ensure that exchange is good. I have been accidentally logged as CO (Colorado) and CT (Connecticut) before. Here the CQ zone is the exchange, also repeated twice. I am using the 6Y4K contest station in Jamaica as an example:

CQ TEST W6AER W6AER CQ

6Y4K 6Y4K

6Y4K *599 03 03 W6AER*

W6AER 599 08 08 6Y4K

6Y4K *TU W6AER CQ*

Alternatively, PSE K can be used instead of the CQ at the end as well as QRZ. Certain contesters formulate the response by using the same call at the start and the end. For example, me responding to 6Y4K would be "6Y4K *599 08 08* 6Y4K". Some contests exchange a QSO "*serial number*". So always check the rules or watch a few QSOs and likely you can figure out what is going on. The above format is not written in stone, though as they say when in Rome, do as the Romans do. If you hear "?", "NR" or "AGN" the call needs to be repeated as the other party did not get your information. Once you see a "TU", "73" or "QSL" you are good to go and log it.

Unless conditions are great, you may get some garbage decodes in RTTY as there is no error correction. There are some tricks to "*head decode*" through the noise. I obviously do not mean that literally. You may get a report back that looks like "TOO". This was 99% likely meant to be a signal report of 599 but the shift did not occur due to noise. Looking at your standard keyboard layout, QWERTY is 123456 when upshifted, so it is easy to reverse engineer numbers for example. T=5 and 9=O, just like that. Your keyboard here makes a great point of reference. If you need some further reading on RTTY, there is a great website at aa5au.com/rtty I would recommend checking out.

The PSK Modes

PSK is short for **P**hase-**S**hift **K**eying. They are further broken down into binary PSK (2-PSK) and quadrature PSK (4-PSK) or just BPSK and QPSK respectively. The more commonly used versions would include PSK31, 63 and 125 and these increase in data speed with the number going up. **PSK-31 is by far the most popular** of the bunch and

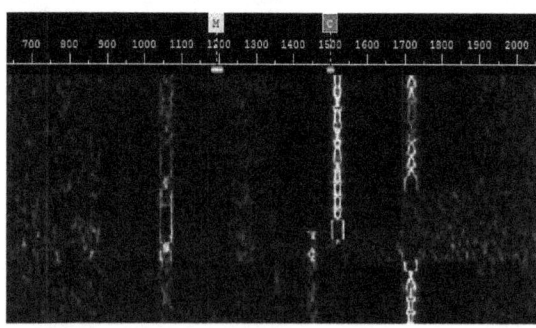

ironically, the slowest. It is used as it is the most likely to work when conditions are less than ideal.

There are some other variants used by hams though even less so such as the QPSK-31, 63 and 125 modes. If there is no Q in front of the mode name, safe to assume it is the more common, binary (2-PSK) version. While there are other variants, you are very unlikely to encounter them in the ham radio world. The main, and lately the only activity for the mode is on the 20m band as pictured above. Specifically, 14.070MHz. Much of the software I had mentioned in The RTTY section will also work for these modes as well as other modes I will be covering from here on.

If you are wondering, the PSK number is not the "*version*" but rather a variant indicating how wide the signal is. In the case of PSK31 is about 31Hz wide. Due to its nature, PSK31 is about 70-80% duty cycle. This means you really need to be careful if you are running your radio at 100W as discussed earlier. Though, this is not nearly as bad as with RTTY and the modes on WJST-X. You will hear your fans going though likely and the shack will certainly get warmer. Ideally you want to run it at 60-80W barefoot. Like how you would operate RTTY without an amplifier.

Do not use CAPS if you can avoid it. Takes longer to send upper case in PSK as they use more bits. Does not matter much when using PSK125 though as it is significantly faster, but in PSK31 you will notice for sure. I frankly do not think it is a big deal but if you ever see someone writing in all lower-case letters, it is likely not due to them being lazy.

If you notice someone ignoring your questions and seem to be sending form a macro, they likely are. In fact, I had someone email me once after a QSO saying they were sorry, they could not respond specifically to what I was asking, likely about power used. They were operating off a tablet in their backyard with only 5 preset macros. They kept the QSO to a bare minimum, and that is fine as well.

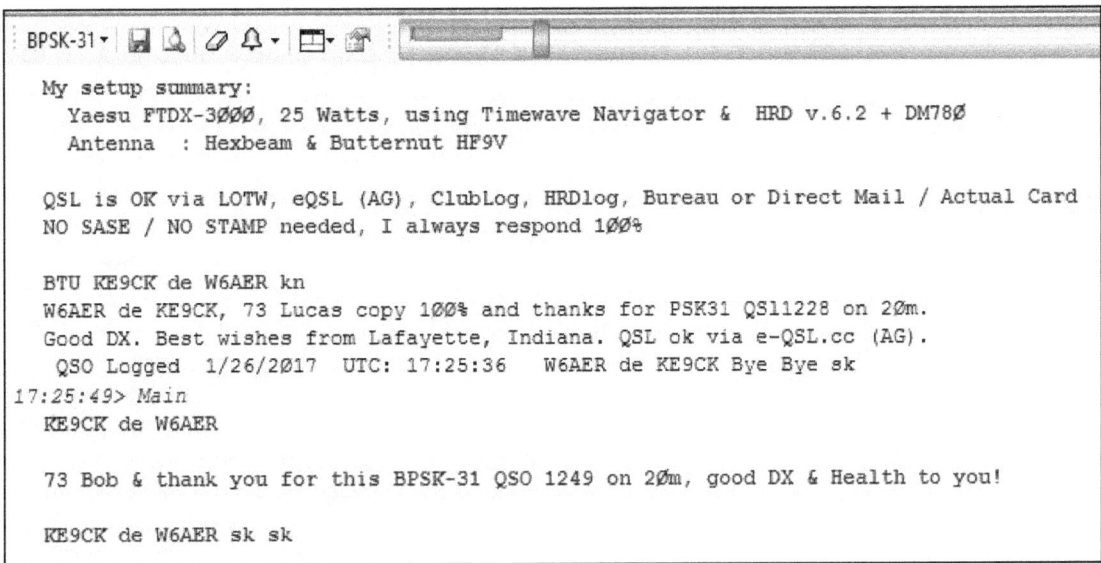

Most hams use macros on PSK modes, kind of like FT8 with the pre-generated messages but significantly longer as seen above. When it is a non-DX contact often the QSOs are longer as well. Above is a PSK-31 QSOs I had with Bob, KE9CK.

I recall working dozens of Japanese **(JA)** stations on 15m in a row on PSK31 before FT8. I was also often working Europe in the mornings via PSK31 from the US West coast during this time. Seems the pattern has mainly shifted to FT8 now. As I had mentioned earlier, my first DX outside the North American continent was using this mode.

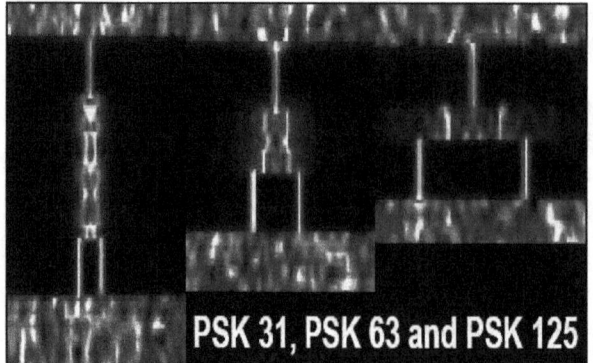
PSK 31, PSK 63 and PSK 125

Pictured here is a comparison of PSK31, 63 and 125. As the signal width increases the transmit time decreases. PSK31 is just under 63Hz wide, PSK63 is twice that at 125Hz and PSK125, again doubles for 250Hz. There are wider modes, but very rarely used and not going to be DX friendly or even ham friendly.

PSK31 can be used as a K2K mode (**K**eyboard **to K**eyboard) assuming you type faster than the text is transmitted. Most of us do. PSK125 however, would require superhuman speed. Why do we not all use PSK125 then? Much wider mode and since PSK31 does not error correct like FT8, you will get some corrupted text or "*garbage*" here and there. You can even see some of this the example I used earlier. Due to this the callsigns are usually sent 2-3 times in a row, as in RTTY, before main text and depending on propagation, QSOs may be kept short. PSK63 is a compromise of accuracy and speed.

Occasionally you still see some DXpeditions use RTTY and PSK31 in addition to FT8 but not so much anymore. This is too bad as it would give a new level of excitement to some activations and may encourage others to branch out a bit. Might even bring back some folks who love these modes and may introduce others to them. There are several contests during the year though when I do see some DX (though not rare DX) show up and work a few stations in these modes. Always nice to see and I usually try and be there too.

Ham Radio Deluxe, covered earlier, even includes a **SuperBrowser** which allows you to simultaneously decode everything in a 3KHz slice on PSK-31. Much like what WSJT-X does now. Very neat for finding a wanted contact.

SSTV

If you are already set up for digital, this is worth trying. SSTV is often listed as digital, though it is technically a voice mode. The main operating frequency is even located in the US voice segment. Oh, and it is also not really TV or even moving. Confused yet? It is just images being exchanged. There are some digital SSTV modes out there too, though 95% heard is still analog and I suspect it may stay that way for a while. **The main watering hole for SSTV is the 20m band at 14.230MHz**. The bandwidth used is about 3kHz, same as SSB and is operated in USB mode. Please try and stay clear of this frequency if you are using phone mode. I see a lot of QRM here at times and with much of the rest of the band underutilized, I could never understand why folks do not spread out more. I can see during contests folks parking here...sort of...but let's be courteous to other operators, modes and let's share the bands and respect all modes and operators.

You will not work any rare DX likely via SSTV mode, however, I have worked many entities in this mode across the world. I use Ham Radio Deluxe's **DM870** and **MMSSTV** interchangeably for this mode. If MMSSTV name looks familiar, it should. It is the same author who brought you MMTTY, Makoto Mori, JE3HHT. You can get MMSSTV from hamsoft.ca/pages/mmsstv.php. It also includes great help files to get you going. You may want to edit some of the template images it comes with, so they are personalized. Does not have to be anything fancy. Use large font and keep it simple and high contrast. Do keep in mind that it may not be received crystal clearly on the other end.

Afreet Software also has SSTV tools which can be rather handy if you get serious about this mode at dxatlas.com/sstvtools. If you only want to receive the above, **RX-SSTV** from ON6MU and can be downloaded from qsl.net/on6mu/rxsstv.htm.

There are also digital variants of this mode as I had mentioned. One software used is **Easypal** by Erik Sundstrup, VK4AES (SK 2015). It has not been updated in a while but if you want to test it out, you can obtain it from vk3evl.com or at g0hwc.com. Digital SSTV (DRM) is much more likely to get your callsign right as it is sent continuously versus analog SSTV where one must receive the image and hope it is legible enough. The use of VOX trigger on external audio interfaces is not recommended for digital SSTV as many have encountered issues. Built in sound cards within transceivers should be fine.

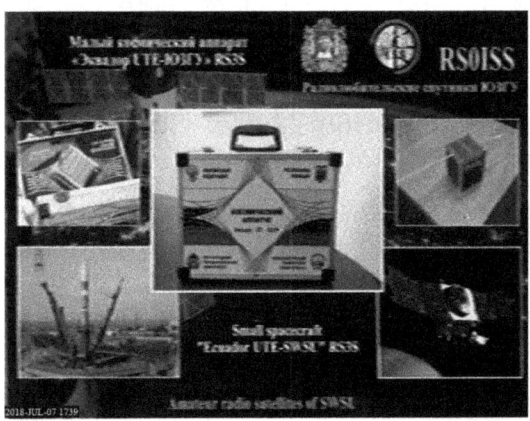

The other digital SSTV software is **KG-STV** developed by JJ0OBZ and can be obtained from www2.plala.or.jp/hikokibiyori/soft/kgstv thought the site is in Japanese. You can get started with these instructions at sgcderek.github.io/blog/kgstv-reception.html. For the Linux users out there, **QSSTV** is a great open-source option and great instructions on how to go about it are at charlesreid1.com/wiki/Qsstv and for the MacIntosh users, software from Black Cat Systems is still your best option at blackcatsystems.com.

You can even receive DX from space, as seen here, though it will not count for DXCC. One popular example is the somewhat regular SSTV broadcasts from the International Space Station (ISS) which operates under the callsigns NA1SS (USA), RS0ISS (Russia) as well as some others when there are astronauts up there from other countries. I only mention this as this is also technically DX, although generally about 250mi (400km) only from earth when directly overhead. It can be surprisingly easy to receive even though it is moving at 4.76mi (7.66km) per second. Does not really count for much other than the fun you have receiving it. Though if you think about it, you are receiving a signal on the 2m band several hundred miles away. It is not going to be a high-resolution image but gets the job done. You can read more about this on my website if interested at w6aer.com/international-space-station-iss-ham-radio-contact/ or alternatively you can go to ariss.org/contact-the-iss.html.

Other Digital Modes

We have now touched on FT8, FT4, PSK modes, SSTV and RTTY. But wait, there is more! Before WSJT-X was used for the "*FTs*" it was used for the "*JTs*" as in JT65 and JT9 mainly. JT9 was very popular a few years back, mainly on the lower bands and is still used on the 630m and 2200m band regularly today. I still see some JT65 and JT9

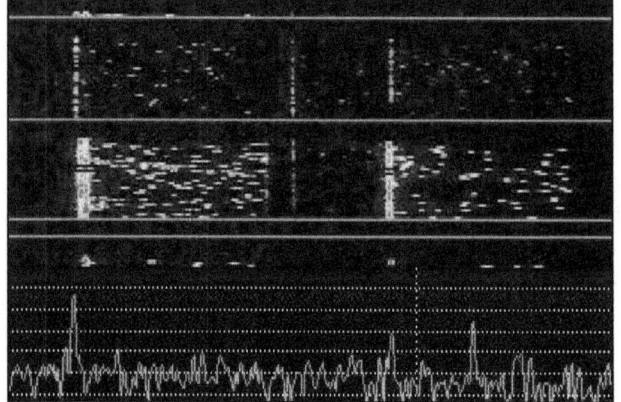

activity occasionally on 160m as these modes do very well with high noise levels. In fact, much better than FT8 as I had pointed out earlier.

These traditionally ran on top of the minute and transmitted for about 50 seconds, though on the 630/2200 bands, longer cycles are sometimes used.

These can be up to several minutes per turn. JT65, pictured here, can still be found but more so via EME (Earth Moon Earth) communications as well as occasionally on the lower, more noisy bands.

Other less used digital modes would include MSK144 (6m Meteor Scatter), JT4, Domino and Olivia, Contestia, MFSK, Hell, Thor, Throb among others.

These are not used much for DX anymore, however this was not the case before the JT and FT modes were introduced and FT8 adaptation skyrocketed. There were a lot more hams experimenting with other digital modes, and it seems to be making a comeback. I have used all the above at least a few times and have made at least a few DX contacts on most of these modes.

They all have their plusses and minuses; some perform better under certain conditions than others. Knowing when to use them is key. Olivia, for example, does very well when conditions are less than ideal, hence nicknamed the "*magic mode*". You can read more about Olivia at oliviamode.com. Olivia will decode much better than PSK31 and RTTY, though not as well as FT8. The tradeoff is speed. Olivia tends to be on the slower end, though there are 40 variations of this mode. So how does one know what variant is used?

Enter **Reed-Solomon Identification. RSID for short** is something you may need to get familiar with for lesser used digital modes, especially if there are variations within the mode, such as speed and signal width. It is used to identify the mode before transmission. Most software will have a pop-up to alert you to this and by clicking on it, you will switch to the correct mode and frequency to respond. Pretty easy from there on. For the modes supported in WSJT-X, RTTY and PSK31, this is not needed, in fact for some might not even be an option. Be sure to enable this if you are going to try and use the more obscure modes, especially for DX.

Contestia is a mode which is very similar to Olivia. In fact, they are related. Contestia is notably faster the Olivia, though not as popular, at least from what I have seen. MFSK dates back a while and you will still find it on the air here and there. This mode is closely related to RTTY, though will work more reliably. MFSK16 will also allow for image transmission. This caught me by surprise once when a fellow hams photo popped up. These will be tiny but relatively good quality. Fun mode!

The software I have covered in the digital software section will also get you going in experimenting using these other modes, as most if not all support these. You can spend months playing around with other digital modes and I know this since I have done just that. Finding a QSO now-a-days though might be a bit harder but a post to the DX cluster or simply calling CQ might surprise you. Additionally, most of these other digital modes have contests. I have been pleasantly surprised at who showed up in the past. **If everyone is waiting for someone else to call CQ there will be no QSOs.** Try creating a schedule with a friend, especially if it is a location, you are still after for an award. One of my goals soon is to get WAS endorsements for some of the rarer modes. If interested in the same, reach out to me.

If you would like to see what digital modes look like on the waterfall and/or want to hear them, look no further. Check out w1hkj.com/modes. I have spent hours here trying to identify "*mystery*" signals. This is an amazing collection with descriptions and technical data. If you would rather have something on your desktop, check out **Artemis** from Aresvalley. The Radio Signal Recognition Manual. Get this exceptional software at aresvalley.com. Personally, I love to mix it up a little. I would recommend you at least give some of the other modes a try. You might be surprised what DX you may dig up.

What is a "LID" and how not to be One

I think this is a good way to close out the modes section as this can be applied to all three modes we have covered. You have heard me use the term already in this book but now time to elaborate a bit.

```
17   0.5   445   ~   K6SAT HK6W -10
------------------------------------ 40m
 2   0.6   446   ~   KP4DVM KG5RJ EM12
11   0.2  2231   ~   R U STUPID?
13   0.5  2493   ~   CQ KL7J BP40
------------------------------------ 40m
-2   0.5  1000   ~   LW6EQG N5SKT EM12
 3   0.5  1747   ~   CQ W9RF EM57
-4   0.3   498   ~   CQ PY5EG GG54
```

Let's start with the basics. You do NOT want to be labeled as one. A "*lid*" refers to a bad operator in ham circles. As I have already mentioned, easy to get on this list but hard to get off. **We all make mistakes; lids are those that keep repeating them and don't take advice from others.**

If you use digital modes long enough, you will see some very odd things occur. These can also be observed especially on FT8 fox and hound mode as many are not comfortable using this mode or forget to switch to it when needed. Great example above from someone who clearly is annoyed at another ham doing something wrong. Likely repeatedly and it happens more than you think. A "*lid*" label might stick!

I have already touched on this, but you never want to send out a signal, no matter how modest, over someone else, especially the DX. This happens when you tune up tube amps (dummy loads are widely available) or adjust antenna tuners even. Move off frequency... always! Or upgrade your gear to solid state if you find this difficult. Be courteous to other hams, please. You would not like being on the receiving end of a super loud tone causing your ears to ring either while wearing headphones trying to listen to a weak DX signal.

If you follow the procedures I have outlined in the past chapters, you will be off to a great start. Will you make mistakes? Of course! We are all human. The key is learning from them and trying not to repeat them. By reading this book, I suspect you are trying to learn and become not just a better DXer but a better operator. There is a very good chance you will never be labeled a *"lid"*.

CHAPTER 10: On the Air – Navigating DX Challenges

In this section, I will give a basic overview of propagation basics, methods, trends as well as anything to help you figure out the "*when and where*" as it relates to DX. There are certainly things I have discovered as a ham which I was not aware of; things nobody told me, though I wish they had. I truly believe this section is the one that separates a ham radio operator from a true DXer.

Some hams are just focused on getting their DXCC, others the 5 band DXCC which only involves the 10m, 15m, 20m, 40m and 80m bands. Some may only work one mode, such as phone. This will limit you greatly! **Work all bands and all modes!**

For example, I have worked many ATNOs on the WARC bands (12m, 17m and 30m) initially. This is in part since many do not even have antennas for these bands, sometimes this means less competition. On the flip side, possibly less DX as well, but not something I generally found to be true.

The above observation is especially true for some serious contesters whose focus is restricted to the contest bands. There are no contests on the WARC or 60m bands. Others may not bother as they are too focused on the 5 band DXCC so they can get that on their walls. Regardless, this means less DXers to compete with and the DX also possibly looking for you on lesser used bands and modes. They may also want endorsements. They may also be after an award other than DXCC. Be ready!

When conditions are less than ideal, especially if very poor, digital is your ticket to DX. However, when the sun is cooperating, many are going to try and go back to operating CW and phone as well. This includes the DX! Some are already starting to suffer from FT8 burnout. Sometimes I am even getting to that point.

By not working CW, you may miss out on some DX where the station does not have the ability to work FT8 or perhaps, does not like the mode. Possibly the operator is technologically challenged. There are those who have bad mouthed FT8 and even all digital modes. Though I find these are usually the people who can't get it to work. Phone only, while I love it, will limit you as well. This is one of the least efficient ways to make a QSO. Though you may find some DX stations are only set up for phone, and will not do CW or digital, so be ready. No other way to work them!

So how do you get up there in numbers? To quote my friend Rusty, W6OAT "*The secret is to outlive everyone.*" There is certainly truth to this. But you can't just stand by and wait. You must keep your count moving up in numbers. When entities you need come on the air, not to state the obvious, but make sure you work them. You never know when you will get a chance again. This is especially true once you get up to over 200-250. You might be waiting for activations at that point.

I have already talked about upgrading your license class to the highest you can. This will help you to be able to work anywhere ham radio is permitted in your country. Many DX stations will be operating on the lower portions of the band when working CW and many times in the US Extra Class portion of the SSB allocation. These are great places to start looking for contacts. Nine out ten times, this is where I stumbled upon great DX. With digital, it is much easier but be aware that many DX operators will not use the standard FT4/8 frequencies as we have covered. Watch your DX cluster, your waterfall, your repeater and tune around.

The Solar Cycle

If you are reading this, you have likely lived through at least one of these solar cycles. They are generally considered as being around 11 years in length. Of course, this is not an exact science. Some say 9-13 and if you look at the data, some will argue that this is more accurate. These cycles have very distinct highs and lows, with a gradual increase and decrease. This rise and fall occur slowly over many years.

Technically you could argue that the cycle is 22 years. Every 11 years the suns poles switch polarity. South becomes North and North becomes South. This reoccurs continually. For DX purposes, this is irrelevant though, but good to know.

So, what is it and why is it important? It has to do with sunspot activity of the sun. The number of sunspots increase during the peak and decrease, or even altogether disappear at the bottom of a cycle. This affects HF propagation dramatically. As I am writing this book, cycle 25 is starting up and we have already noticed an improvement in HF conditions. So... I better hurry up and finish this book. DX is waiting!

Sometimes cycles can start up more slowly than expected and wind down faster than earlier cycles. The reverse can also be true. No two are alike and some are better, at times much better, than others. The past 2 cycles were not all that impressive compared to what radio amateurs experienced a few decades ago when there were close to 24/7 openings for a few years near the peak. Let us cross our fingers and hope for another one like it in our lifetimes. I do suspect that our level of accuracy in counting sunspots has increased since the 1st recorded cycle, now well over 250 years ago. Our ability to predict the size of upcoming ones accurately though, not so much. But there is a lot more to this.

While I will give you what you need to know for DXing purposes, I will not dive deep into propagation as you really do not need to have a deep understanding of it. And frankly, I am also nowhere near an expert on the topic either. It is complicated and the more I learn about it, the more questions I have. I also realize just how much we do not actually know about propagation yet. This section of the book is not intended to be a replacement for a book on propagation. There are many great books written on the topic if you decide to explore it deeper as I did. I will try and keep science to a minimum. Knowing some basics about propagation, however, will really improve your chances of making great DX contacts and covering your walls with awards.

Day vs. Night

The bands which tend to generally perform **better at night are the lower bands**. Basically, the 40m band and below. On the flip side, the **higher bands are going to be daytime bands**. But not always! I have been on 10 and 12m at midnight working DX. So, the key word is "*generally*". 40m, 30m and 20m are bands you can usually find someone on here on the US West coast, day, or night. Sometimes even 17m can be open into the late evening to Europe even when the sunspot counts are low. So, you just never know! Use these as guidelines, there are always exceptions within reason. For example, you are not going to work Austria **(OE)** on 80m or 160m at noon from the West coast or work India **(VU)** on 10m at midnight.

Enter the Ionospheric Layer Alphabet

If you have even a basic ham radio license, chances are you recall this from your study materials. Consider this a highly condensed refresher. **Radio signals are either absorbed or reflected by the ionosphere**. When an element loses an electron, sometimes more than one, it is ionized. These radiated particles collect in the ionosphere at various layers. This is a grossly oversimplified explanation, but for DXing that is all you really need to know.

There are up to 4 layers of ionosphere in the earth's atmosphere, D, E F1 and F2. During the night technically there are 2, E and F, which is a combined F1 and F2 layers.

The **D Layer** is the lowest and therefore closest to us. It is sometimes called the "*daytime layer*" and it sits at around 45-55mi (70-90km). This only exists during the day and disappears during the night. D Layer can act like a sponge for some wavelengths or frequencies if you will. I sometimes jokingly refer to this as the "*death*" layer as it can really kill signals. D layer is responsible for killing signals from 60-160m at night, some would argue even 40m. I would tend to somewhat agree. Though, this is not as consistent. Alternatively, some remember it as D for dense, as it is also our densest layer.

The **E Layer** is, as you guessed it, somewhat in the middle. This layer sits at around 55-75mi (90-120km) though various sources seem to disagree on this a bit. As you can see, it is not a perfect science. This layer gets the most ionized when the sun's rays come in at a higher angle. This may happen more often during the summer solstice.

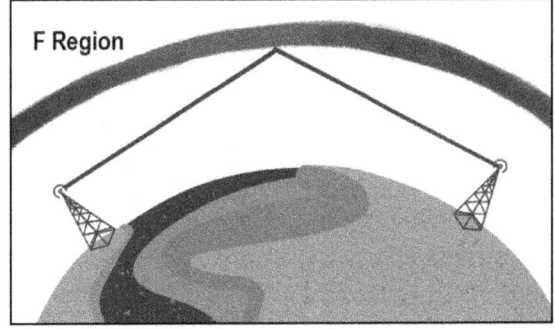

The **F Layer** is the furthest from us. There are two layers during the day, F1 and F2. At night, when the sun stops barraging them, they merge back into one single layer again. This repeats very predictably. There is always going to be some version of the F layer around. These two F layers are found at about 80-125mi (130-200km) and 185-250mi (300-400km) respectively. The **F layer is primarily responsible for ham radio propagation**. The illustration here shows a single skip (obviously not to scale), meaning the signal refracted once off the ionosphere. Though, multiple hops are not only possible but common. The above is an illustration of skywave propagation. Each F layer hop is estimated to be between 2000-2500 mi (3200-4000km).

The higher the frequency, the more likely the radio wave will go through the ionospheric layers (See MUF later in the book).

For a second think of radio waves as just a "*waves*". Light waves bounce, sound waves bounce and guess what….so do radio waves! These are all in the same spectrum, just different ends of it. Radios waves fall between audio waves, lower frequency and light waves, higher frequency. Making some of the physics somewhat predictable.

Most HF radio waves will be bounced off the ionosphere if conditions are just right. This is the reason that we use VHF and UHF frequencies to talk to ham radio satellites as they generally penetrate the ionosphere. But this is not always the case! There are times when ionization is so great there even VHF can get reflected. Alternatively, there are times that all HF communications are limited to short distances. That is propagation for you!

Propagation Reports Demystified

Sometimes you will hear these referred to as the "*Solar Index*". With the aid of modern science, we tend to be able to predict propagation a bit better than we did many years ago. Predictions have been at times dead wrong though, again not a perfect science. But then again, what is? While it helps to know what might be coming, you often get surprises as well as some possible disappointments. If you pay attention to (non-solar) weather reports for your local area, you know that sometimes they predict rain with a percentage chance, or a certain temperature is expected. Occasionally these predictions are wrong. No different here. But we still pay attention to both as they are very valuable guidelines.

One thing to note is that the sun does rotate around its own axis, and it is about a 27 ½ -day cycle. Though this is not a completely accurate number due to its gaseous body, it is beyond the scope of this book. The point is cycles will repeat roughly every 4 weeks during peak times. Therefore, if you missed a good opening, it may just be repeated shortly.

Solar Flux Index

This is usually expressed as SFI for short and is sometimes just called "*Solar Flux*". This is measured using Solar Flux Units or SFUs. And no, it has nothing to do with university. The numbers generally range from the mid 60's (Though could be lower) to 300. For DX what you need to know is that when it hits 125-150 for a few days is when you can start seeing improvements in propagation. Solar flux reports will not be the same from different parts of the globe, in fact can vary quite a bit. The main source for data is currently at the Dominion Radio Observatory near Penticton British Columbia, Canada. Here they measure using 10.7cm or 2.8GHz, which is just above where your kitchen microwave and some Wi-Fi operates, which is around 2.4GHz. There are historical reasons for this. If they change the frequencies now, it would be hard to compare with past data.

Use this information together with sunspot numbers (next topic) to get a better feel for possible propagation. **<u>Higher</u> numbers here are generally good!** Why generally? Solar activity can intensify auroras at the poles, sometimes making the ability of signals to pass through nearly impossible.

Sunspot Count and the Wolf Number

To continue diving deeper into sunspots, this number you will find expressed as SSN usually, but sometimes just as SN. **Higher Sunspot numbers will improve DX propagation on most higher HF bands**. Of course, during the bottom of the solar cycle, you may have many days with zero sunspots. This does not mean you cannot make contacts. You still might be pleasantly surprised at times. **<u>Higher</u> numbers here are generally good!**

These visible sunspots are actually cooler than the sun surface and therefore appear black to us. This is actually what makes them visible. They push out a lot of UV (**U**ltra **V**iolet) rays. This UV radiation is what ionizes the ionosphere as we had discussed earlier.

The Solar Flux and the Sunspot Count together are usually referred to as the Solar Indices. (plural)

This is where the wolf number comes in. This number is indicated by the letter "*R*" if you ever run across it. This indicates the average for a 12-month period and is usually used to analyze a solar cycle. Not something that is useful day to day for DXing but gives you an idea of where we might be in the cycle.

Sunspots: The High, The Low and the Ugly

When times are "*good*" meaning high sunspot count and high SFI, you will find everyone, their neighbor, and their dog (if licensed) on the 10-15m bands. Also, 20m might be open all night and folks will hang out here.

Which is great and fun but sometimes the other bands get forgotten. And this is where I monitor with a second receiver. While everyone is trying to get their fill on those bands, sometimes semi-rare DX moves to a less busy band and calls CQ. I have seen this a lot! This might be because they hate out-of-control pileups, or they need fillers themselves. Might take a few minutes for the crowds to show up, even after a DX cluster post. At times I am pretty sure I was one of first to work some lonely DX and post them on the cluster. It is worth keeping your eyes open where others are not. Don't forget the 6m band! No need to always follow the rest of the herd. Think outside the box!

Things are not always ideal in the ham world. Far from it! So, what can you do when the sun is not on your side and sunspots are low? If you have been through a cycle as a ham, you know that towards the end of the cycle, as propagation declines is when you can get great deals on ham radio gear, especially used items. So, **shopping is**

one thing to do. But real DXers do not hibernate during the lows. We shop for gear and move to the lower bands.

Some hams forget that when the upper bands are weak, often there is less noise on the lower bands and even 80/160m can be a great watering hole. I have worked many ATNOs (**A**ll **T**ime **N**ew **O**nes) during the bottom of the cycle on 80 and 160 meters. Lately 60m has been increasing in activity and this band is also great at low sunspot counts. So, what do you do when the spots are low? Business as usual, with slight operating adjustments. Possibly not as many QSOs, but they are there if you know where to look. As I always say when someone says that it is not even worth turning the radio on, **there is always something to work!**

Additionally, there is always satellite DX work which is independent of the sun. Exception being a few satellites which have dead batteries and only work during the day. Even with LEO (**L**ow **E**arth **O**rbit) satellites, you can work neighboring countries. These tend to pass most places on earth several times a day. The WAS award is also available for working satellites. Something worth checking out! See chapter 12 for operating awards.

When conditions are not so good, it is also a good time to get your HF WAS finished or add WAS band endorsements. There will be US hams who may also not be seeing much DX and are happy to have a QSO with anyone. A good time to get on the more exotic digital modes as well. I often do this when conditions are not the greatest but want to do something other than FT8. Of course, you can always dust off the microphone or the paddle and just call CQ, see what happens. You just never know!

And lastly things can get downright ugly, in fact there are times where this can lead to a complete HF blackout. Which leads us to…

Geomagnetic Storms

Solar or geomagnetic storms can really ruin your day or even days. Solar flares are basically eruptions on the surface of the sun which have increased particle and plasma count. A little can be good, a lot can increase the K index. That can be bad for radio. The above can lead to a CME (**C**oronal **M**ass **E**jection), which in turn releases particles via what is called solar wind. These can even interfere with GPS navigation at times. Afterall, GPS uses radio waves too. The thing to remember is **geomagnetic disturbances would be indicated by high A and K numbers.**

Get to Know your K and A

Commonly referred to as the "*Magnetic Activity*" or Geomagnetic Indices (plural). The K Index numbers are derived from measurements taken and reported every three hours from several locations on the planet. These are then averaged and reported. Sometimes you see these referred to as K or Kp. You may also see A expressed as Ap. P is short for planetary. This is measured similarly to Decibels (though not quite logarithmic) where a single number move is larger on the scale.

K is indicated in levels of 0-9 and A is 0 to 400.

The A index is the current time range, and the K index is for the whole day, so essentially for a 24-hour period. Basically, **lower numbers here are generally good! This is the opposite of the SFI and SSN.** What you really want is just enough but not too much or RF wave absorption will occur. Having high SFI at the same time as high A and K can work against HF propagation.

Maximum and Minimum Usable Frequencies

The **M**aximum **U**sable **F**requency is also known as MUF for short is in refence to the highest (being the maximum) frequency which will support propagation to a given location or path. There are times when this can be rather low and there are times when 10m and up are open. This changes and it is good to keep your eyes on it. When the MUF is favorable for the higher bands, there is a good chance you will experience good propagation. Basically, anything above the MUF frequency will not get refracted by the ionosphere. Going back the layers, these are essentially determined by the D and the F layer. The number you see may not be as good as it looks though as there are other factors involved. I usually look for reliable communication a little below the number given. The number by the way is in reference to the highest usable frequency. So, if the number is 25 for example, I would assume the 15m band at 21MHz will be OK and possibly the 12m band at 24MHz though not likely as reliable.

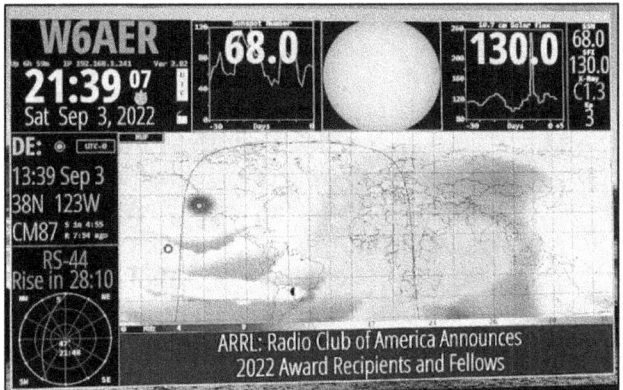

Above is my HamClock, mentioned earlier giving me basic propagation data such as SFI, Sunspots as well as a MUF map. This can be configured to your taste. The map has a color overlay giving me a nice visual as to how the MUF is behaving. As you can see here, some parts of the world are open to some great DX opportunities. As a bonus I even have alerts set for upcoming ham satellite passes, in this case RS-44.

High K and A numbers usually means the MUF is coming down.

On the flip side, **LUF** is the **L**owest **U**sable **F**requency. You will not hear about LUF as much though it can be important especially for those trying to work the lower bands. During daytime the D layer absorbs RF therefore frequencies below the LUF will be very hard to make contacts on. **Your best chance for DX contacts is between the LUF and the MUF.** In the winter your daily maximum is likely midday but, in the summer, it is in the mid- afternoon. However, this does not necessarily correlate to when propagation is going to be good for your location, though helps to know.

Sporadic E

You will see a lot more of this during the summer than any other season and seems most present around the summer solstice. You will likely encounter sporadic E mainly from April to August with some possibility of it during winter, though usually much weaker. Keep in mind this may not be the same for the Southern hemisphere. Some hams report that before lunch and late afternoons are when Es peaks. I have worked many stations early afternoon as well as after sunset. You just never know with sporadic E. This is likely the least understood propagation method.

Sporadic E is used to describe the ionization which forms in the E layer. When formed this helps to refract signals. This is something to look for mainly when operating on the 6m band however, can occur on nearby bands, specifically on 10m and on 2m. For those blessed with the 4m band, I understand it happens there as well. Propagation via Sporadic E can have several hops and are shorter than those via F layer propagation. Some studies have tied this to thunderstorms forming and cloud activity.

Trans Equatorial Propagation

Sometimes you hear this referred to as just TEP. TEP is something that is important to understand if you are serious about 6m. As the name suggests, it involves the equator though the not the geographical one but rather the magnetic one which is slightly different. This propagation phenomenon is mostly present in the spring and the fall. Generally, kicks in approaching early dinner time when the sun has enough time to ionize. This is something which, depending on where you live, may not apply to you. I personally do not think I have ever experienced this.

The Grayline Phenomenon

Sometimes you see this spelled as "grey line" or "greyline" though this is mainly British English. Also, sometimes as "gray line" as two words. These are all the same things, just rather inconsistent. I will go with "grayline" here like most other hams, even though my spell check hates it.

During what is called grayline, there is a notable enhancement in signal strength during twilight. You will probably notice this more on 40m and below and occurs at sunrise and sunset at either your QTH and/or the target DX locale. When both are in grayline simultaneously, there might be 15-20 minutes of pure DX magic, though on 160m for example it can be as short as 5 minutes or less. Some have also reported great success at grayline on the higher bands as well, so do not rule that out.

Sometimes this is referred to as the "*terminator*" line, nothing to do with the movie. This is the area between the dark and the light regions. The in-between region if you will. Northwest and Southwest directions are going to be your best bet for enhanced propagation. A good way to track grayline, among other things related to ham radio using the Geochron if you want a quick visual, more info at geochron.com. You can also use Simon's world map to accomplish this at g4eli.com/world-map. PSK Reporter, which I had already mentioned is also great for viewing grayline.

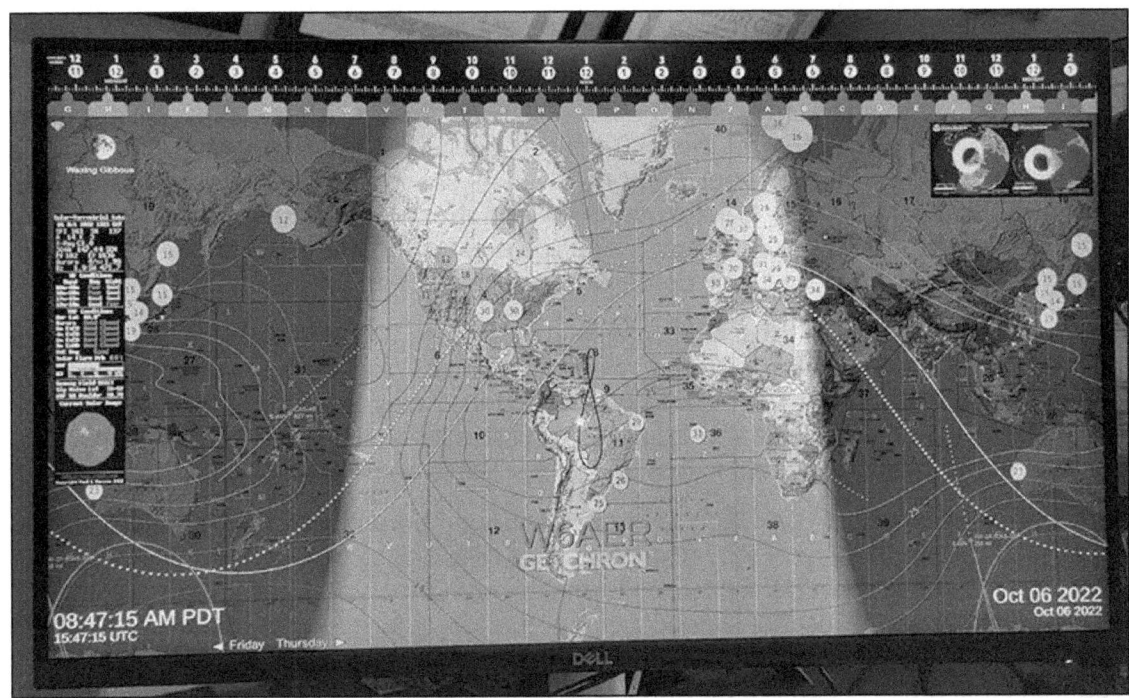

In my experience, using a grayline map is pretty much a must from the West coast to work certain entities on the challenging lower bands. I always keep a grayline map running and you will usually find me on the radio during local sunrises (unless I decide to sleep in on the weekends) and sunsets. Some radio operators live and die by the grayline. You will find them glued to their radios precisely twice a day. Of course, I can think of a few who seem surgically attached to their chairs. If you did not get the Vitamin DX Publishing company tag line, "*Take Twice a Day at Grayline*", now you hopefully do.

The basic concept is the quick dissipation of the D layer along the grayline which is no longer absorbing signals and now taking advantage of the F layer which is now refracting. In some propagation reports, widgets, and other ham related pages you may see reference to **SR and SS**. These are nothing technical and once you know, obvious. These are short for **Sunrise and Sunset**. Important to know if you are going to try and make contacts via grayline. Therefore, they are often next to SN and AFI, as in Sunspot Number and Solar Flux Index.

Long path, Short Path and Weird Path?

There is a good chance that most of your QSOs will be short path, though it depends on when you like to operate and on what bands. Of course, if you are using a vertical then this is not an issue as you are receiving from all directions. Some would say equally badly, but I do not think that is true at all. There are other factors involved.

Long path may work better at times for multiple reasons. These could be the time of day at a location, ionospheric behavior among other factors. At times it is the only thing that will work for you. For example, if you are trying to work a location via a dark path, meaning nighttime, long path might be the way to avoid daylight and D layer absorption.

The opposite is also true. A grayline map here is also very handy as it will show who is in daylight and dark.

It is a good idea to sometimes turn the beam 180 degrees to check, or if you are lucky enough to own a SteppIR, hit the 180-degree button so the elements reverse for a quick check. Saves wear on your rotator as well as saves you several minutes of time. I have found this to be very helpful, especially on the higher bands. There are times I get surprises! The best way to learn to operate and know when to operate long path is simply from experience. This will come with time, likely sooner than you expect.

There are even times that the path makes zero sense. This happens on the 6m band as well, but sometimes for different reasons. For HF, meaning a station's signal is coming in strong and yet appears to be curving in a way that one would not expect. Neither long or short path but a weird path. My first instinct at times like this is to assume it is a pirate station, but often it is just an oddity. There could be many factors involved.

An example of this could be when signals experience bending around the poles due to auroras. If this happens you may also hear what they call "*auroral flutter*" which sounds very distinct not to mention a bit hard to make out. Once you hear it, you will know it for sure next time. If you are working digital modes, you may not even realize when you had encountered this. On CW and SSB you can't miss it. As a reminder, even if you think it is a pirate station, as I had mentioned before, just work it and sort out the details later. You just never know!

Sporadic E signals may not come in the direction you expect it from. Es clouds are irregularly shaped and are not necessarily in the direct path. Therefore, you may find on the 6m band that you are pointing one way and are hearing something unexpected. Then when you turn towards it for a better lock, it gets weaker. You have just experienced a weird path! Sometimes you may see these **abbreviated as LP and SP**, same thing. No WP, though sometimes feels like there should be one!

Hops are not Just for Beer

"*Hops*" is the term sometimes used instead of "*skip*" when referring to skywave propagation, often used interchangeably. To oversimplify it, as in how many times a signal bounces around before reaching the other station. This can be just once (or none in case of ground wave propagation) or 3-4 even in some cases. This is when you get to the really "*good stuff*" as it is referred to by DXers, which you may not always get the chance to work. For example, when I see great propagation from the West coast of the US to parts of the middle East on the high bands, I know there are at least 2 hops involved. Therefore, when you hear a ham say they are "*working skip*" this is what they are talking about.

There are other factors of course, such as radiation angles. Skip via the E layer will be about half the distance of that of the F layer. Meaning E layer will need twice as many hops which is part of the reason 6m DX is such a challenge at times. Remember that the F layer is much higher than the E layer! Incidence angle is what determines your skip distance. Generally, the lower the better!

Ground Wave Propagation

Ionospheric propagation is what we usually work DX with. Tropospheric propagation is also possible, but there is also a third often forgotten one, ground wave. Ground waves involve no skip or reflection of any kind but rather the use of the earth's surface to do their radio wave magic. Vertical antennas will work very well for ground wave propagation. Ther tend to work over shorter distances and range from 40m to 160m bands, mainly the top band, 160m. Range is a few hundred miles at the most under ideal conditions, such as salt water at night, but generally about half that on a good day. And about a fraction of that on land.

Ground wave is a combination of a few propagation methods. It is sometimes used interchangeably with surface wave propagation but incorrectly. Surface waves are one of the methods by which ground wave travels.

So why know about this for DX? For example, if you are in the Caribbean where you have islands very close to each other, some locations will be in your dead zone or skip zone as some call it. You might be able to make contact via ground wave propagation. Ground wave propagations' biggest enemy is attenuation, therefore limiting the travel distance. This will work even during bad solar conditions as it does not involve the ionosphere.

NVIS

This is like ground wave propagation in some ways and is often confused with it. Both are used for shorter distances but work very differently. NVIS stands for **N**ear **V**ertical **I**ncidence **S**kywave. Signals can travel via this method around 3-500mi (5-800 km) and involves a high takeoff angle versus ground. This method works very well in mountainous regions for example and is often used by the military as well as AM broadcast stations in certain areas. Some atmospheric ionization is required for this to work.

Horizontally polarized antenna, such as a dipole that is located lower than ¼ wavelength, are ideal for NVIS. This might be a good way to work DX nearby on the lower bands, therefore as with ground wave propagation, not to be written off. I have heard of hams using this method to finish their WAS in New England where states are located much closer. NVIS antennas by the way are sometimes jokingly referred to as cloud warmers or cloud burners. I think you can figure out why.

Check the CW Beacons

Beacons are transmitters around the world (and maybe even locally to you) where you can check signal strength and propagation from point to point. One point being your receiving location and the other area being the location it is being transmitted from. Before anything else, let me start with the most important thing. **Never transmit on a beacon frequency.** This will get you on a lot of naughty lists you do not want to be on as well as earn you a well-deserved "*lid*" label. Easy to get, hard to shed.

Most beacons use CW, and some transmit only at a set time. Others run 24/7 with only perhaps a few seconds of break. They sometimes transmit with very low power, then medium and then higher power. Though usually the high power is still at a somewhat lower level. If you copy all three power levels, it will be a good time to try make contact with the general locale where you copied the signal from.

The NCDXF, better known as the **N**orthern **C**alifornia **DX** **F**oundation and the IARU aka **I**nternational **A**mateur **R**adio **U**nion, maintain a nice collection of beacons around the world. You can get more information on this at ncdxf.org/beacon. You can also check at benlo.com/ham/AzMap to see where to point your beam for the above beacons. The beacons are on 5 different bands 10,12,15,17 and 20 meters are they are staggered so they are never on at the same time. Max power here is 100W.

NCDXF Beacons are at 14.100, 18.110, 21.150, 24.930 and 28.200 Mhz. Again, please never transmit on these frequencies. People sometimes get carried away, especially during contests, and I have heard hams get chewed out via CW.

If you like what the NCDXF is doing funding DXpeditions as I had covered earlier and are using the beacons, again, please consider donating to them. It will be money well spent.

Tools like **Faros** can be handy if you want to have your radio automatically scan, log or mark open paths for you. If you have a FlexRadio for example, you can set up one of your slices to monitor the propagation beacons while you are decoding FT8 on some other slices. Talk about multitasking! The Faros software can be downloaded from dxatlas.com/faros.

There are also VHF, mainly 6m, beacons which are extremely helpful as band openings on 6m are very short at times. There are even beacons all the way up on 10GHz and higher. Beacons can even be found on ham radio satellites to tell you when the satellite is line of sight. But that is beyond the scope of this book, maybe a future one.

Other Propagation Resources & Alerts

Thanks to the internet, there are many websites to turn to. You can even get propagation updates via regular emails and even phone apps. Almost feels like cheating!

Hamspots I had already touched on and can be found at hamspots.net Similar to PSK Reporter, this site also shows what spots you are reporting back if you have it enabled in your software. You can do spot searches by different modes and by bands. See if DX is hearing you and who the DX is hearing if enabled. Look for patterns here. Using Hamspots is also a great way to "*stalk*" a wanted DX station.

Digimode Automatic Propagation Reporter, usually referred to in ham circles as PSK Reporter comes up a lot in this book as well, and for a good reason. Can be found at pskreporter.info and really complements the above. They also have a map overlay with amazing controls, which is my favorite feature and very useful one! This is also a good place to check who is hearing you as well. You can also spy on some successful DXers to see where they are hanging out!

WSPR is short for Weak Signal Propagation Reporter. This is a one-way transmission at a very low power with the purpose of seeing how far a signal can be received. Why do this? To test antennas, study propagation throughout the day and to catch possible openings. As a DXer, might be worth keeping your eye on these reports. For **WSPRnet** reports, check out wsprnet.org.

The **R**everse **B**eacon **N**etwork or **RBN** for short, is located at reversebeacon.net and is basically a way to see where you are heard on CW. The system uses a set of SDR receivers around the world to gather all the data. Reports will include **w**ords **p**er **m**inute (WPM) and signal strength in dB. They do not use S units for reporting for a multitude of good reasons. There is also a great beacon list here and you can do your browsing by band, so the information is not too overwhelming unlike some sites which can be hard to navigate and often feel like a massive data dump of all ongoings. Please consider contributing to them. Check out their website for more information on this and I encourage you to join in.

You can also play around with the **VOACAP** - **V**oice **o**f **A**merica **C**overage **A**nalysis **P**rogram, which is a propagation prediction tool. This can be found at voacap.com. There is an interesting history behind this US government developed tool, which is free to use. Voice of America is a multi-language radio broadcaster, which is still around on shortwave as well as on other platforms and is the official broadcasting service of the United States. This tool was developed to assist this service when using their shortwave radio system.

PropView is the DXlab suite tool which is one of the best I have seen on a desktop. For sure give this, as well as the suite, a try if you have not done so. Again, visit dxlabsuite.com for more information. There is also a tool called **W6ELProp** by Sheldon C. Shallon, Ex-W6EL (SK) which you can check out at qsl.net/w6elprop. This software has not been updated in a while, but you can likely still make it work with newer versions of windows.

For the MacIntosh there is DX toolbox from Black Cat Systems and this will set you back $24.99 currently when you register it. You can find more information at blackcatsystems.com.

Likely you have already heard of **PropLab**. This is a professional grade prediction software available for $240 from Solar Terrestrial Dispatch. spacew.com is the website if you want to check this out, though this might be overkill for most DXers.

Stu Philips, K6TU runs an amazing website at k6tu.net – **Propagation as a Service** which provides incredible services for those looking to do custom propagation reports to their QTH. There is a subscription-based service as well, and if you are in the market for this, you can't go wrong with it. He even offers antenna performance models over specific terrain.

One thing you will notice is that 160m and 6m prediction models are missing from most of the resources I am mentioning. These bands are rather hard to predict accurately, and in the case of the 6m band, nearly impossible due to its nature. There are more than one propagation modes on 6m, including the most common being Sporadic E. **Your best bet for the 6m band is to just monitor it.** The same goes for 160m. During their corresponding season I tend to monitor them regularly on a secondary receiver. During winters, I always have one receiver on the 160m band.

The NOAA 3-day forecast email service is very popular. You can find this at pss.swpc.noaa.gov/ProductSubscriptionService and this service will bring the 3-day solar weather forecast straight to your inbox.

HamAlert (not to be confused with Hamspots) is a very nice and simple to use tool with not just a clean and easy to use website but also an app for your phone to keep you in the loop as an ATNO comes up. I have this setup for the entities I still need. You can read up on it at hamalert.org.

For solar news there is also a very nice and clean page by **Matt Manjos, W5MMW** at solar.w5mmw.net. This page gets a A+ for the above and excellent readability. **SolarHam** can be found at solarham.net and gives you a nice glace of ongoings as well. For those more visual, **Tamitha Skov, WX6SWW** has a YouTube channel at youtube.com/channel/UCkXjdDQ-db0xz8f4PKgKsag and a big thumbs up for one of the coolest and most fitting callsigns I have ever ran across.

No list can go without spaceweather.com being mentioned. This is a more in-depth page than others and will give you all the data you could possibly want. This is one of my go to pages. The Australian government also maintains a page on solar weather at sws.bom.gov.au.

One you have all likely seen as is often embedded into ham radio websites is from **Paul Hermann, N0NBH**. I think he has the most viewed banners in the ham radio world, and for a good reason. You can visit the main site at hamqsl.com. He has nearly every platform covered. If you decided to use his banners or apps, please support him with a small donation. Of course, I have already mentioned the new Geochron 4K which also uses the above service for its banner and data source.

Lastly, if you are not sure how propagation is, just call CQ. Simple as that. You might be truly shocked who comes back to you. Several times in my ham life now I have received unexpected band fillers from just calling CQ phone. And this number is several dozen via FT8.

Band Characteristics

For simplicity, I would break these down into four sections based on similar characteristics. These are somewhat generalized but there is a reason for it. You will always find exceptions and some weird propagation. You will likely find yourself working mostly the "*middle bands*" as I call them over your ham radio career as these bands are not as dependent on where the solar cycle is at a particular time. Though, when sunspots are good, they will perform even better. **Propagation will shift throughout the day and by season as well for all bands.**

The Magic Band(s) – 6m (and 4m)

For those parts of the world where the **4m band** is available to you, these mostly can be applied as well. The **6m band**, which is of course on the lower end of VHF, combines characteristics of the upper HF and VHF. Sporadic-E is the most common propagation mode for the 6m band. This can happen below and above 6m as well. **I would not expect to see much F2 propagation on the 6m band unless the solar flux was above 200.** But it does happen and when it does, it is great. When working a station via sporadic-E, there is a likelihood of one-way propagation. This might be due to the path signals are taking. The other station is not ignoring you, likely just genuinely cannot hear you. There are also a lot of "weird paths" to be found here.

On the 6m band April to August or so is ideal, though May to July tend to be the peak. I have worked South America and Asia with less than 100W and 3 elements, however for more challenging locations you may want 5 or more elements and a bit more power. Be careful with RF safety though on this band. The WAS award can be obtained barefoot on 6 meters with modest antenna from within the US with some patience. Outside the US your mileage may vary. Once you hit a certain number of entities, moving to a much larger 6m Yagi might be in order.

Sporadic E is not well understood, there are many theories. Some studies have concluded that there is a possible correlation between the formation of thunderstorms and cloud formations. Wind shear has even been suggested as a possible cause. This appears to be especially true with heavy lightning, though this is when you do not want to be operating! Openings come and go and can be just a few minutes or many hours. I have decoded FT8 stations in Asia and Europe with positive dB reports one minute and no trace the next. Almost like someone unplugged my antenna. Welcome to the 6m band!

Auroral propagation as well as Tropospheric propagation can occur when it comes to the 6m band. Tropospheric propagation is more likely to take place in the summer and it is caused by weather, specifically temperature inversion. This will make the radio waves refract. Ideal levels of humidity and temperature layers can create these conditions. High pressure system days will increase the likelihood of these occurring. Auroral propagation is just that, auroras giving the 6m band a helping hand. Or can do the exact opposite, depending on ionization levels.

Trans-Equatorial Propagation is another way to experience the 6m band, as we had covered earlier. Other propagation modes for the 6m band include meteor scatter, which is popular during certain annually reoccurring meteor showers. Here as the meteors pass, their trails ionize. These can be very quick, therefore special modes are often used. These would be faster digital modes or fast CW. You can find out which ones are active at amsmeteors.org/meteor-showers/meteor-shower-calendar.

The 6m band is also popular with some EME (**E**arth **M**oon **E**arth) operators where the signal is bounced off the moon to make contact between two points. This requires a bit more patience, larger antennas and more power in most cases but has been done successfully by many with much less. This is also a good way to add to your entity count on the 6m band.

The Upper Bands – 10m to 15m

This of course also includes the 12m band. **These will be better fall through early spring**, as are most bands in general according to some. The noise here is higher in the summer. These bands do best then the sunspot counts are high or towards the peak of the sunspot cycles. There can be years when they seem almost dead near the bottom of the sunspot cycle. These are also considered to be daytime bands, though do not count them out at night. I have seen great openings on all, including the 10m band several hours into sundown.

The **10m band** is where you will find many newer hams on HF as there are US technician privileges here, though the band is not always open for DX. In fact, during the bottom of the cycle you may only see local traffic. On the flip side, when the sun is cooperating, it will be booming, and jam packed. In fact, can be open to multiple paths at the same time, if not worldwide. Antennas are the smallest for the higher HF bands, therefore easier for many to get on with antenna restrictions. If the band is open, follow the sun on your map to aid with your antenna direction. Likely can work the planet in one single afternoon.

The **15m band** is very popular in Asia and gets very active on the US West coast when open. It does, however, tend to open at least a few hours a day even when conditions are not ideal, so it is worth checking on it. The 15m band is also hopping from the US East coast to the EU when open, sometimes much further. Near the equator, this is likely your best bet for DX as well. Caribbean radio traffic can be second to none on 15m at times.

The **12m band**, as expected, is a lot like the 10m band, but a much narrower slice of the spectrum. This is a good band if you want to stay clear of contests as there are none here. It is a WARC (**W**orld **A**dministrative **R**adio **C**onference) band which refers to the 12m, 17m and the 30m bands. These three bands were made available to hams in the early 1980's and do not allow contesting. On the flip side, since there are no contests filled with DX, you will have to be on a lookout for your wanted DX list. I have often worked DX, sometimes very lonely DX here on major contest weekends.

The Middle Bands – 17m to 30m

This also includes the **20m band**, often called the money band. The reason is, for most folks this is the most likely place that you will work the most entities. Most DXers will have their highest entity counts on this band and for a good reason. It is very active, has propagation even during low sunspots, though not as great. These are mostly considered daytime bands as well. But there are times they are open late into the evening and even all night if the conditions are just right. I have worked Europe at midnight from the West coast before during good openings. This is one band you want to make sure you can always operate on. In other words, whatever antenna(s) you end up with, **be sure at least one antenna always covers the 20m band**.

The **17m band** is another band, much like 12m, where some non-contesting DX stations park during contests weekends, so be sure to check to get your fillers. It is also a WARC band. The difference is that 17m is open a lot more than 12m is. I often find 17m to be open as well, at least to some degree when 20m is doing well. When 15m is open, 17m will also have a lot of activity. I would almost call the 17m band the second most active digital band, or at least a tie with the 40m band for that title.

Both the 20m and the 17m band behave similarly. They tend to follow the sun therefore will be open to the East in the morning and West in the afternoon, often into the evening. For the US this means, Europe/Middle East in the AM with a possible repeat in the evening and Asia/Oceana in the PM.

The **30m band** will be somewhat of a hybrid of the 40 and 20m bands, as expected, though has more attributes of the 20m band. Always remember the 200W power limit here. A big plus is exactly that it has a power limit as you are a lot less likely to be blasted away with stations using way too much power. That is if they follow the rules. This happens a lot with the other bands and seems to be a problem in some instances. It is a little more prone to interference from users outside of ham radio as well as atmospheric noise. The 30m band does not have any contests just like the other WARC bands.

You will want to make sure to stay clear of the 30m WSPR frequency at 10.140MHz. Many hams monitor wspr.net to track propagation.

The Lower Bands – 40m to 160m

Sometimes the 30m band also gets bundled in here. I consider 30m to be a *"Middle Band"*. In the lower bands here also include the channelized 60m band and the 80m band which is sometimes referred to as the 75/80m band do to its actual location and massive span. Antenna sizes go up a lot here, perhaps too much so for many hams with limited real estate. These bands are considered nighttime bands for DX purposes, though you will find lots of local traffic during the day, especially on the 40m band. **These bands also tend to perform better when the days are shorter.** This means winter for the Northern hemisphere, summer for the Southern.

These lower bands can experience severe attenuation from the D Layer. These tend to be best when the sunspot counts are low, or the bottom of the sunspot cycles. But the opposite can also occur! Also, of course they are good nighttime bands. The 40m band is also used by select broadcasters due to its great propagation characteristics and you will often pick up AM shortwave stations at night and sometimes past sunrise. Be careful not to QRM them. As in, do not transmit on top of them.

The **40m band** is the busiest of the bunch and the easiest to operate on. It is also a good band to get your feet wet as is beginner friendly and I would argue the 2nd best band for DX in general. This is a great place to place your focus on when the sunspot cycle is near the bottom. Noise will be lower; traffic will be higher. Those who cannot work 10-15m on a regular basis may move down here. Antennas are smaller than those for 80 and 160m but still can be hefty. The 40m band can also be plagued by atmospheric disturbances, such as thunderstorms which can cause lightning crashes. Long path is always something to be checked on 40m, especially in the winter when there are more dark paths which may be workable.

As I had mentioned earlier, the **60m band** does not currently qualify for many DX awards, including the DXCC. It was first introduced in 2002 and yet many hams have not even operated here. Most DX here will hang out on 5.357MHz which is also where FT8 is. I would not count out this band as there is DX and when more awards are available, you will be that much ahead. Always observe your local power limits on this band.

On the **80m band** you will find a lot of regional traffic during the daytime and conditions for DX here will also improve at night. Sunset and sunrise and for sure key here, think grayline. 40m, 60m and 80m all can be open an hour before and after both sunset and sunrise so be sure to check them. However, some openings to some parts of the world can be as short as a few minutes! This is a good place to check for long path propagation as well if you are fortunate enough to have a directional antenna on these bands, though most use verticals on 80m. Signals will be much weaker than on 40m due to its nature. This is in part due to atmospheric noise. While it is considered a winter band, keep in mind when it is summer for us in the Northern hemisphere, it is winter for the Southern hemisphere. They will have less daylight. I have worked some great South American DX in the middle of summer.

The top band, or the **160m band** is perhaps the most challenging of the bunch when it comes to the lower bands. Sometimes it is also referred to as the gentleman's band. Here you will find more activity from fall to spring, peaking in the winter. Though, I have worked DX, including band filters here in the middle of the summer when there *"wasn't supposed to be any DX"*, including Africa from the West coast which was booming in. You will likely only hear DX phone here during DXpeditions and contests, though there is plenty of local late-night chatter. RTTY was never popular here due to high noise levels. CW used to be DX king here, but now FT8 seems to be the new DX king. This is likely since FT8 deals much better with high noise levels present on this band. Occasionally there is still JT65 and JT9 activity here. There are some who have observed that there is a rise in propagation on 160m at the beginning stages of a geomagnetic storm. This band can also be very negatively affected by auroras.

During the day, this band is for the most part local only. 30 minutes before and after sunset and 30 minutes before and after sunrise is when it is especially worth monitoring it. As with the 80m band openings they can also be as short as a few minutes, and I have seen this with DXpeditions in the Pacific. Trying to hear them from the West coast, they came up from the noise for 10-15 minutes and they went just as fast. Knowing when to listen is a key factor here. This band will be one of the most challenging to work, but also one of the most rewarding. It will take many years for most DXers to even go beyond a few countries, depending on your location. DXCC on the 160m is quite respectable.

Noise will always be a factor on these bands to at least some degree. It will make these bands even more challenging than they already are. This might be why so many shy away from anything lower than the 40m band. In fact, many hams, once they get their 5 band DXCC, never set foot on the 80m band again. I have seen several ads over the years with "*I just finished 5 band DXCC, selling 80m antenna*". Really too bad!

Though there are two other ham bands here we had already covered. These are the **630m and 2200m bands** which are relatively new. No award is given at this time for contacts here. Also, since they are somewhat new, there is practically no information on propagation for these bands outside of them being mostly winter and night bands.

DX Nets

If you tune around the bands, sooner or later you will encounter a "*net*". There are a lot of various nets out there which exist for various reasons. Nets are usually scheduled by time and frequency (plus or minus QRM) and generally do a check-in. Meaning there is a running list of callsigns as well as a net controller. This person is in charge and keeps track of check-ins and maintaining general order. Most nets will be on the phone mode, single side band, but there are exceptions.

There are also what they call DX nets. These can generate some heated debates among hams. Some seasoned DXers look down on these nets, but they really should not. This is something worth exploring as may just be something to assist you in getting your numbers up, if it is your cup of tea.

Once you checked in, wait for your turn. When you get called, try, and make contact with the DX using your callsign and report. Hopefully the DX will hear you and reply and vice versa. The net controller may confirm or may have your try and repeat to make the contact. Once you are good, log it. This process requires time and patience and is not for everyone. You may have to sit through a lot of other hams trying to attempt a DX contact. Some will fail to do so. It is certainly a viable option though if you are interested in it and have the time.

Some DX nets are open to all, others are for clubs, groups, or even certain areas. These nets exist for the sole purpose of getting folks a DX contact. For many newer DXers with limited gear and skill, this is a good way to get your feet wet or get a new one in the log you may not have had the opportunity otherwise.

All the DX nets I know of are on the 20m band, though there might be others. You can find a list of DX nets online, though some are closed per above. These lists change so it is best to do a little homework. One drawback is that if a DX net meets at the same time regularly, this may not be the ideal time for propagation to a certain region on a given band. I do want to note that I was told these nets have been decreasing in popularity over the past few decades. Therefore, lists online may contain a few or possibly many defunct ones. If you are past the 200-entity mark in your count, it is somewhat unlikely that anything would show up on a DX net which you may need.

Holidays and Soccer

No, you did not misread this. If you live in the US, you know that on the weekends there will be more operators because those not yet retired might be on the air. Folks like me for example. The same is true for after hours. 4th of July, which is Independence Day in the US, Christmas, Thanksgiving as well as other holidays also see a jump in operators. Though during the World Series (Baseball) and Superbowl (American football) the bands tend to be less active. Go figure!

The same is true overseas. The difference is soccer (football outside the US) is king pretty much everywhere else. Don't expect the rest of the world to be on the air during championships. I recall traveling in Mexico **(XE)** once during a World Cup. The city I was in, Mazatlán, was pretty much shut down during the game. Waiters were glued to the TV; many stores were closed and on occasion you would hear a loud "*goooooal!*" followed by something resembling a miniature earthquake. I suspect some of these folks were hams generating all the seismic activity. You get the picture.

It also pays to know holidays from other cultures and religions. For the United States for example a good starting point is timeanddate.com/holidays/us especially if you are WAS hunting. To look up any other country, go to timeanddate.com/holidays. Also keep in mind about time differences. In many Muslim countries, such as Jordan **(JY)**, Friday is Sabbath. This is also very important to keep in mind. On my Sunday night, Hungary **(HA)** is already back to work. You see a drop-off in ham activity Sunday afternoon. This is especially true after a major contest weekend. People are wiped and done.

Scheduling a Contact

Sometimes called a "*sked*" for short, though I really hate that abbreviation. This involves contacting a specific DX station you are interested in working via another method first. Nowadays it is usually email. The contact information is generally available via QRZ or other indexes.

Scheduling involves agreeing to meet up on a specific frequency and time to attempt to contact each other. Make sure you **use UTC time and date to schedule** so all parties are clear as to when. Some operators are open to schedules, some not so much. If you are in Delaware **(W3)** or North Dakota **(W0)** in the US for example and the DX is looking for WAS, your chances are better for a schedule. California **(W6)** and Texas **(W5)**, not so much.

Alternatively, if you are in a rare grid in Russian Siberia **(UA)**, you are also more likely to get someone to work you. Some hams are also more likely to schedule on harder to work bands, like 80 or 160m. Additionally, many are interested in different alternative digital modes. I certainly fall into both latter categories.

Some DXers look down on schedules. They feel like you need to "*find*" the DX on your own. Usually these are the same folks who think that DX clusters are the work of the devil. They are entitled to their opinions. Personally, I feel there is absolutely nothing wrong with having a schedule if you wish to occasionally pursue this route to add to your totals.

Scheduling is very common for some parts of the hobby. Notably the EME (**E**arth **M**oon **E**arth) or Moon bounce operators and those experimenting on the newer and more difficult to work 630m and 2200m bands.

This is not something you should or can do for DXpeditions obviously. Similarly, bugging the DX or DXpedition to work you or move bands/modes on the DX cluster is not the way to go. I see this a lot and it is very aggravating not to mention a waste of time.

Frequencies to monitor

There are some blocks of frequencies unofficially designated for DX use. These are referred to specifically as DX windows. These seem to be "*out the window*" as are not observed as much anymore and frankly many newer DXers are not even aware of them. As a DXer, you should be tough!

6m	50.100-50.125MHz	CW/Phone
	51.000-51.100MHz	Pacific
10m	28.000-28.025MHz	CW
15m	21.000-21.025MHz	CW
20m	14.000-14.025MHZ	CW
40m	7.000-7.010MHz	CW
	7.040MHz	RTTY, Digital
80m	3.500-3.510MHz	CW
	3.590MHz	RTTY, Digital
	3.790-3.800MHz	Phone
160m	1.830-1.840MHz	CW
	1.840-1.850MHz	Phone

Generally, you will find these on the lower portion of the band specific mode plan. Meaning the bottom of the SSB or CW allocation of the non-WARC bands. Not found on 12,17 and 30 as these are smaller allocations bands to being with. Something else to note is that the DX windows often fall into the US extra portion of the band, therefore yet another reason to upgrade your license if you still need to.

These are not so much rules as agreements. Some consider 1.825-1.830MHz also the CW window. There are mixed opinions on this. None of the above is written in stone, but the above is the agreed upon list, with some overlaps I found from various sources which I have been referring to over the years.

Regardless, nobody owns the frequencies, so everything is on a first come first serve basis. However, we all need to learn to play nice in the sand box. So, if someone **nicely** asks if you can move up a bit due to a weak DX station nearby or a net starting, be nice. It really is not a big deal to turn the dial. Takes more effort to argue about it.

Finding Information on Active DX

Most DXpeditions will have a QRZ page setup at minimum. These can be looked for by their callsign via the search bar on QRZ or even just by searching for it via your favorite search engine. For larger operations, usually a dedicated website is available. This site will likely list operating frequencies and modes as well as other useful information, such as how to confirm and/or QSL.

I have already mentioned **The Daily DX** at dailydx.com. This amazing resource also often gives information on operating frequencies, modes and even times to look for the DX. Have I mentioned it is worth subscribing? Other DX news sources have already been covered in the book, be sure to bookmark them as well and sign up for their email lists to complement the above if you like.

ClubLog will have some of the expeditions listed at clublog.org/expeditions.php. **ClubLog Livestream** is also an amazing resource. You can find a list of stations currently using it at clublog.org/livestreams.php This site will not only give you data on who was worked last but also on what band a given callsign is active on. The last few callsigns worked will be a good indicator of where propagation is strong and of course you can easily check if you are in the log as well.

DX in Contests

I know some DXers out there hate contests, though I still don't get why. But I know many others, including myself, who love them. **This is a great way to get band fillers**, especially when you are just starting out on your DX journey. There are a lot of DX stations who are not regularly active outside of contest times but start booming in on the dot once a given contest starts. Sometimes even sooner to warm up. This is also a good time to get them! I still find that occasionally I get a filler, especially mode fillers such as on CW and Phone.

For those who do not know what contesting is, simply put, it is a test of capabilities of various hams or groups of hams against their peers. These are based on a point system and often multipliers. Speed, skill, equipment and even a desirable location all come into play. At times even a rare prefix helps. Contests can be phone, digital or CW. Can be DX or localized. If you get on the air and hear a station call "*CQ TEST*" this is short for contest. They are not testing their gear. That would be "*TESTING*". A newly minted ham once asked me this, so figure I will cover it here.

You can pretty much find a contest every weekend of the year and some minor ones, even during the week. For some hams, this is pretty much the only time they get on the air. These hams might be the ones you still need to work! Basically, you can think of it as the opposite of ragchewing if you wish to oversimplify it. Due to their quick pace and number of stations active, contests are something to pay attention to even if you are not going to submit a score. Though it is something to consider.

At times you hear the word "*radiosport*" used to refer to contests. This is not to be confused with the headset brand RadioSport. Contests are generally sponsored by a club or organization. Some of the most popular ones are sponsored by the ARRL, CQ Magazine or even contests clubs like the **N**orthern **C**alifornia **C**ontest **C**lub, NCCC for short. They also sponsor the California QSO party which does bring out a lot of DX as well. They can be found at nccc.cc. Love the domain name and it has many useful resources available and not just to members.

I have already mentioned this but **note that the WARC bands (12,17 and 30m) as well as 60m do not allow contests.** Also, some contests do not extend down to 160m but there are many that do. There are even some 160m only contests, usually during the winter months. There are two VHF contests as well during the year where the 6m band is very active. Because of this, I would recommend trying and spending more time on the WARC bands outside of contests once your numbers start going up. Usually, one can get more fillers during contests for those active DX entities. There are even DXpeditions that will take time out and work a contest. I have seen this and those were some serious contest pileups. Also, as I had mentioned, some DX do "*warm-up*" before the contest to make sure their gear still works fine and to make adjustments to maximize their scores. This is a great time to get in the log before the masses show up. Will not count for contest points but will for DXCC. A DX cluster in conjunction with an early start is very helpful for the above. Though, by the time it's on the cluster, it might be harder to get in the log. Always best to find these yourself first. An hour before starting time is a good time to start looking around a bit.

Where do you find out about contests? I use two resources for this. One is the listing in QST. Every month there is a page dedicated to the listing of all contests for the month. The other place I look is at contestcalendar.com ran by Bruce Horn, WA7BNM. Additionally, ng3k.com/Contest is also a great resource. This is maintained by Bill Feidt, NG3K. Yes, this is the same gentleman I had mentioned earlier who keeps a great list on upcoming DXpeditions at ng3k.com/Misc/adxo.html.

While some refer to parts of the year as "*contest season*" I feel like all year is contest season. In the winter there are several 160m contests as well as a 2nd VHF contest (first one being in the summer). Furthermore, the state QSO parties are spread out over the calendar year. So, if you are outside the US, or even within, and are looking to complete your WAS award, this is a golden opportunity! To read up on these online contesting.com is a good start.

Always remember that contesters will usually not chit chat or deviate much from their script. They will talk fast and often use voice memory. This is not because they are rude, it is a contest. They expect you to be fast as well. It is all about points and number of contacts after all. Know the basic rules and the required exchange before calling. No, you do not need to submit logs, so even if you only make one contact, that is OK. I sometimes only make 5-6 depending on the contest and other times several hundred. Also, no need to repeat your call unless asked or they misheard. Do not work the same station twice on the same band. This is called a "*dupe*" or duplicate and it is not only unnecessary but also slows them down. Could even be a penalty. I know accidents happen, I have even done this accidentally before, but try and avoid it please.

Exchanges will always be posted in the rules and if required the signal report will always be 59 for SSB and 5NN for CW. For RTTY 599 and FT4/8 will be determined by the software or more likely replaced with a grid if in contest mode. Make sure you set the proper contest mode if needed in the applicable digital software.

Strategy is going to be different for a DXer than for a passionate contester. You will not be looking for points or point multipliers, you will be mainly looking for fillers for entities on various modes and bands you do not yet have. Though, you can do both. I have done both at the same time successfully in the past. If you are calling CQ don't be surprised if a DX station you may need calls you. They need the points, and you <u>are</u> one!

Do not mess around though and make up numbers in a serial contest. This would be where the contacts are sequentially numbered. This will not only make you look like a lid but will possibly cause other issues. If you are #342 then #24 an hour later than when logs get compared, you may have just caused a station, or several, a point. Not the time to be funny. Some folks take contesting very seriously and they <u>will</u> remember you! For serial CW contest I usually start with "01". Why? Because many expect more than a single digit. This way it is clear to the other station that I just started, or I am only making a few contacts. But that is just how I do it. It is up to you.

If you start to get more series about contesting, I do recommend **NCJ – National Contest Journal** from the ARRL. You can find out more about this excellent publication at ncjweb.com.

The Sixes – Mark of the Beast?

The mark of the beast is a very appropriate name for the three most difficult bands in my opinion. Many find these <u>devilishly</u> hard to work. These would be **6m, 60m and 160m** but all for very different reasons. The 6m band depends on sporadic E at times and this is not only very seasonal but also harder to come by on the West coast than, let's say in the Southern part of the United States. A good DX contact will require multiple hops. Bad news for the West coast. Outside of sporadic E, the MUF must be very high, and this does not come around too often. The season peak is May to July. I also feel that the 6m band is also often ignored as it does not count towards the 5 band DXCC, though you can get an endorsement for working 100 entities as I had mentioned. Very few seem to do this though.

There are also fewer folks on 6m than on 20m but that is nothing in comparison to the lack of users of the often-confusing 60m band. Some older radios do not handle 60m, and there is a lack of premade antennas for this band as well. So likely you will need an antenna tuner and an 80m antenna short of making a dipole. Also, since it is not recognized for DXCC award purposes, many do not pursue it. It is a shame, as it is one of the most fun bands.

Lastly, 160m has limitations for many due to the large antenna size requirements as well as the fact that during the day, it is pretty much dead. Due to folks sleeping schedules and lack of space for antennas, it is also rather difficult to get to a hundred entities for most folks. And don't even get me started on the horrible noise levels. Many contests also do not extend down to the 160m band. This band, just like the other two will require patience!

On-Air Etiquette

The perfect way to close out this chapter is with the *"DX code of conduct"*. Those who may not be familiar with it, these are a set of guidelines created by hams, though you can almost think of them as rules. I do. You can get an expanded version of these at dx-code.org/english.html. Much of what is covered here, I have also covered throughout this book.

1. I will listen, and listen, and then listen again before calling.
2. I will only call if I can copy the DX station properly.
3. I will not trust the DX cluster and will be sure of the DX station's call sign before calling.
4. I will not interfere with the DX station, nor anyone calling and will never tune up on the DX frequency or in the QSX slot.
5. I will wait for the DX station to end a contact before I call.
6. I will always send my full call sign.
7. I will call and then listen for a reasonable interval. I will not call continuously.
8. I will not transmit when the DX operator calls another call sign, not mine.
9. I will not transmit when the DX operator queries a call sign not like mine.
10. I will not transmit when the DX station requests geographic areas other than mine.
11. When the DX operator calls me, I will not repeat my call sign unless I think he has copied it incorrectly.
12. I will be thankful if and when I do make a contact.
13. I will respect my fellow hams and conduct myself so as to earn their respect.

By following the above code, you will not only be a more successful DXer but are will also contribute to the success of others by making the airwaves enjoyable for all.

CHAPTER 11: After the QSO

Now that you have your hardware, software and techniques in order, you are hopefully ready to work the world. Maybe you already have and that is great! But there are a few things you still should and, in some cases, need to do. This is especially so if you want to get award credit for your contacts. This is an area where new DXers have many questions. I know I did when I started out and I could never find a central place where I could get the answers I needed. I am hoping to change this for others with this section.

Setup your QRZ page

While you may not think this is important, I can guarantee it may save you some hassle. Don't think of this site as social media, though kind of is in a way. I like to think of it as the "*ham directory*" with news and a swap meet. Some use it to showcase their gear or at least humanize the callsign with personal photos. I like that part quite a bit. Yes, some go a little overboard with the animated GIFs and such, or the welcoming users in 27 languages, including Klingon, but that is OK. It is a good place to post your QSL info as well as membership information and so on. Things that other hams looking you

up find useful. Feel free to scan this QR code to visit my QRZ page as an example. You do not have to write anything as elaborate as I did though. Please say hello in the web contact log if you decide to visit.

It can also help the DX to check your mailing address, email you with questions and so on. Increase the likelihood of getting that shiny QSL card in the mail. If you are allergic to HTML or can't even spell it, they offer a design service as well. Although you really do not need to have much, if any knowledge with their basic editor. Alternatively, you can just put some basic text, fill out your correct information and can be done with it. I would recommend doing that much as a minimum. Takes all but two minutes. Less time than a couple of FT8 contacts!

I fully support freedom of speech and expression. You may encounter some political garbage, posted by users, which should always be left out of ham radio. Furthermore, people are not the same as their politicians. I see this internationally and domestically (left vs. right) and so on. Try and refrain from it in your own page and comments. Not the place for it in my opinion. It's a worldwide hobby and should be uniting people, not dividing them more.

Design your own QSL Card

This is not as hard as it sounds. This part scares some hams to death. You can go very simply, or you can fancy it up as much as you like. This is completely your call if you include the basics to be able to verify a valid QSO. Some hams have two sets of cards. One nicer one for "*special*" contacts, however you may define that and one for bureau mailings for example. Sometimes this is to cut costs, both printing and shipping. I just keep one as frankly the cost is only a few dollars more per 500.

The US standard QSL card size is 5 ½" x 3 ½" (14cm x 9cm approximately) but you will see sizes slightly smaller than that as well as larger. This is especially true from overseas. You want to give your card a 1/16" bleed area on all 4 sides when designing it. This region may get trimmed after printing. This is done so you do not end up with white lines around your card. Therefore, consider the safe area 1/16" (1.6mm) away from the sides. Meaning, be sure there is no text or images to close to the sides.

Some cards are monochrome (one color only) to reduce cost, some are full color. Frankly there is not that much difference in cost if you go to the right place. Some cards may also have 4 sides as they open, though these are usually from DXpeditions. Often done so due to the many sponsors.

I use Adobe Photoshop and keep sections in layers so I can adjust on the fly and rearrange it when I am in the mood or need to. However, this is overkill for most users if you will not be using Photoshop regularly, no point in getting it just for your QSL cards. There are many good software alternatives. **Gimp** which can be found at gimp.org and is completely free. There might be a small learning curve but plenty of resources, including great videos out there to help you. The other software I would look at is **Affinity Photo**. It can be found at affinity.serif.com/en-us/photo. The cost is $49 and is rated on top of many lists.

In my option, the best cloud-based solution is **Canva**. As proof, I used it to help design this book's cover, with final touches from Adobe Photoshop. I practice what I preach. You can visit them at canva.com. There is a free account but for pro content there is a subscription fee. If you will use it somewhat often for projects, it is well worth it, or just stick with the free version until you need more features.

I would advise that you only **use an image you own, took, or have rights/permission to use**. You do not want to get into trouble! This includes images from movies, cartoons, photographs, works of art and so on. Think copyright! Even if you are using the image for something non-profit, hobby related, you would be surprised at how uptight some corporations and even individuals can be about this. Better safe the sorry. Royalty free is likely OK as these sometimes only have restrictions for commercial use.

The same goes for including an image of someone else or perhaps immediate family, though the latter is generally fine. I have seen some QSL card photos snapped at a ham convention and while the card owner was the center of attention, there were clearly identifiable folks in the back. Creative cropping can usually solve this problem as well as, if possible. Alternatively, just get their permission to use their likeness. If nothing else as common courtesy. Selfies are something which may stand out for all the wrong reasons. Unless you are very good at taking them, **ask someone else to take your photo**.

Resolution, specifically DPI (**D**ots **P**er **I**nch) comes into play as you do not want to end up with something blurry or pixelated. 300 DPI is the rule of thumb I use for print materials. Sometimes even better. Most images on the web might only be 72 DPI so if you do use something from there, get something very large in size and scale it down correctly.

If you live in an area with something distinct, such as a natural landmark, this can be an option. Though sometimes these get way overused. Again, use your own or a royalty free image. Card collectors, like me, like

to see something unique from the other end of the QSO if possible. Your ham shack, antenna farm, or something else you are into are also good options. Many hams also use their classic cars, motorcycles, pets and so on. Pictured here is a QSL card from Nasir Khan, AP2NK in Pakistan who has done a nice job with an image of his shack, himself and antennas all displayed in a very clean design.

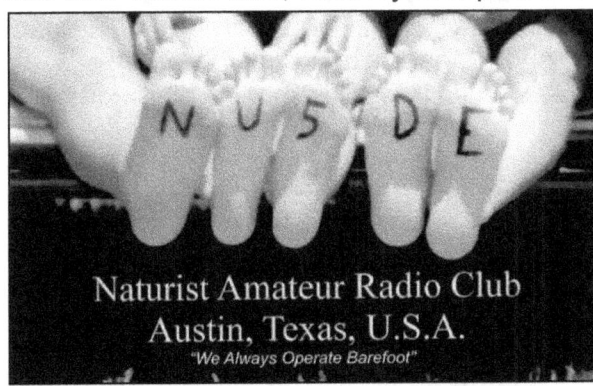

What you do not want to do is use images which may shock or offend some hams. Some may even get hams in legal trouble in certain countries. The biggest example is nudity. I have received several cards over the years with nudity, though done mostly tastefully done. Keep in mind many younger folks are active and some others may also find them offensive. Personally, I could care less, but use caution.

Ironically, NU5DE, the Naturalist Amateur Radio Club in Austin Texas does not have any visual nudity featured on their official QSL card. A+ for the callsign! Be sure to also operate them barefoot (no amp).

So, how did I go about designing my own card? I will go over it in detail so you can take or leave section out as you like when designing your own. There are only really a few rules to follow. On the **front side** of the QSL card I include the callsign in easy to see big letters, QTH information as well as my website, email, and twitter. I also included a small map on the front indicating where I am with my

grid square CM87sp in the black box. The photo was taken by me a few feet from my house. Yes, good take off to Hawaii and Asia! No fancy camera needed; I used my phone for this image.

I also added a few extras like a QR code which when scanned takes you to my website. This can also be your QRZ page. You can get your own QR code at the-qrcode-generator.com or dozens of other sites by doing a search.

The **back side** is where I write the QSO Information. This is the actual one I use, created with photoshop. This seems to work fine for me, and I just write the info in, circle under QSL if I am requesting or returning the card. You can also cross out the one not applicable. Both ways are acceptable if it is clear.

Confirming QSO With	Month	Day	Year	UTC	Band	Report	Mode	QSL
					M			PSE TNX

For ZERO I usually draw a line across, so it does not get confused with an "O" especially in callsigns. Most people figure this out, but it is one extra step that someone may appreciate, especially if sorting your card via the Bureau or whoever. The date can be an issue as **not everyplace writes the dates in the same order**. We like to be different here in the US. We, for example, order things as month, day, and year. Much of the world is day, month and year in increasing size order. Makes a lot more sense to do it this way. I stuck with the US version above and this is since 80% of the cards I get are stateside and mostly towards VUCC award and/or satellite communications. However, I clearly labeled them in my layout so there is no confusion. This also matches the layout of my logbook so when I am filling these out, there is less of a chance of me making a mistake. There is no right or wrong way to do this if it is clear. Figure out what works for you and what is clear for others and go with it.

On the back of the card, I also include my **callsign again** as some card checkers do want your callsign on both sides. Also easier for them and is likely a fraud prevention measure as well. Some could care less, I do it just to be safe to make it easier for the receiving end. I include my **mailing address**, for obvious reasons. If you have an odd address, some of which I come across occasionally such as some territorial islands please make sure you format in a way that those not familiar with it understand it. In other words, complete routing. **County** where you are residing if in the US, If applicable. Many county hunters out there, this is helpful for them. I know some states call them parishes, but this is the same thing. In Canada, these are provinces and territories.

The **grid square** is needed for 6m contacts or anything for the VUCC award, but also good to include for HF. Look it up at levinecentral.com/ham/grid_square.php. Also helpful is you **ARRL section** if in the US, you can find yours at arrl.org/sections. The **ITU Zone** can be looked it up at mapability.com/ei8ic/maps/regions.php. Also, a good idea to include the **CQ zone** at mapability.com/ei8ic/maps/cqzone.php. If you operate QRP, satellite or mobile a lot, might be good to have **checkboxes** for these and I even leave a space to write in the satellite we operated on. It makes it easier for other hams track, not to mention is required for some awards.

Using **club and organization logos**, if it is the approved and an actual approved one, is perfectly fine. Many clubs provide members with a clean copy of the logo for use via their webpage or members only area. You can also just write these as text, or not at all. I also include membership numbers in case folks want to check, verify, or confirm your exchanges for other awards and such.

One thing I started doing for my QSL cards is **version control**. I do this on the very bottom right via both a version and a date when I last edited it. If you make a lot of changes, this will save you a lot of headaches. I have learned this the hard way as I have mixed them up before. This can be in a tiny font as this is just for you.

An interesting option for printing QSL cards is not using paper. I have seen cards printed on thin plastic, engraved on flexible metal but my favorite but far is the one shown here, featuring native Hawaiian Koa wood. Eric Brundage, KH6EB (formerly WH6EYY) had this amazing 2-sided card which I was surprised to receive, now almost 10 years ago for a couple of digital contacts. This was around the time I changed callsigns to reduce CW weight, hence the two callsigns.

Print your QSL Card

This is something you can do yourself if you are keeping it very basic, and there is nothing wrong with that or you can send it out to a professional place to have it printed for you as well. It does not have to be a place specializing in printing QSL cards. If you have the design, pick a card stock, size and you are good to go. Full color cards are cheaper than you think!

Good options for printing your QSL cards are as follows. UX5UP Print can be found at ux5uoqsl.com and has quality cards. Another good resource is at kb3ifhqslcards.com, KB3IFH QSL cards. They also offer other goodies for your shack besides excellence cards. CustomQSL.com advertises itself as the *"Home of the Next Gen Ham QSL cards"*. Owner Tim Hijazi, KB3K pays great attention to design and quality. He is highly recommended by many DXers.

Gotprint.com is where I go to print my QSL cards. And no, it is not a QSL card printing place. Good prices and very nice quality though you will need to upload a ready to go CYMK (**C**yan **Y**ellow **M**agenta **K**ey/Black) file. Blue is B so this makes it less confusing. This basically just means it is formatted for printing. Normally you work in RGB (**R**ed **G**reen **B**lue) when dealing with images at home or on the internet. They offer everything printable, and you might as well get some business cards with your QSL cards if you go this route. I pay about $0.05 per color card with monochrome printed backside. However, if you need more assistance, go with folks who help you with every step or even do the layout for you.

As far as printing the cards yourself at home, unless they are the most basic cards with no color, it will likely cost you more in paper and ink or toner. Not to mention it will not look as nice even on the best printers as if you had it professionally printed. Also, if the card gets wet and was printed on an inkjet, it is ruined! The amount of time you may waste printing it yourself costs money too. You could be hunting DX instead. Take my word, just spend the $30-40 or so and get your 500 attractive looking cards. As a point of reference, if you are paying more than $0.20 per card at 500 volume level, you are getting royally screwed. A ham once told me he was getting about 3 cards for a buck and thought that was good. No, it is not! Always buy in bulk to cut costs and even if you do not, that is way too much to pay for QSL cards. Trust me, you will use them if ordered in bulk. Especially so early on in your DX career.

Proper Card QSL Fill out Procedures

When you get around to sending out your cards, there are a few things to be aware of. Pay special attention to the date order as I had already mentioned. Many, if not most places outside the US have the order….well, in order. In the US when we say 5/6/2022 this means May 6th, 2022, but someone from Europe would say it is June 5th 2022. I feel it is best to write out the name of the month if it is not obvious.

Also, due to the letter "O" and zero being similar, I always put a line through the zero to indicate clearly that it is a number when filling out a card. Just as you should in your callsign when designing the card as we covered.

Lastly, and perhaps most importantly, if you make a mistake, <u>start over</u>. If you need to correct a card, there is a good chance it may get rejected on the other end if it must be checked. Something to keep in mind for your incoming cards as well. If you have the capability to print directly to your QSO cards, that is perfectly acceptable. So are self-adhesive labels and "verified by" stamps, though these are very much optional.

Storing your QSL Card Collection

There are many ways of doing this. Some hams display them on the shack wall. But because there is only so much room for wallpaper, you will need to move or find a new method. There are a variety of holders out there, but they can get expensive if you are rather active and again, you will run out of room fast. Maybe reserve this for your most special ones.

Some use a shoe box or something like this and this works very well for most. I personally like to use photo albums. My preferred one is the one made by Pioneer. No, not the electronics company. These have 4x6" pockets. More pages can be added so these are very expendable, and they come in a variety of colors so I can have one for each continent or organized by other criteria. Look around and see what works for you. This is a good time to think "*outside the box*".

QSL and Getting your Contacts Confirmed

We have covered everything you need to make a few QSOs. Now you have a growing logbook, QRZ page and some cards. But it does not end there.

There are many ways of confirming your contacts nowadays, mainly electronic, but **the "*old fashioned*" QSL card is not dead**. I certainly would not have spent so much time on the topic if this was the case.

I love getting QSL cards, even domestic cards, especially for satellite and VHF contacts. Have many albums full of them and still get a handful every year via the bureau. Additionally, I get about 2-3 a week by US mail alone. If you are a stamp collector, this is like a two for one. Still think they are dead?

As covered before, ARRL's DXCC program uses the LOTW system, also known as the Logbook of the World. But they can also accept physical cards. Though in some cases, this is the only way to confirm a contact realistically! You will need to use a hybrid of the two to get your numbers growing. To acquire a physical card can be done a few ways…so here we go!

Send Some Green Stamps & SASE

One way of doing this is using a SASE (**S**elf **A**ddressed **S**tamped **E**nvelope) with so called green stamps (slang for US dollar bills) where you may or may not need to send your QSL card. Some operators ask that you do not send your card as they do not collect them, although I find this to be rather rare. You still want to make sure they know which QSO you are requesting a QSL card for, even if you do not send a card to them.

In the past some used to send or were requested to send an IRC as an alternative to green stamps. IRC stands for **I**nternational **R**elay **C**oupon. **These are now defunct**, and most places do not know what to do with them, including my own local post office. They stopped selling these in the US around 2013. I used to get them sent to me from overseas as late as 2018. In the past you could trade these in for an international stamp to mail a letter with your QSL card or send them to a DX station for them to use. I still have a small batch which was sent to me by stations which I cannot use any longer, mainly from Japan. If you see a reference to IRC, just know it is no longer a viable way of requesting QSL cards in the United States at least. The world has pretty much switched to "*green stamps*".

Always include the SASE envelope if domestic, or just a SAE (no stamp). This should be large enough to fit their card but not too large. Try not to fold anything, it is best to use properly sized envelopes. I recommend a **#9 envelope for the internal SASE and a #10 security type envelope for the external envelope** containing the QSL card and "*green stamp*" bills to cover the return cost per the DX instructions. This way you do not have to fold your SASE as besides being kind of messy it can also play havoc with some sorting machines on the way back. Folding also makes the envelope thicker, and this will attract more attention from potential mail thieves. Speaking of which, **do not put your or their callsign on the envelope to avoid postal theft.** Not so much an issue in the US and Canada, but frankly beyond that I would not. This includes Europe!

Use paper money only, no coins obviously. Usually, $2-4 will do, but varies and the universally accepted currency is US Dollars per above, though there are exceptions. I have even seen as high as $5 which I question a bit outside of DXpeditions.

Do not send US stamps inside the envelope. These cannot be used to mail back to the US from overseas. Vice versa applies as well. They need to obviously use their own local postage stamps. Include your card with QSOs filled out. Use a security envelope as I had mentioned. These are the ones with an internal liner or pattern, to obscure contents to reduce the chance for postal theft. Mail it and wait. It may take a while to hear back due to various factors, so patience is key here.

Be aware that some in countries owning US Dollar is illegal and even get the operator in into trouble so always ask if not sure. I also recommend that you **put extra tape on the envelope flap**. They can come open and items like cash can fall out so sealing it well is important. Also, it is less likely to be tempered with. For some regions where mail theft might be common, or the postal service is less reliable, registered mail might be a good idea. However, check with the recipient first as this may not be feasible or there might be an easier way of going about getting your card. In addition, this can get expensive! Operators' information is generally current on QRZ, but always check with them first.

The Bureau System

You may hear "*buro*" especially via CW. Same thing, shorter name and frankly less likely to be misspelled. You can send your QSL card via the Bureau as well as receive them via the bureau system if the other party uses the system as well. They will usually say so on QRZ or just ask them. In the United States, this system is operated by the ARRL. Over 200 entities are serviced by the bureau system usually via their equivalent organizations.

The way the system works is via mass shipments. Your outgoing cards get sorted into bulk shipments to be sent to another country along with others from your region. They then get distributed locally once they arrive at the destination. The incoming Bureau is exactly the reverse of this. Simple, but **can be very slow**. The cost savings of using this method is very appealing if you are not impatient. I find that I get cards very quickly from certain places in Europe and Asia. Other entities can vary greatly. I had a card once which took seven years to get a response back to, but I have heard of longer. The bureau is all volunteers, which makes me appreciate this even more. If you see your local sorters, thank them for performing this service for the ham community.

If you live in the US, you can find more information about the outgoing services at arrl.org/outgoing-qsl-service and here for incoming at arrl.org/incoming-qsl-service. Note that it is organized in the US by call area, so if you have a 6 in your call, you will claim your cards with the "*ARRL Sixth District DX QSL Bureau*" even if you reside on the East coast now. The exceptions are Hawaii, Alaska, Puerto Rico, Guam and US Virgin Islands which have their own distribution due to geography. Visit the corresponding website for procedures and local practices based on your number. Generally, you will need to have an account with some money on deposit to cover the cost of sending you the large envelopes (hopefully) with all your confirmation QSL cards. This does differ, so check with your local bureau as they may have their own system.

I generally get my bureau packages about once a year and spend the next week sorting them as time allows. I am always looking forward to these and there is always a nice surprise or two in there. To save on postage, generally they wait to send until there are enough cards to make it worthwhile sending.

While I think 99% know this, please **never** send a domestic card to your domestic bureau. Meaning if you live in the US, do not send cards to other US operators. This system is not intended for that, it is to be used for DX only. All folks who sort cards are volunteers, let's not make their lives harder. There are some smaller clubs and organization based bureau services you may want to look into for the above.

If you move, be sure to update your information with them. Also, if you do not want your cards, let them know. Unclaimed cards are an issue and these take up space to store.

Upload your Logs Regularly

While this might be obvious, make sure you upload your logs and do so often. It might be a good time to upload when you do your computer backups, which you should also do often. In fact, uploading is a type of backup. If there are issues, such as a mismatch, or other mistakes, this is a good way to catch and correct them as well. It is a lot harder to correct an issue many years later, so it is best to find these early on and address them.

So, where to upload you ask? Below is an overview of the most popular sites or methods. The pros and cons are mentioned if applicable. Some may not be a factor for you but might be for some hams, so please take that into consideration. A con for some could be a pro for you and vice versa.

LOTW

The Logbook of the World (LOTW for short) is by far the most popular site to upload contact logs to, at least when it comes to DXers. This site is maintained by the ARRL. At the time this book is being written, it has been around for nearly 20 years; started in 2003. Also, it is by far the most complicated to set up and for some to use according to many hams. I have already touched on this a little as well as the accompanying TQSL software. It does require the TQSL software for upload as well as a valid digital certificate. This is what many hams seem to struggle with, at least the first time. However, once you get it configured, you can expect it to do just do its thing and be relatively maintenance free. Be sure to back up your settings and your password for TQSL once you get it configured and working. Screen captures here are also a good idea, just in case! This has saved my bacon before.

To sign up for LOTW can be a little tricky. It does involve confirmation of a mailing address by receiving a postcard with a code. In some entities, this can be a little troublesome.

After installing TQSL, start to configure it. This will be a multi-step process. You will need to request a callsign certificate. You will also be mailed a postcard. This card will have a password to be used when you complete setting up your account. This will <u>not</u> be your permanent password though. Once you enter this password, ARRL will send you a TQ6 file. This is what needs to be loaded into your TQSL software. You will need to configure your station location (QTH) and enter the information requested, such as your grid, state, and additional station data.

As I have mentioned earlier, but worth repeating, there is very well written guide by Gary Hinson, ZL2IFB which you can read at g4ifb.com/LoTW_New_User_Guide.pdf.

This document is about 40 pages and does a wonderful job of explaining the process. I highly recommend reading it as well as **setting up your own LOTW account if you have not yet done so**.

There is no cost to upload, but there is a small fee per QSO when applying for awards as well as some for printed awards, certificates, and endorsement fees. The system is usually relatively fast! Matching takes place all in the cloud. The only negatives for some are no physical cards and the lack of incomplete utilization by some DX stations. Hence the cards being still needed. Some also do not use it due to being tech challenged, but the above guide can solve that for you hopefully.

eQSL

It is gaining a lot of popularity and relatively easy to use. Also offers awards like the ARRL although not always as well recognized. This might change with time. There are a lot more awards available, however, than from the ARRL. To take advantage of their awards you need to be at least a bronze member. The basic eQSL service is free to use and just as LOTW it does require an upload. Website is at eqsl.cc. You likely want to go through the process of getting the "*Authenticity Guaranteed*" for your callsign. This involves uploading a copy of your license. They also offer a life membership option; this is called platinum. Might be worth it, especially for younger hams. I went for this one a few years back to show my support. There are also bronze, silver and gold membership levels, and each unlocks additional benefits.

However, no physical cards are issued, though printing can be requested for a small fee. These are then mailed to your address. This can take time, however. You are not able to use eQSLs for LOTW awards. It is still not as widely adapted as LOTW and many DXers do not use it. Not as accurate in some instances and there have been some technical issues, some of which I have personally experienced. Support is handled by volunteers.

ClubLog

ClubLog can be found at clublog.org and was started by Michael Wells, G7VJR and Marios Nicolaou, 5B4WN. This site is a bit different. It allows one to upload logs, analyze them, tracks statistics and trends. It then displays them in a very useful fashion. In addition, it offers the ability to request cards from some DXpeditions as well as users via an OQRS (**O**nline **Q**SL **R**equest **S**ystem) system. There are other OQRS systems which I will cover further down. You will likely get your card from these rarer DXpeditions via OQRS as well as anyone who sets up an account to do so. This will be mentioned in the DX announcement, their website or QRZ page.

This service, like the ones above, also does require an upload for proper matching of QSOs. Alternatively, you can manually enter your data and time for each QSO when requesting cards. But this is the least efficient way of going about it. **Uploading your logs and/or connecting it to your LOTW account is highly recommended.**

Cards can be requested via the OQRS system directly for a fee or via the Buro. The checkout uses the PayPal system. If you do not yet have an account with PayPal, I do recommend setting one up as you will likely be using it a lot for card requests. Very fast ordering, card delivery speed varies by the actual sender. Amazing data analysis and statistics. No cost to use except for cards requested. Not all DX uses it for card requests, though it is still very popular and growing. This is almost a must for the DXer as it allows for fantastic visualization of your band fillers among other features. I find this site extremely useful in my DX hunting.

HRDlog

This is not as widely used but some logging applications have upload capability to this, notably Ham Radio Deluxe. Website can be visited at hrdlog.net. While visiting the online log, check out their sister site ham365.net as well which offers a very nice interface to give you a quick glance at what is going on in the ham world as far as action is concerned. They share the same database. There is no cost to use either site. No physical cards, electronic only. Both sites are very informative!

QRZ

The de facto "*white pages*" for hams also has awards available as well as another online log which is used often by hams. I get regular requests asking me to make sure I upload to QRZ as well. I do this automatically, but this tells me it is gaining traction, which is great! Don't forget to set up your own page! Website is at qrz.com. Very widely used and respected. Offers its own awards. Some would say a tie for 2nd place with eQSL for award seekers. No cost to use basic features, though well worth upgrading. No are no physical cards exchanged via QRZ, though user pages will likely have information on how to go about it if interested.

QSL Managers

Some DX (and even some stateside stations) use a QSL manager. You can usually find out if a station has a QSL manager by checking on QRZ or their website, which will give more information.

A QSL manager is basically a person who volunteers to manage the outgoing QSL requests for a station. If you live stateside and the DX manager is also stateside you may be able to just send a SASE and your card. You do want to confirm his with them though to be sure as this could vary slightly manager to manager. QSL managers outside your country will likely ask for green stamps (USD), possibly money via PayPal or they may have an OQRS system. Which leads me to…

OQRS Systems

For DXpeditions refer to the official website for QSL information, though likely this will be OQRS. The fee charged is to cover QSL card expenses, postage, and supplies. Usually very little if anything is left over as a "*donation*". Most OQRS systems use the PayPal system. Some of the notable OQRS systems are:

- Tim Beaumont, M0URX can be found at m0urx.com
- Charles Wilmott, M0OXO can be found at m0oxo.com
- Silvano Borsa, I2YSB can be found at win.i2ysb.com/logonline

- Bob Schenck, N2OO can be found at oqrs.net/n2oo

Both M0URX and M0OXO use The Bespoke OQRS systems by Istvan "Pista" Gaspar, HA5AO. This is an excellent system which serves many hams. It seems to be working consistently and flawlessly as do most OQRS systems.

You can receive a QSL card for a small fee in most cases without the need for sending your card, money and/or a SASE. The cards may still need to be checked for LOTW credit as not all DXpeditions upload to LOTW. Can be slow in some rare cases but be patient.

SWL Reports

If you been a ham for a while, you may have noticed that you occasionally receive QSLs for contacts you did not make. These are generally accompanied by "*callsigns*" such as HA234 which do not conform to the usual callsign system in place internationally. If you have, they are more than likely SWL reports from those who do not have a ham radio license (yet in most cases) and are just listening to QSOs. You are likely to see this on eQSL as well as get actual cards, usually via the bureau system. You should and can easily respond to both with confirmation. They might be future ham operators, possibly future DXers in the making. Be supportive of them! These cards will usually indicate details of the exchange, date and time and the station you had been observed working. Check your logbook and if the data matches, just send a "*Thank you for the SWL report, 73!*" back with a card or eQSL confirmation. If the card came via the bureau, you can send it back via the bureau as well. This will be really appreciated by the recipient.

Tips for Fixing Logs, Missing QSOs

I have a few tricks I use to double check my QSO data as well as to recover "*missing ones*" as in ones I did not log for some odd reason. It could be that something went wrong with the software, and I was not aware of this immediately or just needed another cup of coffee. It does happen.

Since I submit to multiple online systems, there are times I use these to cross reference. If something shows up in a couple of systems, likely I goofed! I will check however if I was even active then or if it could be a mistake on the other end. That also does happen.

Every now and then I am not 100% sure I am in the log or may have gotten a callsign wrong. This happens and if it has never happened to you, it will. There are some stations who may not use LOTW, for example but are active on eQSL or upload to ClubLog. I can check if I showed up and got a confirmation via eQSL or if I find myself on ClubLog QSO search before I send out a card or do an OQRS request. At this point, if all good, you can also add this now corrected or updated QSO to your logbook and upload to LOTW once everything matches up nicely.

Additionally, there are times my LOTW does not get confirmed even though I got a card via eQSL or show up elsewhere. Often, it is the time difference or an error on the indicated band. This can be easily cleared up once the discrepancy is identified, and the other station is altered to this. I find that at times, it is easier for me to just adjust my logs a few minutes to match the DX data and be done with it. However, when it is a band issue or the date is off, I will reach out to them, usually via email, and resolve it. This is generally easily resolved if enough data, and in some cases, proof is provided to the other stations. Though this is rather rare.

The most stressful errors may come up during rare DXpeditions. You think you are it the log, but turns out you are in fact, not. This could be due to many things. Maybe you heard wrong, or the other side wrote down your callsign wrong. Or perhaps it was just a dream about a P5 (North Korean) QSO. Though, likely not.

First, check the online logs if they are available. Sometimes these are delayed, so patience is key here. If needed, reach out to the person in charge. You can usually find this on the website, QRZ page and so on. Some DXpeditions will be happy to correct this error, some, maybe not so much. Regardless, always approach them in a friendly manner. You are more likely to get assistance than if you approach them all worked up and angry. This happened once when I nicely asked if they could double check and another ham sent an email asking them *"if they know what they are doing?"*. Guess who had their log corrected. Be nice! We all make mistakes.

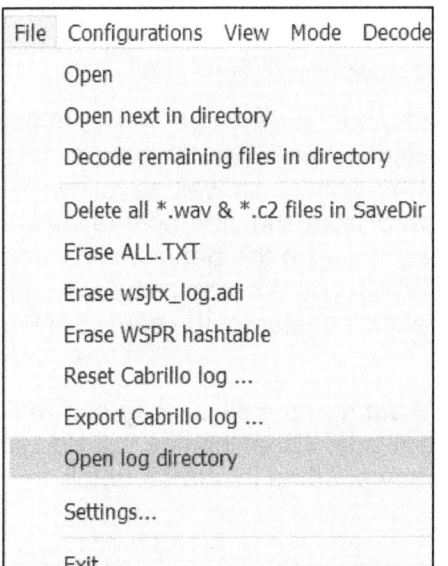

If you had your QSO via FT8 or a similar mode in WSJT-X, you may want to send along this as proof. You can possibly recover the date and time by looking at your ALL.TXT file. This can be found under the file and open log directory as seen here. Also, if you have a QSO that shows up which you are not sure when it happened and you happen to use WSJT-X, the above also applies. A directory will open and the files you are interested in are below. Since these are rather large files, using notepad is not the best choice here. I had mentioned Notepad++ earlier and here this or other alternatives might be a must. You can find it at notepad-plus-plus.org. Notepad may not even open this file if it is too large or at best it will take a very long time as it was not intended to be used for this purpose. Generally opening the ALL.TXT file and searching for the callsign in question will give you the QSO details and will save your bacon. If not, try the other two files but since these files are *"live"* as in being written to actively. Therefore, it is best to stop decodes of WSJT-X. You can also just park on a dead frequency for the time being, so you do not get the *"file has been altered alerts"*.

wsjtx_syslog.log	10/10/2022 8:18 AM	LOG File	1,044,078 KB
ALL.TXT	10/10/2022 8:13 AM	TXT File	250,961 KB
wsjtx.log	10/10/2022 7:56 AM	LOG File	1,007 KB

One last tip. After a successful digital QSO with a rare DX, I make it a habit to screen capture QSOs, just in case! I only had to resort to using this as proof once. Obviously, this is not an option for CW and phone contacts, but I know hams who regularly record their DX contact audio as well. I also do this occasionally too.

Card Checkers

Once you have received a stack of QSL cards from either sending away for them directly, via the bureau or requesting them via OQRS services, time to get credit for them. Some credits you will already see in LOTW or other electronically coordinated logbooks where they get matched automatically. You do not need to submit these again for DXCC credit if they have shown up in LOTW already.

Next, figure out who your local ARRL or CQ card checkers are. You can find a list of all the ARRL card checkers at arrl.org/dxcc-card-checker-search.

For the CQ card checker list visit:

cq-amateur-radio.com/cq_awards/cq_waz_awards/cq_waz_checkpoint.html.

You can likely find card checking at your local ham radio conferences as well as at the International DX Convention, IDXC. Check the convention websites for details. It is important that you fill out the required paperwork ahead of time as well as follow the instructions provided with them. These can be fetched from the appropriate award website above. The cards should be listed in the same order on the paperwork as the stack of cards to make your card checkers' life easier. What I do is organize by callsign, alphabetically starting with numbers if applicable. It makes it easier for the card checker and the order makes sense.

Some card checkers may require an envelope with a stamp as well so they can mail on your behalf. Check with them if not sure, just be ready. Other places will do the mailing for you, such as the IDXC since they are already sending in a large batch.

Meet Other Hams / Conferences

There are countless amateur radio conferences taking place throughout the world at regular intervals throughout the year. There are great places to learn from the knowledgeable presenters, talk to other hams, get exposed to new ideas, meet the vendors and even ask questions. I enjoy it all and always look forward to attending them. All the above can be incredibly valuable resources to enhance your general ham radio knowledge, which includes DXing. Sometimes just talking to other hams, you find that we all seem to have many overlapping issues, problems and certainly similar goals. This is especially true for DXers. As I had mentioned, this was one of the triggers for me to write this book.

One of my favorite things about ham conventions is to have the chance to play with the newest gear and see old friends. Scott Esters, AF6RT is seen here checking out the all new Elecraft K4 Transceiver.

There are some major conventions that are well known, some not so much but do not let that turn you off from attending them. These can also be very rewarding. In many ways they maybe even better! Several have a very specific focus and others are more local ham oriented. Try and find one that fits your interests and/or is near you. Some even plan vacations around Hamventions. If you can, I think this is a super idea. I can certainly see this in my future once I hang up my hat from my day job. The big conferences that come to mind are the Dayton Hamvention, Friedrichschafen (yes, that is the real name) and of course the International DX Convention in central California.

You will also find that you will likely make new friends, see old friends and at times get to meet "*The DX*". This is especially true for the **International DX Convention**. The IDXC takes place every year in April in central California, halfway between Los Angeles and San Francisco. You will always find me in attendance. Hope to have an eyeball QSO one day with readers of this book. Come and say hi, let me know your thoughts. Good or bad, I am sure I can always improve something. You can learn more about IDXC at dxconvention.com.

The **Dayton Hamvention** is considered the "*big one*" for the United States. While not DX focused, lots of goodies here for the DXer and guaranteed you will not be bored. In fact, due to its size, you may even have a hard time getting to everything you want to see and do. More information on this can be found at hamvention.org.

For the biggest conference in Europe, the **Ham Radio International Amateur Radio Exhibition** in Friedrichshafen Germany takes the cake. You can find out more about it at hamradio-friedrichshafen.com. For the largest conference in Asia, the **Ham Fair** in Tokyo, Japan is one to attend. You can find more info here as it becomes available at jarl.org/English. Both are on my bucket list.

Lastly, my local convention here in Northern California is Pacificon. This takes place in October and brings in folks from not just all over the West coast but also some out of state and even international visitors looking for an excuse to visit the San Francisco Bay Area. Not that one really needs a reason. The focus is more general but there is something for everyone, including a separate antenna forum. You can find more information at pacificon.org. No matter where you live, there is one near you at some point during the year. I have four at last count within a few hours of drive from me, besides the IDXC.

Keep Learning for Life

If for no other reasons for you to keep learning here are two. First, there might be an aspect of the hobby maybe you want to try. Be open-minded! At times these might be something you never thought you would be interested in and turns out you are. This happened to me at least twice already during my career as a ham. Some hams get interested in ham satellites, ATV, DMR, EME and so on. FT8 is a recent example where many hams who have been active for decades decided to embrace it and some even got into digital modes for the first time ever. Good for them! Obviously, DX is one, perhaps that is why you picked up this book.

Second reason, you do not want to be left in the dust. There are many hams who have been licensed for decades. Many are very active and current. They keep up where needed and are lifelong learners. However, in some rare cases they will refuse to change their ways even if the rest of society has moved on. There are still hams who will not use a computer or trust anything without tubes. A less extreme example is someone who refuses to try digital modes. Which is their decision, but they are really missing out.

With some, there is a fear of technology. SDRs are a perfect example where some seasoned hams initially turned their nose up to. Remote operating and WSJT-X are others which come to mind.

As I am writing this section at the start of 2022, Windows 10 is the standard and Windows 11 is being rolled out. Several hams I know are still using Windows 7 which is now considered obsolete and frankly a dangerously unpatched operating system. Hope they don't do any banking on it. When asked, the biggest reason given is they do not want to learn anything new. This is too bad, and I am sad to hear it. The benefit of the very little time it takes to educate yourself far outweighs the positive benefits one will get for being more current. New doors will open, and life will likely also be easier.

There are many great resources out there for continued education. One example is **YouTube**, where you can find tutorials on pretty much anything related to ham radio. I often referred you to this resource in this book.

Ham Nation is a weekly show covering many topics relating to ham radio with nearly 500 episodes. I recommend revisiting the older shows as there are many educational moments. They briefly stopped production at the end of 2020 but episodes are available on Twit.tv specifically twit.tv/shows/ham-nation. Ham Nation is currently available at its new home at youtube.com/@HRCC.

Amateur Logic can be found at amateurlogic.tv but it also available on youtube.com/c/AmateurlogicTv as well as many podcast providers including a Roku channel. There are monthly episodes on various topics for all interests. I have been a loyal viewer for many years.

Ham College is brought to you by the fine folks at Amateur Logic and is designed to help people get their ham radio license or upgrade. You can locate these videos on the links provided above. If you still need to upgrade your license, this is an excellent place to start.

Dave Cassler, KE0OG has a website at ke0og.com as well as an incredible YouTube channel at youtube.com/c/DavidCasler both of which are well worth checking out. Many varied topics relating to ham radio with countless videos. You can pretty much find videos on any given topic, and he also answers questions sent to him by loyal viewers.

Many clubs have a YouTube channel as well. The **Northern California DX Club** (NCDXC), for example, posts some of their club presentations as do many other DX and local clubs. Even if you are not near one, or near one which shares your passions, there are no excuses. You can always join a DX club as an associate member, even if you are outside their region. Some DX clubs even allow hams outside their regional zone to join as a full member!

There are many magazines out there worth reading to learn new things and many also provide you with DX news as well as a nice plus. **QST** from ARRL is something you get automatically with your ARRL membership. You can find out more about it at arrl.org/qst. ARRL also offers **QEX** which is an experimenter's magazine. This is something requiring an additional subscription as does **NCJ** (**N**ational **C**ontest **J**ournal) which I had already mentioned. QEX Information is at arrl.org/qex and NCJ information is at arrl.org/ncj.

CQ Magazine is also a magazine with quite a history and covers various topics of interest to all hams, including DXers. You can find out more about CQ magazine at cq-amateur-radio.com. I am a long-time subscriber.

Spectrum Monitor Magazine is an online only magazine which is well worth getting. It covers a variety of things related to radios. Sometimes I get some great project ideas from reading it. Somewhat different and a nice complement to the two magazines I had mentioned above. You can find it at thespectrummonitor.com.

The actual websites of some manufacturers can be true gems. At times packed with great information where one can learn to use something even before it arrives, for the most part.

Give Back if You Can

Finally, per the title, give back if you can! Not everyone can, I understand that. One way to give back is to become a VE (**V**olunteer **E**xaminer) for the ARRL or other VE teams out there. I have met some great future hams doing this over the years. Not just on the VE teams but also new hams and those upgrading. This is a great place to start, and you may just create more DXers. To find out more about the VE program, visit arrl.org/become-an-arrl-ve.

Share the knowledge! You might think you have nothing to add but you might just be surprised. I am often surprised! 5-10 years ago, I would have laughed if you told me I would be writing a book on DXing. Guess what, you are reading one now. Back then my money would have been on a book on satellite communications or radio modifications. Who knows, those may still happen.

You may have had experiences which can be used as teaching moments for others. Perhaps you have tried and failed or hopefully succeeded. Perhaps unexpectedly at either. That happens too and more often than you may think. You can share these experiences by joining support groups, helping young hams, beginner DXers or even those just looking to do something new and possibly "*scary*" to them. Trust me, they will appreciate the time you take by giving them a lending hand.

Another way to give back is by doing presentations for a local ham radio club. Though the magic of the internet, pretty much anywhere really. You do not have to be an expert in a topic. As I had mentioned, very few if any are.

Lastly, **always be positive and welcoming!** If you do not like a certain mode, don't use it. No need to bash it. If you do not like something about a local club, get on board and help improve it. If you see a young ham attend a club meeting, go and say hello! **They are the future of DX.**

I get a lot of email via my ham radio website. Many are from overseas hams, some from those with DX questions. A lot of the questions are the same or at least similar. When I see this, generally I update the website with more details. As I had mentioned in the beginning, much of it made it into this book. If I ever do a second edition, the same will be true. So, I welcome and look forward to your direct feedback!

Patience is key to achieving DXing goals you have, and courtesy should be a given and should come first. Let's all play in the sandbox together nicely. The most Important thing is to enjoy this wonderful hobby and have fun!

CHAPTER 12: Available DX Awards

One may ask, so why DX? One reason is awards, though for me it is secondary. Some DXers do not even bother to apply for most awards, if any. They find enjoyment in the chase and the reward in making the contact. And I get that. My feeling is, while you are at it, you might as well apply for at least some of them. Afterall, you need to put something on your ham shack wall, right? Maybe time to cover the ugly wallpaper or the wood paneling?

Many others, including me, eventually go on and become "*wallpaper*" collectors rather quickly. Meaning, if there is a meaningful award earned, I am likely going to apply for it. Why not! There are some very nice-looking awards out there, including some rather nice plaques. Sometimes these are referred to as "*lumber*" for historical reasons as they were made of wood. However, many of the new awards and some of the older ones which are still being issued, are now made of acrylic. I have mixed feelings about this, but they are still rather nice in most cases.

There are a wide range of awards out there, but since the focus of this book is DX, I will concentrate this section on HF DX related awards mainly.

The Prestigious DXCC Award

The best-known award for DX is, of course, the ARRL DXCC award. Which basically means that you have worked 100 entities and have confirmed them. This can be any combination of phone, digital or CW modes. I will cover later in the book on how you go about confirming. You can read up on this award in more detail at arrl.org/dxcc. I also encourage you to read up on the rules at arrl.org/dxcc-rules. This award has been around since 1937 and had to make various changes over the years. Wars and border changes are the biggest. Countries reunite like Germany **(D)** or split, like the former Yugoslavia **(YU)** and Czechoslovakia **(OK-OM)**. Of course, the biggest change was the collapse of the Soviet Union **(R)**, which now has 15 different entities. And these are just the more memorable ones in relatively recent history.

When it comes to entities, there are other places similarly split up into multiple entities for DXCC purposes, much like the US with Hawaii **(KH6)** and Alaska **(KL7)**. Another example is Greece **(SV)**. Dodecanese **(SV5)** and the hard to work Mount Athos **(SV2A)**, at least from the US West Coast, are counted as separate entities.

You likely already picked up on the fact that I use the word entities, not countries, as this is how the ARRL, who administers the award, counts them. This is also true for many other organizations which offer awards. Currently (as of 2022), there are **340 official entities**. This goes up and down as political boundaries change, or mother nature takes its course as in the case of some locations which had to be removed as they are no longer above water.

To apply for the DXCC award, if you reside in the United States, Puerto Rico or any US territorial possession, you need to be an ARRL member. If you live outside the United States, you do not have to be a member, however you still can be one if you like. There are some benefits to being a member.

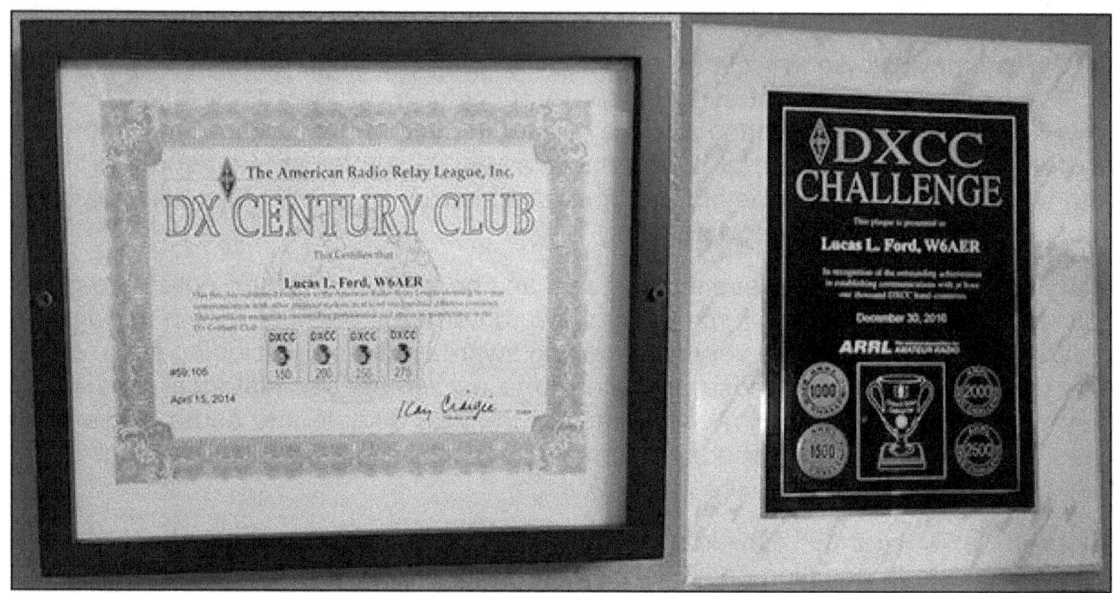

Above are two examples of the awards we are covering. On the left is the DXCC award, in this case mixed. The right is the DXCC challenge award, with a 1500 endorsement.

Those selected few who get to the super high challenge numbers compete once a year for the DeSoto Cup. This would be the place holder in the middle of the award (right picture). This is awarded to the person with the highest challenge number count. Only one is awarded per year and it is considered to be the pinnacle of band filler achievement. I will discuss the DXCC challenge in more detail shortly.

In addition to the mixed variety, the DXCC award is also available for specific modes such as CW, phone and digital modes. Meaning, you had worked 100 entities in one of those modes exclusively. Currently, the digital award would be the easiest to obtain, due to the popularity of FT8. I will cover this in the digital section in more detail. During good propagation and/or heavy contesting, the other ones should not pose too much of a challenge either with the right gear and knowledge.

The goal is to get on the "*Honor Roll*" and eventually if you work hard enough, live long enough and with some luck **"Honor Roll #1" meaning you have worked them all**. This can take a lifetime for many, and most hams never get there.

To get on the DXCC Honor Roll, you must be within the top 10 entities. Currently, our entity count (as of 2022) is 340, therefore if you have 331 worked and confirmed, you are on the honor roll. A total of **340 is Honor Roll #1**. Some may have more than this number and that just means that their count includes deleted entities. The 340 for honor roll must consist of entities <u>currently</u> counted towards it, meaning no deleted ones.

To apply for the DXCC award, you will need to prove either via LOTW (more on this later) and/or with physical cards that you have worked at least 100 entities. These do not have to be on the same band or mode. It is perfectly OK to have a combined total of at least 100 in any variation. You will however need at least 100 for the initial application. I will cover card checkers and parts of this process also in more detail later in the book.

There are endorsements for working more entities, and these are usually issued in a block of 50, such as 150, 200, and then in smaller increments such as 310 and so on as you get to a point close to the working them all. You can add to your total at any time though once you got the first 100 credited, meaning you have earned the basic DXCC award.

There are other DXCC related awards, such as the 5 band DXCC which means you worked 100 entities (do not have to be the same ones) on 10,15,20,40 and 80 meter bands. Endorsements are available for 6, 12, 17, 30, and 160 making it a 10 band DXCC.

On the left is the 5 Band DXCC award from the ARRL. In this case, endorsed for 12, 17 and 30m bands. On the right is the DXCC equivalent called eDX from eQSL.

You can find out more about eQSL at eqsl.cc which is gaining momentum in popularity. LOTW still seems to be king of the hill, but more and more hams are embracing eQSL as well.

The DXCC Challenge

We have already touched on this somewhat but let me further clarify what this is about. This challenge program is based on the DXCC award, as the name implies, and the same exact criteria is used to determine an entity count. The ARRL also administers this award, as you likely guessed from the name. To put it simply, it uses *"band points"* for determining progress. **Mode count does not matter.** Many hams, even seasoned ones, get a bit confused by what counts and what does not.

Basically, one point is scored per entity, per band. It does not matter which mode it was, and no double dipping here. If you worked Hungary (**HA**), on CW and again via Digital on 20m, it is still just one point total. Even if it was a different station. Hungary on 20m counts once. Period. No matter how hard you may have worked to get that contact. Sorry!

However, if you worked Hungary on 40m (any mode) in addition to the 20m contact, then it is now considered two points. One per band.

Let's say you had worked 100 entities on 15m, 100 on 20m, and 100 on 40m bands, all ignoring modes and duplicates. You now have 300 points in total. In just 700 more, you can get a plaque that can be endorsed in increments of 500 as seen earlier. Getting to 1000 is not as hard as it sounds, getting past 1500 is rather hard for most. I certainly find it a *"challenge"* to hit 2000 as pickings get slim and things sure slow down. 2500...yeah...might be a while! Think retirement.

When Contacts Don't Count?

Sadly, sometimes some contacts do not count towards certain awards. This varies by award, so always check. Rules can and do change. Personally, I just work all bands, modes, stations, and sort it out later. Although I do make a conscious effort not to work the same station twice via the same band and mode, other than contests of course. More on these late. Contacts which did not count for anything at some point, might just earn you additional wallpaper or timber down the road. Here is a little overview with some examples of what I am talking about.

The Odd Bands

There are some bands that we are allowed to operate on in the United States and in many other countries. **Some of these bands are not eligible for the DXCC award**, at least not yet. As I had mentioned, my take on this is, let's work DX there too and if it counts one day, I am that much ahead in the game.

The 60m band (5MHz) is the one you likely notice right away, is not on the list for DXCC endorsement. This band became available to us in 2002 but adaptation in other countries has been slow until recently. Initially, we had 50 watts PEP (**P**eak **E**nvelope **P**ower) and this has been raised by 3dB to 100 Watts as of now for the 60m band. **Don't forget the 60m band power limit just as on the 30m band, so leave those amps off!** Let's not annoy anyone and ruin the chances of having this band added to the DXCC program.

Channel 1	5330.5 kHz
Channel 2	5346.5 kHz
Channel 3	5357.0 kHz
Channel 4	5371.5 kHz
Channel 5	5403.5 kHz

All modern HF rigs support this band, yet many hams do not operate here. DXCC rules hopefully, this will change soon as there certainly is DX on this band and frankly a surprising amount. I have worked quite a few now myself, although 95% on digital. Specifically, the FT8 mode. As a side note, if you are not familiar with this, the United States 60m band (5351.5-5366.3kHz) is channelized. In the US 5 distinct frequencies are used, shown on the left. The **60m channels may**

not match up region to region. Be sure to check if your country permits operating on this band. If not yet, I suspect it will be coming soon.

The US channel 3 allocation is mainly FT8 at 5.357MHz and this is where you will find most action on 60m currently. Most 80m antennas such as verticals, with the help of an antenna tuner, work fine here. Alternatively, you can make your own dipole if you have the space. I have even managed to tune up my 160m antenna for the 60m band. If you have a 43ft vertical, it should be a breeze to use for 60m as is a near match.

Those in the US and elsewhere may not be familiar with the "*other magic band*" which is the 4m band, located at around 70MHz. The main magic band being the 6m band 50-54MHz in most entities. In most parts of the world, the 4m band is not available to ham radio operators and therefore not recognized by many organizations giving out awards. Last count, under 50 entities are allowed operating here. I would not hold my breath!

There are two additional bands which are rather new and are not heavily used yet. They are also not eligible for the DXCC award. I suspect it might be a while before they are, or possibly never. Though I hope this is not the case. **These are the 630m and 2200m bands.** One thing to note is if you are planning on operating on this band in the United States, you will need to clear it and register with the Utilities Technology Council, also known as the UTC for shot. You can do so at utc.org/plc-database-amateur-notification-process. If they do not voice any objections within the set amount of time, as in you will not interfere with anything nearby, you are ready. Just be sure to carefully read the current requirements on the website as these may change. I really hope to see more activity here in the future.

Going Fishing? Will not Count for DXCC!

The **Maritime Mobile** designator, for example W6AER/MM while at sea, **does not count for an entity as it is at sea, not on land**. However, more than likely if you are far out at sea, it is a wanted grid square. So, if you hear a Maritime Mobile station and may one day want to go after a grid award, go for it! The same applies also to aeronautical contacts /AM or Aviation Mobile. For the entity to count for DXCC and similar award, the contact must be made from a landmass, even if just a "*rock*" in the middle of the sea. And while this sounds funny, this has occurred before. Many might recall BS7H, Scarbrough Reef 2007. It is worth a quick search on the internet. You will understand my rock reference.

There are a few other rare instances where the DXCC desk will not count a contact for DXCC credit. This issue could be, for example, not being on land completely. This happens more than you think. One cannot operate from a ship docked next to a landmass or a harbor for credit, this is still Maritime Mobile per above. Recently I had a few contacts which were not eligible for DXCC credit as the person was on a ship docked at a harbor. Luckily, I did not need this entity or the band filler, but they should have been identified as /MM. Rules are rules for a reason!

No Permission Granted

For many lesser activated locations on the planet, **the entity or the governing country must give permission to the operation and/or operator.** Therefore, one can't just sneak into a rare county (like North Korea if you are daring enough), or sail to a protected island, operate and have it count for those who worked it. Or technically even operating legally for that matter. Not as simple as it sounds, at least in some cases.

Depending on where the entity is, you may even need permission from other agencies, such as the USFWS (**US F**ish and **W**ildlife **S**ervice) or international equivalents to be able to land or dock. This is in addition to a ham radio license in cases where reciprocal agreements do not exist.

Furthermore, the DXCC desk, as well as other award programs, may ask for proof that the DXpedition actually took place. This may not only involve copies of permits but also proof that the operation took place such as photographic evidence. Why you ask? Well, if you talk to some DXers who have been around for a while, they can tell you stories of a few "*faked*" DXpeditions in the past. This is a topic beyond the scope of this book and frankly could take up another book.

What Else is Out?

Calls made using the aid of a **repeater rarely count. The notable exception is satellite communications**, where that is the entire point. Yes, a ham radio satellite technically is a repeater! This "*not allowed list*" basically also includes any VOIP (**V**oice **O**ver **IP**) service such as DMR, Echolink, D-Star, C4FM (Yaesu Fusion), or any other system that involves the internet for passing data instead of actual radio signals. There are some awards for these as well, not just via the ARRL or awards in relation to working DX per se.

One-way contacts, such as WSPR, while great for propagation studies, do not count either as it must be a 2-way contact. There are some awards for reception reports, but they do not count towards your DXCC totals.

Of course, if the other station decides that it is not a complete QSO, for example if no 73 was exchanged (even if let's say signal reports are), it is invalid. Simply does not count. I know this is silly in some ways, but both parties must confirm for it to be valid or at least agree for it to be valid. Some folks are very "*old school*" when it comes to this, and I tend to fall into this category myself. I would like to know that the other party got the report, and I can safely log them.

And lastly, there is a sad example of what does not count. One of which we got reminded of recently (as of 2022) when Russia **(R)** invaded Ukraine **(UR)**. During wartime, many countries may suspend amateur radio operations. While this is in place, contacts <u>may</u> not count. This varies, so always check! Not only have I seen Ukrainian hams still on the air as I am writing this. Tt seems to be business as usual with stations active in the occupied and free zones. Now, whether all these contacts will be considered valid remains to be seen.

Worked All Continents Award

This award is commonly called or abbreviated as **WAC**. Administered by the IARU also known as the **I**nternational **A**mateur **R**adio **U**nion and can be applied for via the ARRL since they administer it. If you use LOTW, anything matched there will be counted. Antarctica does not count as a continent here though. This award can be endorsed for bands once you get your five band WAC award. You can see the rules at arrl.org/wac.

This award is not as tough to obtain as the DXCC award since technically it only requires one entity from each continent. One of which you will be operating from, so that should be a breeze. You can find more information about the IARU at iaru.org. This is likely an award you will naturally be eligible for once you start to DX more seriously. If you have been at it for a while, I guarantee you already have this or could have been hanging above your shack door.

CQ DX Award

This, as you may have guessed, is offered by CQ Magazine. It is basically the equivalent to the ARRL DXCC award. These awards are awarded by modes only, meaning SSB, CW and Digital. You can visit CQ Magazine at cq-amateur-radio.com for more information on this award.

Worked All Zones Award

This one is known as the **WAZ** and is administered by the friendly folks at CQ Magazine as well. There are 40 zones total to collect, and this can be rather tough. Some feel this is much harder to obtain than the DXCC. I agree fully. There are many well-seasoned hams who are yet to obtain it. Some zones can be very elusive, especially from tough DX spots. This award can be endorsed by band as well. There is also a 5-Band WAZ award for working all zones on 10,15,20,40 and 80 bands. That is one big accomplishment!

You can look up the WAZ zones at cq-amateur-radio.com/cq_awards/cq_waz_awards/cq_waz_map.html or at mapability.com/ei8ic/maps/cqzone.php.

Just as with the DXCC, Maritime Mobile, Aeronautical Mobile, Cross-band or mode contacts are not eligible.

Collecting Prefixes

CQ Magazine also has awards for collecting prefixes called **WPX**. It is worth looking into and chasing. Many DXers, including me, do just that. This award considers the prefix only and therefore it is easy to get over 1000 rather quickly. If you worked me as W6AER, W6 is the prefix. If I travel to Oregon and use the W6AER/7 indicator (Though I do not have to) then I count for a W7 towards your prefixes. Internationally, for example in Mexico I am XE2/W6AER and it would count for XE2. Some hams use CALL/XE2 instead. Both technically work, you will encounter both orders at times. When you are the DX, generally you want to start with the country you are operating from, then your home callsign.

These can be applied for via your LOTW account online as well and therefore is super convenient. You can get more information at arrl.org/cq-awards. Just please be aware that there will likely be additional charges from CQ magazine before you get the award.

This award has a very nice, classy look but does take a while to receive once initially ordered as they involve hand calligraphy and therefore are not mass produced. There is separate "*wallpaper*" for mixed, phone, CW and digital. There are endorsement stickers given, much as for the DXCC award. Information on this award can be found at cq-amateur-radio.com/cq_awards/cq_wpx_awards/cq_wpx_awards.html.

Hunting for Islands

Islands on the Air (IOTA) is another award available for DXers and is administered by the **R**adio **S**ociety of **G**reat **B**ritain (RSGB). This award counts islands (or groups of islands) worked as entities only and there is a list available on their website. It is one of the most popular awards DXers chase after the DXCC.

I rather enjoy chasing this award. There are 1200 Island groups and, in most cases, multiple islands within these. Chasing this award can keep you busy for a while if not forever. There are still islands out there which have never been activated. The endorsements are in blocks of 100 as well as regional endorsements. Visit rsgbiota.org or iota-world.org for more information.

Other Notable Awards

Worked all Europe, also known as the **WAE** award, is from DARC. **DARC** is short for the **D**eutscher (German) **A**mateur **R**adio **C**lub. You can get more information at darc.de/en/der-club/referate/committee-dx/diplome/wae-award and no, you do not have to work all of Europe to get the basic award. This should be relatively easy if you are already in Europe or on the US East coast.

The **AJD** (**A**ll **J**apan **D**istricts) and the **WAJA** (**W**orked **A**ll **J**apan **P**refectures) awards are given for working Japanese stations. These are much easier to obtain from the US West coast than other parts of the US. Band and mode endorsements are available. While these are the two most popular, there are other awards offered by JARL. JARL is the **J**apanese **A**mateur **R**adio **L**eague. You can read up on their awards at jarl.org/English/4_Library/A-4-2_Awards/Award_Main.htm

Worldwide Antarctic Program (WAP) is also an interest of mine. The **W**orked **A**ntarctic **C**allsigns **A**ward (WACA) is obtained by working 10 different callsign from this polar region. You will need to collect physical cards for this award. I especially enjoy getting cards for my arctic contacts, perhaps even more than tropical photos on QSL cards. Everyone seems to like to put palm trees and nice beaches on their cards, very few can use icebergs and penguins. For more information visit waponline.it.

There are many non-DX specific awards out there also, some of which I will go over very briefly. I will do this, as not all readers are in the US and this might be of interest to those outside of the United States. Afterall, for those outside the US, we <u>are</u> DX!

Awards such as **Worked All States (WAS)** are earned for working all 50 states (48 mainland, Hawaii and Alaska). This can also be endorsed by band and mode as well as the digital award can be endorsed with additional digital modes, which keeps things exiting. If you are interested, you can read up on it at arrl.org/was. This award can be quite a challenge on some of the bands, specifically 160m and 6m. Some have also accomplished getting this on the 2m band. Hats off to them. Seriously! That is no small accomplishment.

There is also something called the **Triple Play Award** for working all 50 states in 3 modes. That is 50 states each on CW, phone, and digital, for a total of 150 contacts towards this award. Many International operators go for this award and is even challenging to get from within the US due to some of the smaller or less active states. Delaware and Rhode Island come to mind first. North Dakota can also be tough. California and Texas, not so much!

There is even an award called the **WUST (W**orked all **US T**erritories) if you work them all and can submit proof. These US are territories, which are <u>not</u> to be confused with US states. Better known territories include Puerto Rico **WP4**, US Virgin Islands **KP2**, Guam **KH2**, American Samoa **KH8**, Northern Mariana Islands **KH0,** and Guantanamo Bay **KG4** which is an odd one as it is on the main island of Cuba **CO**.

Lesser-known territories are Baker and Howland **KH1**, Desecheo Island **KP5**, Johnston Island **KH3**, Midway Island **KH4**, Palmyra & Jarvis **KH5**, Kingman Reef **KH5K** (now deleted as underwater), Kure Island **KH7K**, Swains Island **KH8S**, Navassa Island **KP1** (used K1N last activation, pictured) and of course the famous battleground from WWII, Wake Island **KH9**. You can get more information about this at metrodxclub.com/award-application. *Photo courtesy John Miller, K6MM.*

The **VUCC (VHF UHF Century Club)** uses grid squares instead of entities. You can find out more at arrl.org/vucc. This is a very popular award on the 6m band as well as for amateur satellite operators. My other ham radio passion. A very select few have even accomplished this on the 2m band as well! Again, hats off!

Some operators take this a step further and try and go for the relatively new **Fred Fish Memorial Award** which was established in 2008. This award is named after Fred Fish, W5FF (SK) and is received only if you have worked all 488 maidenhead grid squares in the lower 48 on the 6m band. Meaning all grid squares in the US states outside of Hawaii and Alaska. When I say a step, I mean a huge step further. Fred was the first to do so and this award is named in his honor. Some of these grids can be rather challenging and without rover stations, this could never be accomplished. Rover stations are those which are set up temporarily from a grid to operate where there a few or at times no operators. You can find more information on this award at arrl.org/ffma.

There is even an award for **Worked all Counties** in the United States which you can find information about at countyhunter.com. There is a total of 3077 of them. The program is run by CQ Magazine, and the initial award is given once reached 500 then endorsed in increments. It is a tall order to get them all, or to even get above the 2000 mark. I know this since I am stuck there myself. Some of these counties have very few operators as I had mentioned before. There are many, especially in the Western US, that have none! Therefore, some counties require activation from a rover or visitor. I know many international operators outside the US who chase this award as well.

There is a version of the **DXCC award for QRP operators**. QRP is used in reference to low power operating, usually 5W or less. I must say I am very impressed with those who have managed to get this award and blown away by those with an over 200 entities count using QRP. That is an amazing accomplishment. The award is based on an honor system, but I suspect most obey the rules. Hams for the most part tend to be very honest folk. I also suspect all these operators have amazing patience, not to mention very efficient antenna systems. Of course, if you ask a *"true DXer"*, they may jokingly say QRP is using only 100W. Don't take it literally. Not true!

Additionally, there is a **satellite version of the DXCC award**, but due to the current satellites out there at the time of the writing of this book, it is nearly impossible to obtain. At least from the United States. On the other side of the planet, the QO-100 geostationary satellite is keeping hams busy, and I will admit, I am very jealous! For those who may not know, a geostationary satellite means it appears to stay in place when observed from earth. Meaning, it orbits at the same speed as our planet. Much like the ones used for the reception of satellite television, hence your dish never having to move. A geostationary satellite or a HEO (**H**igh **E**arth **O**rbit) satellite versus the current LEO (**L**ow **E**arth **O**rbit) satellites would be a true game-changer for us. I know AMSAT has got something in the works. This might be a topic for another book one day. It is very hard to get above even 10 entities currently from California.

Lastly, if you still need more wallpaper and lumber on your wall, check out the DX awards website by Theodore Melinosky, K1BV at dxawards.com where he details everything else available out there. There might just be one not covered here you are interested in.

Final Thoughts

Let me start off by thanking you for reading this book. Hopefully you have enjoyed your journey from karaoke to rockstar. Perhaps you were already there but were curious or wanted a refresher. I hope this book helped make you a better and more successful DXer. If nothing else, at least made you think about a few things and perhaps you have even learned something new. Maybe even tried something new!

It took a long time to get this book finished. In fact, much longer than I expected it to take. Nevertheless, I had a blast writing it. I even learned some new things as I started fact checking and researching topics further. This book also became much larger than I expected. I even had to change the format size for the paperback to accommodate for this. I decided to do that instead of editing it down to make it fit a smaller format. Afterall, I promised a "*complete*" guide to DXing and I wanted to live up to my word as much as possible. Frankly I still feel some things could have been included. Maybe next time?

If you did enjoy the book, I would love to ask you to please leave a positive review at your place of purchase. Your positive rating and if you have the time a brief written review would be greatly appreciated. This will also help other hams and potential new DXers in the making to discover it.

If you would like to stay informed about other publications or upcoming books, please sign up for the VitaminDX publishing newsletter at vitamindx.com to be notified. And if you are a visual person, check out my new YouTube channel @MaximumRF, alternatively maximumrf.com. You will even find a video there on how this book came about if you are interested, amongst other topics related to this book and even other aspects of amateur radio.

Lastly, if I decide to do a 2nd edition, I would love to get your direct feedback. You are more than welcome to reach out to me about what you would like to read more about, what you feel might be missing and whatever else comes to your mind.

You can send me email via **Lucas@w6aer.com**

Hope to hear you on the air, and most importantly, good DX!

73 de Lucas, W6AER

Glossary

59 – Signal report used on SSB for DX and contest exchanges.

599 – Signal report, usually sent as 5NN during CW DX exchanges.

73 – Best Regards/Wishes.

88 – Love and kisses, not often used.

AC – Alternating Current, used in household power.

AF – Audio Frequency, 20Hz to 20KHz generally for human hearing.

AGC – Automatic Gain Control.

ALC – Automatic Level Control.

Alligator – Ham using too much power who cannot hear well.

Amplifier – Used to increase power output of your rig.

Anderson Powerpole – Type of connector used for power in ham radio.

Antenna Farm – Requires no water, is a collection of antennas by a ham.

APRS - Automatic Packet Reporting System.

ATNO – All Time New One, an entity not worked before.

ATU – Antenna Tuning Unit.

ARRL – American Radio Relay League.

AWG – American Wire Gauge, measurement of wire diameter. Smaller number refers to larger diameter wire. Sometimes just # is used.

Balun – BALanced to UNbalanced.

Band – A range of frequencies.

Band points – Used to count entities per band worked for the DXCC Challenge points.

Bandwidth – How much spectrum space a given signal takes up.

Barefoot – Not using an amplifier. Generally, means 100W or less.

Beacon – Used to check propagation, transmitted at given intervals and usually at different strengths to help determine conditions.

Beam – Generally an antenna with directionality and gain, such as a Yagi.

Big gun – A large, powerful station.

BNC – A type of connector short for Bayonet Neill-Concelman but sometimes defined incorrectly as British Naval Connector.

Boat Anchor – Vintage gear, generally vacuum tube.

Brick – Used in reference to an AC or DC adapter which plugs in the wall.

BTU – Short for "*Back to you*" in CW and some conversational digital modes.

Bug – A Semi-Automatic, mechanical CW key.

Bureau – sometimes "*buro*" for short. Used to send and receive QSL cards between entities and then forward to operators.

Cabrillo – A log format usually used in contesting.

Callsign – Amateur Radio operators ID with this, a combination of assigned letters and numbers.

Cans – Ham slang for headphones.

Capacitance Hat – Found on top of verticals to increase efficiency and to reduce resonant frequency.

CAT – Computer Aided Tuning/Transceiver.

CI-V – A Version of CAT control use by Icom. See CAT.

Cloud warmer – Slang for antennas, sometimes referring to antennas radiating inefficiently or NVIS antennas.

ClubLog – A web-based service with many features, including OQRS requests

Coax – An unbalanced transmission line with an inner conductor surrounded by a shielded outer conductor.

Contest – Also known as "*radiosport*" not to be confused with the headset of the same name. Operators compete against each other for contacts. Contests are scheduled ahead of time and have specific rules and exchanges to be followed.

Counterpoise – Sometimes called an elevated radial system. Used when ground radials are not practical or preferred to give an antenna something to work against.

CQ – A request for a QSO (contact) to any station. Can be directional such as CQ EU or CQ DX for non-domestic contact requests.

CQ Magazine – Ham radio publication, also offers a couple of DX awards.

CQ Zone – Consists of 40 zones, used for contesting and award purposes mainly.

CW – Continuous Wave, also see Morse Code.

EIRP – Equivalent/Effective Isotopically Radiated Power. The actual power leaving your antenna accounting for losses or gains in antenna design, coax loss after initial power output.

EMI – Electromagnetic Interference.

D Layer – Also sometimes called the D region. Closes region to the earth. It is a daytime layer and is responsible for lower frequency absorption.

dB – Decibel. Logarithmic scale measurement unit.

dBd – dB measurement based on ½ wave dipole, 2.15 dBi.

dBi – dB measurement based on isotropic radiator

DC – Direct Current, used in batteries, many lower power electronics.

Dipole – A bidirectional antenna cut to ½ wavelength.

DSP – Digital Signal Processing, could be hardware or software driven.

D-Star – Digital Smart Technology for Amateur Radio.

Duty Cycle –The time spent in receive versus transmit cycles. Full duty cycle means 100% power out.

DQRM – Deliberate interference, QRM is manmade Interference. It is the combination of the two.

DX – Distant, refers to a station outside a given geographical area.

DX Cluster – A list of reportedly active ham radio stations.

DX Net – An HF net with the specific purpose of working DX.

DX Window – A section on certain bands unofficially designated for DX contacts only.

DXCC – DX Century Club, an award offered by the ARRL for working at least 100 entities.

DXer – A person who enjoys DXing. DXer is also the name of the monthly publication of the Northern California DX Club. See NCDXC

DXpedition – The activation of an entity for amateur radio contact purposes.

E Layer - Also sometimes called the E region. It is the middle layer between D and F. This is also the layer where sporadic E occurs. It can absorb and reflect radio signals.

Elmer – Someone who assists other hams, especially new ones.

EME – Earth Moon Earth, also known as Moonbounce.

Entity – A country or territory as counted towards the DXCC award for example.

ESP – Sometimes jokingly used to reference barely audible signals.

eQSL – A web-based QSO confirmation system.

F Layer - Also sometimes called the F region. These recombine F1 and F2 at night as is the layer most responsible for ham radio propagation. It is the outermost layer of the 3 layers.

FB – Used in CW, short for *"fine business"* used on sideband. In other words, good or all received.

FCC – Federal Communications Commission.

Feedline – A way of connecting antennas to equipment. This can be coax or ladder line for example.

Fine Business – See FB.

Fist – Sending style in CW. A *"nice fist"* refers to a good sending style.

Free DV – Digital voice communication mode.

FSK – Frequency Shift Keying.

FSTV – Fast Scan Television, moving video transmission.

FT8 - Franke-Taylor 8-FSK modulation, a popular digital mode.

Gentleman's Agreement – Not an actual rule or regulation, but something that is unofficially agreed on. Such as a band plan or even a format.

Gentleman's Band – The 160m Band, also see Top Band.

GA – Good Afternoon, CW shorthand.

GLONASS – The Russian GPS Network.

GM – Good Morning, CW shorthand.

GND – Often used as abbreviation for Ground.

GPS – Global Positioning System, normally used in navigation and military applications using Satellites to calculate positions. In ham radio often used to set clocks and in a GPSDO.

GPSDO – GPS Disciplined Oscillator. A unit which locks to a satellite signal to provide an accurate reference frequency. Usually, 10MHz.

Green Stamp – US currency, used for requesting QSL cards.

Grayline – Increased propagation at dawn and dusk where darkness and light meet.

Grid Squares – Maidenhead locator system used to indicate location.

Ground – Connecting to earth.

Harmonic – Dual meaning. Can be a multiple of a frequency. Also, can be used in reference to one's children when on the air.

Heterodynes – Created by the mixing of two other frequencies. Used to shift frequencies in some receivers.

Hertz – Unit or measurement and refers to one cycle per second. This unit of measurement was named after Heinrich Rudolf Hertz who studied electromagnetic waves and proved their existence.

HF – High Frequency, specifically 3 to 30MHz though often and incorrectly includes the 160m and 6m bands.

HOA – Homeowners Association.

Homebrew – Something made at home, not purchased, or made by a manufacturer.

HRD – Ham Radio Deluxe. A software suite.

Hz – See Hertz.

Iambic Keyer – Uses a paddle with side-to-side contact to send code.

IARU – International Amateur Radio Union.

IF – Intermediate Frequency.

IMD – Intermodulation Distortion, usually caused by mixers.

IOTA – Islands on the Air. An RSGB sponsored award.

IP – Internet protocol. Used in IT in reference to networks.

IPO – Intercept Point Optimization. Used by Yaesu in reference to RF preamplification.

IQ – Incident and Quadrature data streams.

IRC – International Relay Coupon. No longer used in most places.

ISP – Internet Service Provider, basically where your internet service comes from.

JT Alert – Popular add-on software for WSJT-X and JTDX.

JT65 – A popular digital mode before FT8, still used in EME circles.

JT9 – A narrow digital mode, still used on some lower bands.

JTDX – A WSJT-X Alternative preferred by some hams.

Key – is a device used to send CW by hand. This is likely to be a paddle today, but many still use a bug or a straight key.

Keyer – Is an electronic device or circuit inside a radio to allow CW communications. Not to be confused with a Key.

kHz – Kilo Hertz, 1000 cycles per second. Lower case k, capital K is used in Kelvin.

λ – Lambda, measure of wavelength.

LAN – Local Area Network, a term used in IT.

LED – Light Emitting Diode, used in illumination, indicators.

Lid – A ham using poor operating practices or is unprofessional on the air. Something you do not want to be.

Little pistol – A down to earth, simple ham radio station.

Loading Coil - A system where an inductor coil (or coils) is used to electronically lengthen the antenna while allowing it to remain shorter.

LOTW – Logbook of the World, web-based service from the ARRL used to match contacts and apply for awards such as the DXCC.

LSB – Lower Side Band, SSB mode.

Lumber – Refers to plaques made of wood, such as awards for DX accomplishments and for contests.

Magic Band – Refers to the 6m band, sometimes also the 4m band available.

MHz – Mega Hertz, 1 million cycles per second. M is capital for Mega, Lover case m is short for milli.

Moonbounce – See EME.

Morse Code – Used in CW communications, named after the inventor Samuel Morse. It is a set of dots and dashes sent as a set of tones used for communications. Basically, this makes up an alphabet.

MSHV – Software by LZ2HV used to send multi-stream digital, mainly used for FT8 mode.

Net – An organized on-air gathering.

NB – Noise Blanker, reduces pulsing noises.

NCCC – Northern California Contest Club.

NCDXC – Northern California DX Club.

NCDXF – Northern California DX Foundation.

NR – Noise Reduction.

NTP – Network Time Protocol, used to set computer clocks via the internet.

NVIS - Near Vertical Incidence Skywave.

Ω - Ohm, measure of electrical resistance.

OM – Old Man, no matter the age, male hams are referred to as such.

Paddle – This is a device used in CW communications to send code.

Panadapter – Display of waterfall and/or spectrum display, short for panoramic adapter.

Pileup – Sometimes written as "pile-up". A number of stations calling the DX.

Phone – Voice communication modes.

Phonetic Alphabet – Words assigned to letter to ease understanding on the air.

PL259 – UHF connector, mates with the SO239.

Pounding Brass – Slang for operating in CW mode.

POTA – Parks on the Air.

Propagation – Indicates conditions on HF for making contacts.

PSE – Short for please during CW and some digital modes.

PTT – Push/Press to Talk.

Q-Codes – Abbreviations used in ham radio. See section Q-codes on page 145.

QSL manager – Someone who handles QSL card request.

Radiosport – Contesting. Also a brand of headset.

Rag Chewing – Friendly conversation instead of just a report exchange.

RBN – Reverse Beacon Network.

RF – Radio Frequency.

RFI – Radio Frequency Interference, not a good thing.

Rig – Slang for one's ham radio they are using.

RIT – Receiver Incremental Tuning. Sometimes called a clarifier.

RR – Roger Roger, used instead of QSL sometimes in CW and digital.

RSGB – Radio Society of Great Britain.

RST – Signal reporting system, Readability Strength Tone.

RTTY – Radio Teletype.

S-Meter – Signal strength meter. S1-S9 scale above which a +dB system is used.

Simplex – Using the same frequency for reception and transmission.

SDR – Software Defined Radio.

SFI – Solar Flux Index.

Shack – Is the name for the place of operating for the ham.

S/N – Signal to Noise Ratio, also serial number.

SNR – see S/N.

SK – Silent Key has a dual meaning. End of transmission and also refers to a ham radio operator who passed away

Skimmer – An automatic online spotting network, can be local software also.

Skip – Also known as hop, signal reflected by the ionosphere.

Solid State – Electronics which uses Semiconductors.

SO239 – UHF connector socket, mates with the PL259.

SOTA – Summits on the Air, Portable operating from an elevated point.

Split – Using different transmit frequency from the receive frequency

Sporadic E – Clouds of high ionization which form in the E layer/region.

SSTV – Slow Scan Television, a mode used for sending single images.

Squelch – Mutes audio below a certain level of reception as determined by the setting.

SWR – Standing Wave Ratio.

Tail-ending - Transmitting out of sync with other callers.

Tailgating – Calling immediately after the DX is ready for new callers.

TEP - Trans Equatorial Propagation.

TFT – Thin-Film Transistor, a type of display.

Ticket – This is your license to operate.

Top Band – Same as Gentlemen's band, the 160m band.

Transverter – A way of operating a radio on a frequency it was not designed for. Commonly VHF/UHF transverters are used with HF rigs, generally via the 10m band allocation. They allow both to receive and transmit in most cases.

Traps – Contain an LC (Inductor/Capacitor) network which electronically isolate one part of the radiating element from another allowing it to function (and resonate) on more than one band.

TU – Short for "Thank you" used in CW and digital modes.

TX - Abbreviation for transmit.

UHF Connector – A PL259 connector type.

UPS - Uninterruptable Power Supply, used in case of power failure as well as to compensate for low power fluctuations.

USB – Upper Side Band, SSB mode.

UTC – Universal Time Coordinated, always used for logging contacts. Sometimes referred to as Greenwich Mean Time (GMT).

Vanity Callsign – Callsign chosen by the operator.

VFO – Variable Frequency Oscillator.

VOX – Voice Activated Switch. Transmit without using a PTT, speech activated.

VUCC – VHF UHF Century Club, similar DXCC for stations worked above HF but based on grids not entities.

W – Watt, unit of measure for power.

WAC – Worked All Continents, an award.

Wallpaper – Refers to awards of the paper kind, but also sometimes to QSL card collections displayed on the wall.

Wall Wart – An AC or DC adapter that plugs in the wall. Called this as they stick out of the wall and are ugly.

WARC - World Administrative Radio Conference, sometimes used in reference to the allocation of the 12m, 17m and the 30m bands.

WAS – Worked All States award.

Waterfall – Historical data on the spectrum display. A visual log of the panadapter.

Wavelength – Refers to length of a specific wave.

Work / Worked – Made a contact with another station.

WSJT-X – Digital software used for various digital modes.

WSPR - Weak Signal Propagation Reporter.

XIT – Offset Transmit Frequency.

XYL – Wife.

Yagi – An antenna named after Hidetsugu Yagi and Shintaro Uda. Sometimes called a "*beam antenna*" is a direction antenna used on a wide range of frequencies, including HF.

YL – Young Lady, though above 18 usually.

Zero Beat – Matching the receive frequency of the other station.

Disclaimers, Warnings, and other Legalese

The author and publisher have made every possible effort to ensure that the information contained in this book was correct as of press time. **The author and publisher hereby do not assume liability for any injury, loss or damage caused by errors or omissions.** This is regardless of whether any errors or omissions result from negligence, accident, or any other cause. Readers are encouraged to verify any information contained in this book prior to taking any action on the information presented here.

The above does not mean everything in this book applies to you or even will work in your given situation. Always use good engineering practice, follow local laws and regulations, and use common sense as with everything else. The author and the publisher are not responsible for misuse of information presented in this book.

This book will often provide web links to other resources. Readers using the eBook version can go directly to these resources. Those reading printed versions are obviously not so fortunate. Some web links are shorter and some longer. URL shorteners were not used for several reasons beyond the scope of this book. **While at the writing of this book all links used tested as working, this may change with time.** The author always welcomes corrections. Due to the print, you may see what appears to be a space in some web links, it is likely an underscore. There are no spaces used in web addresses. Sometimes the same link may appear in more than one location as there is some unavoidable information overlap. This is not accidental but for the convenience of the reader.

Certain products, services and brands are mentioned in this book which the author uses or has used, in instances he has helped someone install and/or sometimes actually recommends. **These are not paid endorsements.** The author's opinions are based on his experiences as a ham and in his profession to possibly help guide you and aid your enjoyment of this wonderful hobby even more. The products that fit the author's needs may not fit yours. So do your own research and use this book as a guideline only.

All items in this book that are known to be trademarks or service marks have been appropriately capitalized.

Microsoft, Windows and Notepad are registered trademarks of Microsoft Corporation. MacOS, OSX, iPad, iPhone and MacIntosh are registered trademarks of Apple Corporation. Android, Chrome, YouTube, and Google are registered trademarks of Alphabet Corporation. FlexRadio, PowerGenius XL, TunerGenius XL and SmartSDR are registered trademarks of FlexRadio Systems. Elecraft K3S, P3, K4D, KPA100 and KPA1500 are registered trademarks of Elecraft. Yaesu is a registered trademark of Yaesu Musen Co., Ltd. Icom is a registered trademark of Icom Incorporated. Kenwood is a registered trademark of JVC Kenwood Corporation. MFJ and Ameritron, Hy-Gain, Cushcraft, Mirage and Vectronics are registered trademarks of MFJ Enterprises. SteppIR and Urban Beam are registered trademarks of SteppIR Communication Systems. ARRL is a registered trademark of the American Radio Relay League.

The Daily DX and *The Weekly DX* publication names are used with permission from **Bernie McClenny, W3UR.**

All other brand names used or mentioned are registered trademarks or service marks of their respective owners. Including but not limited to those related to hardware, software, services and/or ham radio specifically.

www.ingramcontent.com/pod-product-compliance
Lightning Source LLC
Chambersburg PA
CBHW081218170426
43198CB00017B/2644